Also from Cold Spring Harbor Laboratory Press

Related Manuals

Methods in Yeast Genetics
Molecular Cloning: A Laboratory Manual

Other Available Titles

*From **a** to α: Yeast as a Model for Cellular Differentiation*
Cold Spring Harbor Laboratory: 100 Years of Science (VHS)
*Lab Math: A Handbook of Measurements, Calculations, and Other Quantitative Skills
  for Use at the Bench*
Lab Ref: A Handbook of Recipes, Reagents, and Other Reference Tools for Use at the Bench
Lab Dynamics: Management Skills for Scientists
At the Bench: A Laboratory Navigator
At the Helm: A Laboratory Navigator
Landmark Papers in Cell Biology

LANDMARK PAPERS IN YEAST BIOLOGY

EDITED BY

Patrick Linder
University of Geneva, Switzerland

David Shore
University of Geneva, Switzerland

Michael N. Hall
Biozentrum of the University of Basel, Switzerland

COLD SPRING HARBOR LABORATORY PRESS
Cold Spring Harbor, New York

Landmark Papers in Yeast Biology

©2006 by Cold Spring Harbor Laboratory Press
All rights reserved
Printed in the United States of America

Publisher	John Inglis
Acquisition Editor	Kaaren Janssen
Development Manager	Jan Argentine
Project Coordinator	Joan Ebert
Permissions Coordinator	Maria Fairchild
Production Manager	Denise Weiss
Production Editors	Kathy Bubbeo and Mala Mazzullo
Desktop Editor	Susan Schaefer
Interior Book Designer	Denise Weiss
Cover Designer	Ed Atkeson

Cover: Budding yeast. (*Image courtesy of Claudio De Virgilio.*)

Library of Congress Cataloging-in-Publication Data

Landmark papers in yeast biology / edited by Patrick Linder, David Shore, Michael N. Hall.
   p. cm.
 Includes bibliographical references and index.
 ISBN 0-87969-643-5 (hardcover : alk. paper)
 1. Yeast. I. Linder, Patrick. II. Shore, David, 1955- III. Hall, Michael N.
 QR151.L26 2005
 579.5'63--dc22

                            2005025092

10 9 8 7 6 5 4 3 2 1

Authorization to photocopy items for internal or personal use, or the internal or personal use of specific clients, is granted by Cold Spring Harbor Laboratory Press, provided that the appropriate fee is paid directly to the Copyright Clearance Center (CCC). Write or call CCC at 222 Rosewood Drive, Danvers, MA 01923 (508-750-8400) for information about fees and regulations. Prior to photocopying items for educational classroom use, contact CCC at the above address. Additional information on CCC can be obtained at CCC Online at http://www.copyright.com/

All Cold Spring Harbor Laboratory Press publications may be ordered directly from Cold Spring Harbor Laboratory Press, 500 Sunnyside Boulevard, Woodbury, New York 11797-2924. Phone: 1-800-843-4388 in Continental U.S. and Canada. All other locations: (516) 422-4100. FAX: (516) 422-4097. E-mail: cshpress@cshl.edu. For a complete catalog of Cold Spring Harbor Laboratory Press publications, visit our World Wide Web Site http://www.cshlpress.com/

*This book is dedicated
to the memory
of Ira Herskowitz (1946–2003)*

Contents

Preface, xiii
Reprint Credits, xiv

Introduction, 1
Jasper Rine

1 **Cytoplasmic Inheritance, 11**
Susan W. Liebman and Fred Sherman

   Ephrussi B., Hottinguer H., and Tavlitzki J. 1949. Action de l'acriflavine sur les levures. II. Étude génétique du mutant "petite colonie"

   Thomas D.Y. and Wilkie D. 1968. Recombination of mitochondrial drug-resistance factors in *Saccharomyces cerevisiae*

   Tzagoloff A., Akai A., Needleman R.B., and Zulch G. 1975. Assembly of the mitochondrial membrane system. Cytoplasmic mutants of *Saccharomyces cerevisiae* with lesions in enzymes of the respiratory chain and in the mitochondrial ATPase

   Cox B.S. 1965. Ψ, a cytoplasmic suppressor of super-suppressor in yeast

   Wickner R.B. 1994. [URE3] as an altered *URE2* protein: Evidence for a prion analog in *Saccharomyces cerevisiae*

2 **Homologous Recombination, 33**
Lorraine Symington

   Hurst D.D., Fogel S., and Mortimer R.K. 1972. Conversion-associated recombination in yeast (hybrids/meiosis/tetrads/marker loci/models)

   Nicolas A., Treco D., Schultes N.P., and Szostak J.W. 1989. An initiation site for meiotic gene conversion in the yeast *Saccharomyces cerevisiae*

   Cao L., Alani E., and Kleckner N. 1990. A pathway for generation and processing of double-strand breaks during meiotic recombination in *S. cerevisiae*

   Allers T. and Lichten M. 2001. Differential timing and control of noncrossover and crossover recombination during meiosis

3 **Chromosome Replication and Segregation, 49**
Carol S. Newlon

   Stinchcomb D.T., Struhl K., and Davis R.W. 1979. Isolation and characterisation of a yeast chromosomal replicator

   Brewer B.J. and Fangman W.L. 1987. The localization of replication origins on ARS plasmids in *S. cerevisiae*

Bell S.P. and Stillman B. 1992. ATP-dependent recognition of eukaryotic origins of DNA replication by a multiprotein complex

Lundblad V. and Szostak J.W. 1989. A mutant with a defect in telomere elongation leads to senescence in yeast

Clarke L. and Carbon J. 1980. Isolation of a yeast centromere and construction of functional small circular chromosomes

4 Transcription, 67
Fred Winston

Matsumoto K., Toh-e A., and Oshima Y. 1978. Genetic control of galactokinase synthesis in *Saccharomyces cerevisiae:* Evidence for constitutive expression of the positive regulatory gene *gal4*

Guarente L., Yocum R.R., and Gifford P. 1982. A *GAL10-CYC1* hybrid yeast promoter identifies the *GAL4* regulatory region as an upstream site

Brent R. and Ptashne M. 1985. A eukaryotic transcriptional activator bearing the DNA specificity of a prokaryotic repressor

Hirschhorn J.N., Brown S.A., Clark C.D., and Winston F. 1992. Evidence that SCF2/SWI2 and SNF5 activate transcription in yeast by altering chromatin structure

Thompson C.M., Koleske A.J., Chao D.M., and Young R.A. 1993. A multisubunit complex associated with the RNA polymerase II CTD and TATA-binding protein in yeast

5 Translation, 85
Alan G. Hinnebusch

Sherman F., Stewart J.W., and Schweingruber A.M. 1980. Mutants of yeast initiating translation of iso-1-cytochrome c within a region spanning 37 nucleotides

Donahue T.F., Cigan A.M., Pabich E.K., and Castilho Valavicius B. 1988. Mutations at a Zn(II) finger motif in the yeast eIF-2β gene alter ribosomal start-site selection during the scanning process

Mueller P.P. and Hinnebusch A.G. 1986. Multiple upstream AUG codons mediate translational control of *GCN4*, 94

Altmann M., Sonenberg N., and Trachsel H. 1989. Translation in *Saccharomyces cerevisiae:* Initiation factor 4E-dependent cell-free system

Tarun S.Z., Jr., Wells S.E., Deardorff J.A., and Sachs A.B. 1997. Translation initiation factor eIF4G mediates *in vitro* poly(A) tail-dependent translation

6 Cell Division, 109
Kim Nasmyth

Diffley J.F.X., Cocker J.H., Dowell S.J., and Rowley A. 1994. Two steps in the assembly of complexes at yeast replication origins in vivo

Irniger S., Piatti S., Michaelis C., and Nasmyth K. 1995. Genes involved in sister chromatid separation are needed for B-type cyclin proteolysis in budding yeast

Nurse P. and Thuriaux P. 1980. Regulatory genes controlling mitosis in the fission yeast *Schizosaccharomyces pombe*

Weinert T.A. and Hartwell L.H. 1988. The *RAD9* gene controls the cell cycle response to DNA damage in *Saccharomyces cerevisiae*

7 Cell Growth, 127
James R. Broach

> Johnston G.C., Pringle J.R., and Hartwell L.H. 1977. Coordination of growth with cell division in the yeast *Saccharomyces cerevisiae*
>
> Toda T., Uno I., Ishikawa T., Powers S., Kataoka T., Broek D., Cameron S., Broach J., Matsumoto K., and Wigler M. 1985. In yeast, *RAS* proteins are controlling elements of adenylate cyclase
>
> Cameron S., Levin L., Zoller M., and Wigler M. 1988. cAMP-independent control of sporulation, glycogen metabolism, and heat shock resistance in *S. cerevisiae*
>
> Barbet N.C., Schneider U., Helliwell S.B., Stansfield I., Tuite M.F., and Hall M.N. 1996. TOR controls translation initiation and early G1 progression in yeast

8 Differentiation: Mating and Filamentation, 141
George F. Sprague, Jr.

> Strathern J., Hicks J., and Herskowitz I. 1981. Control of cell type in yeast by the mating type locus: The α1-α2 hypothesis
>
> Bender A. and Sprague G.F., Jr. 1987. MATα1 protein, a yeast transcription activator, binds synergistically with a second protein to a set of cell-type-specific genes
>
> Keleher C.A., Redd M.J., Schultz J., Carlson M., and Johnson A.D. 1992. Ssn6-Tup1 is a general repressor of transcription in yeast
>
> Hicks J.B. and Herskowitz I. 1977. Interconversion of yeast mating types. II. Restoration of mating ability to sterile mutants in homothallic and heterothallic strains
>
> Gimeno C.J., Ljungdahl P.O., Styles C.A., and Fink G.R. 1992. Unipolar cell divisions in the yeast *S. cerevisiae* lead to filamentous growth: Regulation by starvation and *RAS*

9 Meiosis and Spore Development, 157
Rochelle E. Esposito

> Esposito R.E., Frink N., Bernstein P., and Esposito M.S. 1972. The genetic control of sporulation in *Saccharomyces*. II. Dominance and complementation of mutants of meiosis and spore formation
>
> Kassir Y., Granot D., and Simchen G. 1988. *IME1*, a positive regulator gene of meiosis in *S. cerevisiae*
>
> Sym M., Engebrecht J.A., and Roeder G.S. 1993. ZIP1 is a synaptonemal complex protein required for meiotic chromosome synapsis
>
> Watanabe Y. and Nurse P. 1999. Cohesin Rec8 is required for reductional chromosome segregation at meiosis
>
> Hepworth S.R., Friesen H., and Segall J. 1998. *NDT80* and the meiotic recombination checkpoint regulate expression of middle sporulation-specific genes in *Saccharomyces cerevisiae*

10 Signal Transduction, 193
Jeremy Thorner

> Hartwell L.H. 1980. Mutants of *Saccharomyces cerevisiae* unresponsive to cell division control by polypeptide mating hormone

Whiteway M., Hougan L., Dignard D., Thomas D.Y., Bell L., Saari G.C., Grant F.J., O'Hara P., and MacKay V.L. 1989. The *STE4* and *STE18* genes of yeast encode potential β and γ subunits of the mating factor receptor-coupled G protein

Stevenson B.J., Rhodes N., Errede B., and Sprague G.F., Jr. 1992. Constitutive mutants of the protein kinase STE11 activate the yeast pheromone response pathway in the absence of the G protein

Choi K-Y., Satterberg B., Lyons D.M., and Elion E.A. 1994. Ste5 tethers multiple protein kinases in the MAP kinase cascade required for mating in *S. cerevisiae*

11 Cytoskeleton and Morphogenesis, 211
John R. Pringle

Byers B. and Goetsch L. 1975. Behavior of spindles and spindle plaques in the cell cycle and conjugation of *Saccharomyces cerevisiae*

Rout M.P. and Kilmartin J.V. 1990. Components of the yeast spindle and spindle pole body

Jacobs C.W., Adams A.E.M., Szaniszlo P.J., and Pringle J.R. 1988. Functions of microtubules in the *Saccharomyces cerevisiae* cell cycle

Kilmartin J.V. and Adams A.E.M. 1984. Structural rearrangements of tubulin and actin during the cell cycle of the yeast *Saccharomyces*

Sloat B.F., Adams A., and Pringle J.R. 1981. Roles of the *CDC24* gene product in cellular morphogenesis during the *Saccharomyces cerevisiae* cell cycle

Chant J. and Herskowitz I. 1991. Genetic control of bud site selection in yeast by a set of gene products that constitute a morphogenetic pathway

Lew D.J. and Reed S.I. 1993. Morphogenesis in the yeast cell cycle: Regulation by Cdc28 and cyclins

12 Membrane Traffic, 243
Randy Schekman

Novick P., Field C., and Schekman R. 1980. Identification of 23 complementation groups required for post-translational events in the yeast secretory pathway

Salminen A. and Novick P.J. 1987. A *ras*-like protein is required for a post-Golgi event in yeast secretion

Schu P.V., Takegawa K., Fry M.J., Stack J.H., Waterfield M.D., and Emr S.D. 1993. Phosphatidylinositol 3-kinase encoded by yeast *VPS*34 gene essential for protein sorting

Lewis M.J., Sweet D.J., and Pelham H.R.B. 1990. The *ERD2* gene determines the specificity of the luminal ER protein retention system

13 Protein Translocation, 253
Howard Riezman

Hall M.N., Hereford L., and Herskowitz I. 1984. Targeting of *E. coli* β-galactosidase to the nucleus in yeast

Deshaies R.J. and Schekman R. 1987. A yeast mutant defective at an early stage in import of secretory protein precursors into the endoplasmic reticulum

Schleyer M. and Neupert W. 1985. Transport of proteins into mitochondria: Translocational intermediates spanning contact sites between outer and inner membranes

Eilers M. and Schatz G. 1986. Binding of a specific ligand inhibits import of a purified precursor protein into mitochondria

14 Ubiquitination and Protein Turnover, 267
Mark Hochstrasser

Bachmair A., Finley D., and Varshavsky A. 1986. In vivo half-life of a protein is a function of its amino-terminal residue

Hiller M.M., Finger A., Schweiger M., and Wolf D.H. 1996. ER degradation of a misfolded luminal protein by the cytosolic ubiquitin-proteasome pathway

Kölling R. and Hollenberg C.P. 1994. The ABC-transporter Ste6 accumulates in the plasma membrane in a ubiquitinated form in endocytosis mutants

Schwob E., Böhm T., Mendenhall M.D., and Nasmyth K. 1994. The B-type cyclin kinase inhibitor p40^{SIC1} controls the G1 to S transition in *S. cerevisiae*

Mizushima N., Noda T., Yoshimori T., Tanaka Y., Ishii T., George M.D., Klionsky D.J., Ohsumi M., and Ohsumi Y. 1998. A protein conjugation system essential for autophagy

15 Genomics, 285
Mark Johnston and Philip Hieter

Petes T.D. and Botstein D. 1977. Simple Mendelian inheritance of the reiterated ribosomal DNA of yeast

Olson M.V., Dutchik J.E., Graham M.Y., Brodeur G.M., Helms C., Frank M., MacCollin M., Scheinman R., and Frank T. 1986. Random-clone strategy for genomic restriction mapping in yeast

Oliver S.G., van der Aart Q.J., Agostoni-Carbone M.L., Aigle M., Alberghina L., Alexandraki D., Antoine G., Anwar R., Ballesta J.P.M., Benit P., et al. 1992. The complete DNA sequence of yeast chromosome III

Winzeler E.A., Richards D.R., Conway A.R., Goldstein A.L., Kalman S., McCullough M.J., McCusker J.H., Stevens D.A., Wodicka L., Lockhart D.J., and Davis R.W. 1998. Direct allelic variation scanning of the yeast genome

Giaever G., Chu A.M., Ni L., Connelly C., Riles L., Véronneau S., Dow S., Lucau-Danila A., Anderson K., André B., et al. 2002. Functional profiling of the *Saccharomyces cerevisiae* genome

Index, 301

Preface

LANDMARKS DEFINE A LANDSCAPE. Here, we present landmark papers that define fifteen different fields within the yeast biology landscape. The handful of papers for each field were chosen and then placed in historical context by an expert in that field. The reader will immediately ask the same question posed by the authors, "How does one pick landmark papers?" A landscape and thus its landmarks differ depending on the perspective of the viewer. Consequently, there is no truly satisfactory answer to this question. Yet, as the authors put pen to paper, they had to make hard decisions. We ask the reader to be sympathetic with the authors' plight. As editors, our most difficult task was reducing the number of landmark papers originally proposed by the authors. If your favorite paper is not a landmark, it is we, the editors, who take responsibility. Although there are surely errors of omission, we are nonetheless confident that the selected papers, reprinted on the accompanying CD, are deserving of landmark status.

The idea to produce this volume grew out of a literature-based yeast biology course. Indeed, the book is designed as a teaching aid for such a course. Each chapter provides not only insight into a particular field by an expert, but also a set of practical study questions. We hope that the selected papers and their accompanying chapters will stimulate critical reading of the scientific literature and promote awareness of how scientific breakthroughs are made.

We would like to express our deep gratitude to the authors—not only for providing stimulating overviews of key events in the history of their fields, but also for making the hard decisions required to keep this book within reasonable limits of size. We also offer our thanks to colleagues who contributed their thoughts and comments to individual chapters. Finally, we are also grateful to our colleagues at Cold Spring Harbor Laboratory Press (with special thanks to Joan Ebert and Kathy Bubbeo) for their patience and guidance and to Nicolas Roggli (University of Geneva) for his expert artwork.

PATRICK LINDER, DAVID SHORE, AND MICHAEL N. HALL
Geneva and Basel, August 2005

Reprint Credits

Articles previously published in *Cell, Biochemical and Biophysical Research Communications, Experimental Cell Research,* and the *Journal of Molecular Biology* are reproduced with permission from Elsevier.

Articles previously published in *Nature* and the *EMBO (European Molecular Biology Organization) Journal* are reproduced with permission from Macmillan Publishers Ltd.

Articles previously published in *Science* are reproduced with permission from the American Association for the Advancement of Science. Readers may view, browse, and/or download material for temporary copying purposes only, provided these uses are for noncommercial personal purposes. Except as provided by law, this material may not be further reproduced, distributed, transmitted, modified, adapted, performed, displayed, published, or sold in whole or in part, without prior written permission from the publisher.

Articles previously published in *Molecular and Cellular Biology* and the *Journal of Bacteriology* are reproduced with permission from the American Society of Microbiology.

Articles previously published in the *Proceedings of the National Academy of Sciences of the United States of America* are reproduced with permission from the National Academy of Sciences.

Articles previously published in *Journal of Cell Biology* are reproduced with permission from The Rockefeller University Press.

Articles previously published in *Genetics* are reproduced with permission from The Genetics Society of America.

Articles previously published in *Genes and Development* are reproduced with permission from the Cold Spring Harbor Laboratory Press.

Article previously published in *Heredity* is reproduced with permission from The Genetics Society.

Article previously published in the *Journal of Biological Chemistry* is reproduced with permission from The American Society for Biochemistry and Molecular Biology.

Article previously published in *Molecular and General Genetics* is reproduced with permission from Springer Science and Business Media.

Article previously published in *Molecular Biology of the Cell* is reproduced with permission from The American Society for Cell Biology.

Article previously published in the *Annales de l'Institut Pasteur* is reproduced courtesy of Anne Ephrussi.

Introduction

Jasper Rine

Department of Molecular and Cellular Biology
University of California
Berkeley, California 94720

I AM PLEASED TO HAVE THE OPPORTUNITY TO PROVIDE THE INTRODUCTORY chapter for *Landmark Papers in Yeast Biology*. Some of the papers highlighted in this volume were obviously destined at the time of their publication to become landmarks in our field. For others, it was not at all clear at the time of their publication, at least to me, that they would emerge as landmarks. In those cases, subsequent work by the authors or others revealed the singular significance and landmark status of the earlier work.

The selection of papers as the landmarks of a field requires a perspective on the role of that field in biology and biomedical research defined in the broadest sense. Let us take a moment to reflect on this challenge. For anyone who has grown up in a field to judge that field objectively against all others requires at least a *modicum of hubris*. Nevertheless, the contributors to this volume will agree that we have much to be proud of as a field for the impact yeast has made on biology writ large. There is abundant evidence to support this assertion. Certainly, the recognition of Leland Hartwell and Paul Nurse by the Nobel Prize, and of Randy Schekman and Alex Varshavsky by the Lasker Award, and of Ira Herskowitz by the Rosenstiel Award reflected widespread appreciation of studies of both *Saccharomyces* and *Schizosaccharomyces* on fundamental processes common to all cells. However, the lasting contribution of the field goes much deeper and broader than that. Progress from our field has provided the conceptual framework that has driven experiments in many areas of biology. Take, for example, the revelation by Terri Orr-Weaver, Jack Szostak, and Rodney Rothstein that double-stranded DNA ends are capable of elevating the frequency of homologous recombination in mitotic cells by orders of magnitude (Orr-Weaver et al. 1981). This discovery was a clear factor in the design of the substrates that were used to achieve the first homologous mouse gene knockout by Kirk Thomas and Mario Capecchi (Thomas et al. 1986). The work by Maynard Olson, beginning in 1977, to produce the first physical map of a eukaryotic genome (Link and Olson 1991), laid the foundation for so much of what was achieved in the bloated genomes of other eukaryotes in later years, including completion of the human genome sequence.

Similarly, the field has accomplished amazing feats by organized cooperation among labs that had no prior scientific links. Under the organizational prowess primarily of Andre Goffeau and Steve Oliver, 96 different labs worldwide, with 92 of them being in Europe, combined efforts to produce the first complete genome sequence of *Saccharomyces* in 1996 (Goffeau et al. 1996), and then of *Schizosaccharomyces* shortly thereafter. This model of cooperation and its success certainly paved the way for newer European Union-distributed collaborations such as the Epigenome Network of Excellence (http://www.epigenome-noe.net/aboutus/noemem.php).

NEW METHODOLOGIES HAVE BEEN A CRITICAL CATALYST FOR THIS PROGRESS

There is something mystifying about predicting progress in science. Asked to imagine how much progress we expect to make in the next 5 years, nearly all of us will overestimate how far we will go, which can be embarrassing when applying for grant renewals in which we have not accomplished all of our specific aims. If instead we are asked to project where we or the field will be in 10 years time, we will nearly always substantially underestimate our progress. The striking difference in the 5- and 10-year perspectives probably reflects the time of adapting new methods and technologies to our research. It seems to take about 7 years for a new potentially transforming technology to move from the benches of its creators to the orthodoxy of common usage. We can usually see the methods that are likely to be available in a 5-year window, so our ambition tends to outpace our progress. But in no 10-year period has anyone that I know been able to predict the new methods that are created and reduced to common practice. The sections of this volume are sensibly organized around biological concepts that experiments in yeast have appreciably advanced. However, the impact of our field is greater when measured against the technologies developed in yeast that have driven progress across biology. We can better appreciate this point by looking at the technologies available in yeast research just a few years ago.

To appreciate how far we have come, consider for a moment the technology that went into a typical Ph.D. dissertation in yeast genetics 25 years ago: toothpicks, Petri plates, a microscope, a micromanipulator, a replica block and velvets, and glass pipettes, which were used to—dare I say it?—mouth-pipette everything. No proteins were purified (with or without tagging), no gene was cloned or sequenced, indeed no gel was run, and the PCR had not even been spelled at that time. From today's perspective, it is difficult to imagine that such a thesis would ever garner the support of a 21st century thesis committee.

Clearly, the two advances that catalyzed progress in this field were the development of recombinant libraries of yeast genes, largely through the efforts of Ron Davis and his colleagues (Cameron et al. 1977; Struhl et al. 1979), and the discovery of methods for the transformation of yeast with recombinant DNA by Hinnen, Hicks, and Fink (1978) and by Jean Beggs (1978). These advances made it obvious that the gene corresponding to any recessive mutation could be identified by transforming pools of recombinant clones into the mutants and screening the transformants for complementing clones. The backlog of mutations identified and characterized in the previous decades would, in a short time, provide an entrée into the molecular basis of processes common to all eukaryotic cells. These advances were certainly aided by the discovery of ARS (autonomous replication sequences) elements and centromeres and the contributions of these elements to better vector design (e.g., Clarke and Carbon 1980; Stinchcomb et al. 1980). Indeed, ARS elements and centromeres were very useful long before their mechanisms were understood.

To my mind, the singular technical advance that gave yeast studies a manipulative power over any other eukaryote for years to come was the harnessing of homologous recombination, coupled with the newly developed transformation method, to replace, with perfect precision, a gene in the chromosome with any form of that gene altered in whatever way an experimenter wanted (Scherer and Davis 1979). The paper describing this method is, by any criterion, a classic, but its acceptance by the community illustrates an interesting point regarding the adoption of new technologies. Despite the obvious utility of being able to manipulate genes, it was a full 2 years before the first paper appeared that actually used this method (Kolodrubetz et al. 1982). It took several additional years before one-step gene replacement (Rothstein 1983) became commonplace, yet a decade passed before gene knockouts of all yeast genes were first contemplated. I think the take-home message here is that there is an inertia in the way we do science that causes breakthroughs in technology to have a longer time to adoption than is necessary or desirable. As a community, we would do well to remember that message if yeast is to remain a preeminent organism for genetic research.

More recent technical advances that have propelled our field include the first physical map of a eukaryotic genome (Link and Olson 1991), methods for using an antibody to clone the gene for a protein (Young and Davis 1983), the two-hybrid interaction method for revealing protein interactions (Fields and Song 1989; Chien et al. 1991), gene filters for mapping cloned genes, the first whole-genome expression arrays (Lashkari et al. 1997), identification of protein complexes using tagged proteins and mass spectrometry (Gavin et al. 2002; Ho et al. 2002), and the first systematic analysis of null alleles of all genes (Giaever et al. 2002). It is humbling to imagine where each of us would be in our understanding of our favorite subject had any of these advances not been applied to yeast.

OTHER LANDMARKS

The choices involved in making any collection of landmark papers involve trading off the specific interest of a paper against the impact that paper made in the field. Organizing papers into categories always results in certain gems being overlooked when they do not fall neatly into categories. Hence, I present here an addition to the 15 categories of landmark papers found in this volume. I call this category "Papers That Make Me Proud to Work on Yeast," which includes papers of two varieties. One class makes me feel good by underscoring in dramatic ways the unity in biology that can pop up in the most unexpected ways. The second class includes papers so clever and insightful as to represent the highest form of our art, leaving workers on other organisms to lament the pedestrian limits imposed by the vagaries of chance and circumstance in the evolution of their organisms.

The first of these personal landmarks involves the genetic test of the Cassette Hypothesis by Ira Herskowitz and colleagues (Kushner et al. 1979). This paper presents as genetically sophisticated a series of experiments as I have ever encountered. The essential elements of its creativity lie in testing whether the *HML* and *HMR* loci of *Saccharomyces* are the source of transposable copies of mating-type alleles. Conceptually, the test of the hypothesis was to isolate mutations that map at one position in the genome based on the phenotypes they present when transposed to a different position in the genome, amid a background of many *STE* genes whose mutant phenotypes would look similar. I use this paper in a graduate class as it is the best way to identify those students capable of the most sophisticated genetic reasoning. Amar Klar's papers on the mating-type interconversion of *S. pombe* are also of this class (e.g., Dalgaard and Klar 1999). If, as a student, you find yourself dazzled by the beauty and elegance of these studies, then you should clearly give up your space in the cold room for you will never be satisfied by anything other than the most sophisticated of genetics.

A second paper in this category, from Jef Boeke, David Garfinkel, Cora Styles, and Gerry Fink, tests the mechanism by which Ty elements transpose (Boeke et al. 1985). Today, it is difficult to grasp the excitement surrounding the mystery of how transposable elements move circa 1980. Although reverse transcription was well established at that time as a principle for propagating RNA tumor viruses, the long terminal repeats (LTRs) of Ty elements could have been analogs of the IS elements of DNA-based transposons or analogs of the LTRs of retrotransposons. The concept in this paper is simple to grasp. The authors reasoned that if Ty elements transposed through an RNA intermediate, then an artificially inserted intron should be removed from all Ty elements descended from that marked parental Ty element. The conceptual beauty of this experiment is matched by its flawless execution in what, for me, is another high point in our field.

A third paper deserves special mention for the powerful simplification it provided to a field that was staggering from too many genes. From the beginning, studies of mating type were fueled by the isolation of mutations that rendered cells sterile, defining so-called *STE* genes. Some mutant *STE* genes affected **a** cells specifically, some affected α cells specifically, and some affected both **a** cells and α cells. We knew that mating required the production of and response to mating-type-specific pheromones, which were chemically dissimilar in the two sexes of yeast and were encoded by genes that followed different organizational principles. The central question was to what extent the response to the two mating pheromones of yeast occurred by similar or different processes. In an elegant experiment of conceptual simplicity, Alan Bender and George Sprague (1986) engineered the expression of the receptor for a particular mating pheromone in the same cell type that produced that pheromone and found that those cells would respond to their own pheromone. In other words, the difference between the two mating types was largely due to the particular pheromone and pheromone receptor made. The rest of the mating process is essentially identical in the two cell types. We now know the mating response pathway in great detail as described by the landmark papers in Chapter 8, but the Bender and Sprague study stands out in my mind for the way it set the stage for all that has followed.

In the category of papers that thrill us by their identification of unity in biology, certainly the work on Ras, described in Chapter 7, is a powerful example. The discovery by Michael Wigler, Jeff Strathern, James Broach, and their colleagues that human Ras proteins could complement yeast *ras1 ras2* double mutants and that activating mutations in yeast Ras proteins modeled after human oncogenic Ras mutations altered growth control was breathtaking (see, e.g., Katoaka et al. 1984). More than any other single discovery, this result convinced people that studies on yeast could provide more than just an understanding of fundamental molecular, cellular, and genetic mechanisms. This result established that yeast studies could have direct applications to human health and disease. A wonderful account of the circumstances surrounding these studies and the discovery of human oncogenes is provided in Natalie Angier's book *A Natural Obsession* (1988), which, in addition to being scientifically accurate, is a real page-turner.

Other examples of outstanding "unification" papers include the study by Melanie Lee and Paul Nurse in which the human *Cdc2* gene was cloned from a cDNA library by its ability to complement the *cdc2* mutation of *S. pombe* (Lee and Nurse 1987). Although this paper is not included in the selection of papers in Chapter 6, it was very influential in the award of the first Nobel Prize for work done in yeast. Logically, we should deduce that the landmark papers which do appear in this chapter will garner even more recognition in the coming years.

With the clarity of hindsight, we can now justify that fundamental molecular mechanisms involving cell growth and division are conserved from yeast to humans, but I do not remember a lot of people betting so at the time of these experiments. To me, the most surprising of these unification papers involves the link between human and yeast aging. It is true that aging and death have been with us as long as growth and division so there is no conceptual reason why the mechanisms could not have commonalities. However, human aging, especially the premature aging brought on by diseases such as Werner's syndrome, seems so different from the aging associated with the accumulation of yeast bud scars that the potential connection went unexplored for decades. In my all-time favorite line from any movie, in *Indiana Jones and the Temple of Doom*, Harrison Ford attempts to impress his would-be paramour with his declaration that, "Nothing shocks me, I am a scientist." I suppose as a scientist, I was not really shocked, but I was mighty surprised by the discovery of Lenny Guarente and his colleagues that the yeast ortholog of the gene responsible for the premature aging of Werner's syndrome, when mutated, considerably shortened the life span of yeast (Sinclair et al. 1997). The jury is still out on how good a model this type of yeast aging—involving the accumulation of replicating but nonsegregating DNA molecules—is for human aging. Perhaps the real question should be: How good a model is Werner's syndrome for conventional aging? Nevertheless, this work opened up yeast aging as an exciting subject for study. The subsequent discoveries of *SIR2*'s role in aging of both yeast (Sinclair and Guarente 1997; Sinclair et al. 1997) and *Caenorhabditis elegans* (Tissenbaum and Guarente 2001) and the multiple paralogs of *SIR2* in mice and humans have many of us on the edge of our chairs in excitement. The recent discovery that caloric restriction extends yeast longevity by a *SIR2*-independent mechanism will surely keep the yeast-aging field young and vibrant for the foreseeable future (Kaeberlein et al. 2004).

THE LANDMARK PAPERS OF TOMORROW

Any thoughtful review of the papers included in this volume will conclude that identifying the landmark papers of today will be much more easily done in distant tomorrows than in the confusion of today. My favorite example of this challenge comes from the first chapter, which shows how 30 years would pass between what we now recognize as Brian Cox's discovery of Psi (Cox 1965) and Reed Wickner's discovery that it is prion (Wickner 1994). Despite this difficulty, it seems that certain recent papers have great potential to be recognized as the landmark papers of our present age in some future landmarks volume. I will bet heavily that the comparative genome sequence papers from Mark Johnston's and Eric Lander's groups (Cliften et al. 2003; Kellis et al. 2003) are among the landmarks, but not because of any particular insight offered by either paper. Instead, these two papers mark the first opportunity for any scientific community to have access to the admixture of sequence conservation and divergence in the genomes of closely related species and have led Ken Wolfe and colleagues to fascinating insights into the evolution of the yeast genome and speciation (see, e.g., Wolfe 2004). Likewise, the sequence of *Ashbya gossypii* by Peter Philippsen and colleagues (Dietrich et al. 2004), the smallest free-living eukaryotic genome sequenced, gives us a sense of just what it takes to be a functional eukaryotic cell and has put this organism on the fast track for bringing genetics to bear on the wide range of fascinating opportunities in such a tractable fungus with a filamentous life style. This opportunity for our community to bring serious function-driven discovery to the "squishiness" of evo-devo studies is too important to pass up. I hope that all yeast labs will spend a considerable portion of the coming years determining how we can most effectively capitalize on this opportunity.

In the last few years, our field has been the beneficiary of powerful studies using various physical methods, such as mass spectrometry, or "quasi" genetic methods, such as two-hybrid interaction mapping, to potentially link the functions of various genes. As a field, we have been conflicted on how to use the data from these studies. At one extreme, we sometimes indiscriminately incorporate any interaction between two proteins as detected by such methods into new lines in the models of our publications, with no further vetting of the interaction. At the other extreme, we dismiss the importance of some interactions because of the artifacts that accompany any method.

Becoming more sophisticated consumers of this type of information is crucial to appreciate another of what I believe will be a landmark paper of the future. The use of the yeast knockout collection and simple robotics has let Charlie Boone and colleagues systematically assess the phenotypes of all possible double mutants among mutations that are individually viable (Tong et al. 2001, 2004). The networks of genetic interactions that emerge from a typical such study are four- to eightfold more extensive than what has emerged from mass spectrometry analysis of complexes copurifying with tagged proteins or with the network of two-hybrid interactions emerging from ongoing work of Stan Fields and colleagues (Uetz et al. 2000). This bounty of interactions from the double-mutant interactions reflects the many different ways that two mutations in combination can affect phenotype. There is probably no such thing as an "in vivo artifact," so all of these interactions reflect something important that has happened in the cell. I think we will need a few years of collective effort before a consensus emerges on the range of interpretations that underlie these myriad interactions. It seems to me that our field is better positioned to deal with the clues from these broad-based methods than most other fields as long as we treat the data with both optimism and skepticism—optimism that some of the interactions from the broad-based studies will be fundamentally and directly important to the genes and proteins in question and skepticism that leads to more stringent assessments of interactions.

Other landmarks of this time will surely include those papers that offer comprehensive approaches to yeast biology. Certainly, the construction of the set of knockout mutants of all genes in the genome, led by Mark Johnston (Winzeler et al. 1999), was a conceptual catalyst for many such efforts, including the recent collection of green fluorescent protein (GFP)-tagged forms of all yeast proteins and the compilation of their abundance and localization (Ghaemmaghami et al. 2003). I have a similar enthusiasm for papers offering compendia of expression profiles and distributions of proteins over the genome. However, I think that we are still missing something important: much better ways of defining phenotype. Too often, we dismiss a mutant's phenotype as being slightly sick or slow growing. At our present level of sophistication, phenotype is what we observe after the cell has exhausted its ability to compensate for the loss of some gene. If we can go beyond our present and often superficial phenotyping and develop better ways of asking a cell, "Where does it hurt?," we will create studies that will be the landmarks of biology and not just of our field.

THE DEFENSE OF WHAT WE HOLD DEAR

I close with musing on the nature of the environment that has fostered so many landmark papers and comment on important steps that we need to take to maintain that environment. First and foremost, our field has been blessed with widespread and near universal belief in the fundamental value of sharing resources and reagents. The founders of modern yeast molecular genetics—Fink, Hartwell, Botstein, Herskowitz, Sherman, Nurse, Mortimer, and others—have vigilantly insisted that any published stain or reagent be freely shared with anyone who requests it, no matter whether that person is your chief competitor. The challenge to this attitude comes most frequently not from those who have grown up in this culture, but from those who now work with yeast but come from other disciplines where reagents are more commonly "sold for co-authorship." We are not going to change attitudes throughout all fields, but we must not let the undesirable attitudes of other fields corrupt ours.

The national and international yeast meetings have tied our field together in many important ways, and Cold Spring Harbor Laboratory deserves special credit for helping foster such interactions with their support of the early yeast meetings. The success of these meetings has outstripped the capacity of CSHL to accommodate such a crowd, and indeed successful specialty meetings have evolved that help communication among our subfields such as the semi-annual CSH Yeast Cell Biology Meeting and the Chromosome Replication and Segregation Meeting organized through FASEB. Organism-based meetings such as the yeast, fly, and worm meetings have a special niche in science by being the best venues in which scientists working on one aspect of biology can recognize the chance connection of their work with work in a different subfield. These meetings need our continued support. The papers in this volume document many examples of how progress in one field of yeast biology stimulated progress in another.

Communication has become one of the most fundamental challenges each of us face in our professional careers because of the sheer abundance of information at our disposal. A PubMed search for *Saccharomyces* returns approximately 68,000 citations and for *Schizosaccharomyces* another 6,000. There are many more papers published each week that are of relevance to each of us than any of us will ever read. How are we to cope with this flood of experimental information? The *Saccharomyces* Genome Database (SGD) has emerged as a lifeline for many of us, allowing ready access to the data deluge that no longer fits in our personal mental RAM. The curators at SGD have done masterful work in annotation, summarization, and presentation of enormously complex data sets and none of us would consider living without it. However, despite the skill applied at SGD, no database can adequately substitute for the knowledge gained by immersing oneself deeply into the primary literature. Doing so gives us a personal audience with authors, seeing firsthand how they thought about a problem and interpreted the data, rightly or wrongly. So given the daily flood of new important work, how do we justify sending our students back to study these landmark papers of yesteryear? For many of us, the answer lies not in finding a way to keep up with all the literature, but in learning how to recognize the small subset of papers, the landmarks of tomorrow, that are the most worthy of our time today. I am sure the editors and contributors to this work share my belief that deep study of the agreed-on classics is the best training for learning how to recognize those contemporary papers worthy of our personal time and limited powers of memory.

ACKNOWLEDGMENTS

Research in my lab has been generously supported primarily by grants from the National Institutes of Health (GM31105 and GM35827). I thank Erin Osborne for helpful comments on the manuscript.

REFERENCES

Angier N. 1988. *Natural obsessions: The search for the oncogene.* Houghton Mifflin, Boston.
Beggs J. 1978 Transformation of yeast by a replicating hybrid plasmid. *Nature* **275:** 104–109.
Bender A. and Sprague G.F., Jr. 1986. Yeast peptide pheromones, a-factor and α-factor, activate a common response mechanism in their target cells. *Cell* **47:** 929–937.
Boeke J.D., Garfinkel D.J., Styles C.A., and Fink G.R. 1985. Ty elements transpose through an RNA intermediate. *Cell* **40:** 491–500.
Cameron J.R. and Davis R.W. 1977. The effects of *Escherichia coli* and yeast DNA insertions on the growth of lambda bacteriophage. *Science* **196:** 212–215.
Chien C.T., Bartel P.L., Sternglanz R., and Fields S. 1991. The two-hybrid system: A method to identify and clone genes for proteins that interact with a protein of interest. *Proc. Natl. Acad. Sci.* **88:** 9578–9582.
Clarke L. and Carbon J. 1980. Isolation of a yeast centromere and construction of functional small circular chromosomes. *Nature* **287:** 504–509.
Cliften P., Sudarsanam P., Desikan A., Fulton L., Fulton B., Majors J., Waterston R., Cohen B.A., and Johnston M. 2003. Finding functional features in *Saccharomyces* genomes by phylogenetic footprinting. *Science* **301:** 71–76.
Cox B.S. 1965. Ψ, a cytoplasmic suppressor of super-suppressor in yeast. *Heredity* **20:** 505–521.
Dalgaard J.Z. and Klar A.J. 1999. Orientation of DNA replication establishes mating-type switching pattern in *S. pombe. Nature* **400:** 181–184.
Dietrich F.S., Voegeli S., Brachat S., Lerch A., Gates K., Steiner S., Mohr C., Pohlmann R., Luedi P., Choi S., Wing R.A., Flavier A., Gaffney T.D., and Philippsen P. 2004. The *Ashbya gossypii* genome as a tool for mapping the ancient *Saccharomyces cerevisiae* genome. *Science* **304:** 304–307.
Fields S. and Song O. 1989. A novel genetic system to detect protein-protein interactions. *Nature* **340:** 245–246.
Gavin A.C., Bosche M., Krause R., Grandi P., Marzioch M., Bauer A., Schultz J., Rick J.M., Michon A.M., Cruciat C.M., Remor M., Hofert C., Schelder M., Brajenovic M., Ruffner H., Merino A., Klein K., Hudak M., Dickson D., Rudi T., Gnau V., Bauch A., Bastuck S., Huhse B., Leutwein C., et al. 2002. Functional organization of the yeast proteome by systematic analysis of protein complexes. *Nature* **415:** 141–147.
Ghaemmaghami S., Huh W.K., Bower K., Howson R.W., Belle A., Dephoure N., O'Shea E.K., and Weissman J.S. 2003. Global analysis of protein expression in yeast. *Nature* **425:** 737–741.
Giaever G., Chu A.M., Ni L., Connelly C., Riles L., Veronneau S., Dow S., Lucau-Danila A., Anderson K., Andre B., Arkin A.P., Astromoff A., El-Bakkoury M., Bangham R., Benito R., Brachat S., Campanaro S., Curtiss M., Davis K., Deutschbauer A., Entian K.D., Flaherty P., Foury F., Garfinkel D.J., Gerstein M., et al. 2002. Functional profiling of the *Saccharomyces cerevisiae* genome. *Nature* **418:** 387–391.
Goffeau A., Barrell B.G., Bussey H., Davis R.W., Dujon B., Feldmann H., Galibert F., Hoheisel J.D., Jacq C., Johnston M., Louis E.J., Mewes H.W., Murakami Y., Philippsen P., Tettelin H., and Oliver S.G. 1996. Life

with 6000 genes. *Science* **274:** 546, 563–567.

Hinnen A., Hicks J.B., and Fink G.R. 1978. Transformation of yeast. *Proc. Natl. Acad. Sci.* **75:** 1929–1933.

Ho Y., Gruhler A., Heilbut A., Bader G.D., Moore L., Adams S.L., Millar A., Taylor P., Bennett K., Boutilier K., Yang L., Wolting C., Donaldson I., Schandorff S., Shewnarane J., Vo M., Taggart J., Goudreault M., Muskat B., Alfarano C., Dewar D., Lin Z., Michalickova K., Willems A.R., Sassi H., et al. 2002. Systematic identification of protein complexes in *Saccharomyces cerevisiae* by mass spectrometry. *Nature* **415:** 180–183.

Kaeberlein M., Kirkland K.T., Fields S., and Kennedy B.K. 2004. Sir2-independent life span extension by calorie restriction in yeast. *PLoS Biol.* **2:** E296.

Kataoka T., Powers S., McGill C., Fasano O., Strathern J., Broach J., and Wigler M. 1984. Genetic analysis of yeast *RAS1* and *RAS2* genes. *Cell* **37:** 437–445.

Kellis M., Patterson N., Endrizzi M., Birren B., and Lander E.S. 2003. Sequencing and comparison of yeast species to identify genes and regulatory elements. *Nature* **423:** 241–254.

Kolodrubetz D., Rykowski M.C., and Grunstein M. 1982. Histone H2A subtypes associate interchangeably in vivo with histone H2B subtypes. *Proc. Natl. Acad. Sci.* **79:** 7814–7818.

Kushner P.J., Blair L.C., and Herskowitz I. 1979. Control of yeast cell types by mobile genes: A test. *Proc. Natl. Acad. Sci.* **76:** 5264–5268.

Lashkari D.A., DeRisi J.L., McCusker J.H., Namath A.F., Gentile C., Hwang S.Y., Brown P.O., and Davis R.W. 1997. Yeast microarrays for genome wide parallel genetic and gene expression analysis. *Proc. Natl. Acad. Sci.* **94:** 13057–13062.

Lee M.G. and Nurse P. 1987. Complementation used to clone a human homologue of the fission yeast cell cycle control gene *cdc2*. *Nature* **327:** 31–35.

Link A.J. and Olson M.V. 1991. Physical map of the *Saccharomyces cerevisiae* genome at 110-kilobase resolution. *Genetics* **127:** 681–698.

Orr-Weaver T.L., Szostak J.W., and Rothstein R.J. 1981. Yeast transformation: A model system for the study of recombination. *Proc. Natl. Acad. Sci.* **78:** 6354–6358.

Rothstein R. 1983. One-step gene disruption in yeast. *Methods Enzymol.* **101:** 202–211.

Scherer S. and Davis R.W. 1979. Replacement of chromosome segments with altered DNA sequences constructed. *Proc. Natl. Acad. Sci.* **76:** 4951–4955.

Sinclair D.A. and Guarente L. 1997. Extrachromosomal rDNA circles—A cause of aging in yeast. *Cell* **91:** 1033–1042.

Sinclair D.A., Mills K., and Guarente L. 1997. Accelerated aging and nucleolar fragmentation in yeast *sgs1* mutants. *Science* **277:** 1313–1316.

Stinchcomb D.T., Thomas M., Kelly J., Selker E., and Davis R.W. 1980. Eukaryotic DNA segments capable of autonomous replication in yeast. *Proc. Natl. Acad. Sci.* **77:** 4559–4563.

Struhl K., Stinchcomb D.T., Scherer S., and Davis R.W. 1979. High-frequency transformation of yeast: Autonomous replication of hybrid DNA molecules. *Proc. Natl. Acad. Sci.* **76:** 1035–1039.

Thomas K.R., Folger K.R., and Capecchi M.R. 1986. High frequency targeting of genes to specific sites in the mammalian genome. *Cell* **14:** 419–428.

Tissenbaum H.A. and Guarente L. 2001. Increased dosage of a *sir-2* gene extends lifespan in *Caenorhabditis elegans*. *Nature* **410:** 227–230.

Tong A.H., Evangelista M., Parsons A.B., Xu H., Bader G.D., Page N., Robinson M., Raghibizadeh S., Hogue C.W., Bussey H., Andrews B., Tyers M., and Boone C. 2001. Systematic genetic analysis with ordered arrays of yeast deletion mutants. *Science* **294:** 2364–2368.

Tong A.H., Lesage G., Bader G.D., Ding H., Xu H., Xin X., Young J., Berriz G.F., Brost R.L., Chang M., Chen Y., Cheng X., Chua G., Friesen H., Goldberg D.S., Haynes J., Humphries C., He G., Hussein S., Ke L., Krogan N., Li Z., Levinson J.N., Lu H., Menard P., et al. 2004. Global mapping of the Yeast Genetic Interaction Network. *Science* **303:** 808–813.

Uetz P., Giot L., Cagney G., Mansfield T.A., Judson R.S., Knight J.R., Lockshon D., Narayan V., Srinivasan M., Pochart P., Qureshi-Emili A., Li Y., Godwin B., Conover D., Kalbfleisch T., Vijayadamodar G., Yang M., Johnston M., Fields S., and Rothberg J.M. 2000. A comprehensive analysis of protein-protein interactions in *Saccharomyces cerevisiae*. *Nature* **403:** 623–627.

Wickner R.B. 1994. [URE3] as an altered URE2 protein: Evidence for a prion analog in *Saccharomyces cerevisiae*. *Science* **264:** 566–569.

Winzeler E.A., Shoemaker D.D., Astromoff A., Liang H., Anderson K., Andre B., Bangham R., Benito R., Boeke J.D., Bussey H., Chu A.M., Connelly C., Davis K., Dietrich F., Dow S.W., El Bakkoury M., Foury F., Friend S.H., Gentalen E., Giaever G., Hegemann J.H., Jones T., Laub M., Liao H., Davis R.W., et al. 1999. Functional characterization of the *S. cerevisiae* genome by gene deletion and parallel analysis. *Science* **285:** 901–906.

Wolfe K. 2004. Evolutionary genomics. Yeasts accelerate beyond BLAST. *Curr. Biol.* **14:** R392–R394.

Young R.A. and Davis R.W. 1983. Efficient isolation of genes by using antibody probes. *Proc. Natl. Acad. Sci.* **80:** 1194–1198.

1

Cytoplasmic Inheritance

Susan W. Liebman
Department of Biological Sciences
Laboratory for Molecular Biology
University of Illinois at Chicago
Chicago, Illinois 60607

Fred Sherman
Department of Biochemistry and Biophysics
University of Rochester Medical Center
Rochester, New York 14642

The lesson of "petites", 15

Ephrussi B., Hottinguer H., and Tavlitzki J. 1949b. Action de l'acriflavine sur les levures. II. Étude génétique du mutant "petite colonie." *Ann. Inst. Pasteur* **76**: 419–442.

Recombination of mitochondrial markers, 16

Thomas D.Y. and Wilkie D. 1968. Recombination of mitochondrial drug-resistance factors in *Saccharomyces cerevisiae*. *Biochem. Biophys. Res. Commun.* **30**: 368–372.

Respiratory chain and oxidative phosphorylation mutants, 17

Tzagoloff A., Akai A., Needleman R.B., and Zulch G. 1975c. Assembly of the mitochondrial membrane system. Cytoplasmic mutants of *Saccharomyces cerevisiae* with lesions in enzymes of the respiratory chain and in the mitochondrial ATPase. *J. Biol. Chem.* **250**: 8236–8242.

Ψ, a cytoplasmic suppressor of super-suppression, 20

Cox B.S. 1965. Ψ, a cytoplasmic suppressor of super-suppressor in yeast. *Heredity* **20**: 505–521.

Yeast prions, 22

Wickner R.B. 1994. [URE3] as an altered *URE2* protein: Evidence for a prion analog in *Saccharomyces cerevisiae*. *Science* **264**: 566–569.

Note: The landmark papers listed above are those discussed in this chapter. Each landmark paper is preceded by the name of the section (with starting page number) where the paper is first discussed in detail.

CYTOPLASMIC INHERITANCE WAS DISCOVERED MORE THAN 90 YEARS ago, when certain plastid phenotypes of higher plants were found to exhibit non-Mendelian inheritance. Subsequently, the fields of chloroplast and mitochondrial genetics developed and became recognized disciplines, whose basis could be related to organelle DNA, which was discovered and characterized during the 1950s and 1960s. As described below, the yeast *Saccharomyces cerevisiae* had a critical role in defining mitochondrial genetics.

The extrachromosomal traits of *S. cerevisiae* that can be attributed to nucleic acids include not only those of mitochondria, but also those of the killer RNA viruses and the 2-μm nuclear DNA plasmid (Table 1). Most importantly, yeast was also instrumental in the discovery and characterization of extrachromosomal traits associated with prions. In this chapter, we emphasize extrachromosomal inheritance associated with two distinct mechanisms: altered mitochondrial DNA and formation of prions.

The major evidence for cytoplasmic rather than chromosomal inheritance is the absence of 2:2 Mendelian segregation as determined by tetrad analysis. Before Mendelian inheritance can be discounted, however, the possibility of multigene inheritance, which would also alter the 2:2 ratio, must be eliminated. This has often been done by a series of backcrosses, as described below for the mitochondrial ρ determinant. Another clue suggestive of a mechanism of inheritance distinct from classical Mendelian genes is a high mutation or reversion rate causing an unstable phenotype. Such instability can also be caused by chromosomal epigenetic phenomena (Pillus and Rine 1989). The discovery of a new species of nucleic acid in the cytoplasm also suggests the possibility of associated cytoplasmically inherited traits.

A simple test, called cytoduction, has been developed to identify a cytoplasmically inherited trait in yeast. This test involves donating just the cytoplasm from a donor to a recipient strain, and then asking if the genetic trait associated with the donor has been transmitted to the recipient (Wright and Lederberg 1957). Mutations in various Mendelian genes, e.g., *KAR1*, enhance the frequency of cytoduction by inhibiting the fusion of the haploid nuclei in zygotes (Conde and Fink 1976). When a haploid containing a mutation in *KAR1* is mated, a transient heterokaryon is formed. As schematically presented in Figure 1, cells arising from the zygote of this heterokaryon contain a mixture of cytoplasms and one of the nuclei. To obtain the cytoductant with the desired haploid nucleus, one can select or screen for a cytoplasmic marker from the donor strain together with a recessive nuclear marker from the recipient. This test is not foolproof, however. Indeed, the indigenous yeast 2-μm DNA plasmid was originally believed to be cytoplasmic because it segregated 4:0 and could be transferred by cytoduction (Livingston and Klein 1977). However, it later became clear that although the 2-μm plasmid was replicated and by and large maintained in the nucleus, it was small enough to occasionally leak through the nuclear membrane into the cytoplasm, a fact that was also true of small chromosomes (Sigurdson et al. 1981).

Winge and Lausten (1940) first suggested the occurrence of non-Mendelian inheritance in yeast by demonstrating a lower spore germination of diploids formed by homothallic diploidization, in contrast to diploids formed by cell fusion between two presumably identical haploid cells. Furthermore, they suggested that the lower spore germination was due to a decreased number of "chondrisomes," an early term for mitochondria. They suggested that the number of chondrisomes is related to the number of cellular divisions, which would be less if homothallic diploids were formed before cytokinesis. The findings of Winge and Lausten (1940) are not easily explained by our current knowledge

TABLE 1. Genome of a diploid *Saccharomyces cerevisiae* cell

Inheritance	Mendelian	Non-Mendelian						
Nucleic acid		Double-stranded DNA		Double-stranded RNA				
Location	Nucleus		Cytoplasm					
Genetic determinant	Chromosomes	2-μm plasmid	Mitochondrial DNA		RNA virus			
				L-A	M	L-BC	T	W
Relative amount (%)	85	5	10	80	10	9	0.5	0.5
Number of copies	2 sets of 16	60–100	~50 (8–130)	103	170	150	10	10
Size (kb)	13,500 (200–2200)	6.318	70–76	4.576	1.8	4.6	2.7	2.25
Deficiencies in mutants	All kinds	None	Cyto. $a \cdot a_3$, b	Killer toxin			None	
Wild type	$YFG1^+$	cir^+	ρ^+	$KIL\text{-}k_1$				
Mutant or variant	*yfg1-1*	cir^o	ρ^-	KIL-o				

A wild-type chromosomal gene is designated as $YFG1^+$ (Your Favorite Gene) and the mutation as *yfg1-1*. Although cir^+ and cir^o strains are phenotypically the same in most strains, the presence and absence of the 2-μm plasmid are inherited in a non-Mendelian manner. T and W RNA viruses, also designated 20S and 23S dsRNAs, encode RNA polymerases (Wesolowski and Wickner 1984). The table does not include the Ty1–Ty5 retroviruses, which are generally inherited as integrated Mendelian elements, nor does it include extrachromosomal circular rDNAs, which arise from chromosomal rDNAs.

Modified from Sherman (2002).

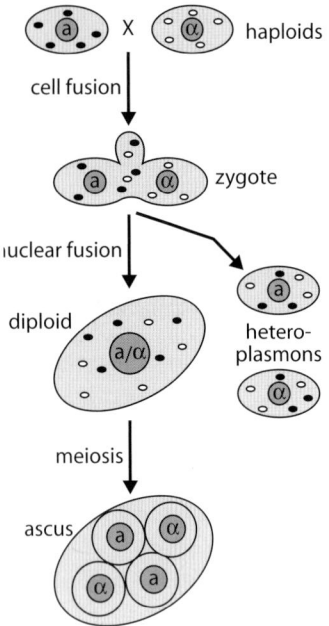

FIGURE 1. The sexual cycle in *Saccharomyces cervisiae*. The two alleles of the mating-type locus are **a** and α. The large circle represents cells, and the small circles, which encircle the mating-type genotypes, represent nuclei. The black and white dots represent mitochondria from one or the other of the two haploid parents. The arrow going from the zygote to the heteroplasmons is thinner as an indication of the low frequency of this event in wild type. (Redrawn, with permission, from Conde and Fink 1976.)

of mitochondrial inheritance and cellular biology of yeast. Although the work of Winge and Lausten (1940) was cited in numerous monographs on cytoplasmic inheritance, including the influential book by Ephrussi (1953), their findings did not have a major impact on the development of the field. In contrast, other works by Winge during this same time period laid the foundation of yeast Mendelian genetics (Westergaard 1963; Mortimer 1993).

The first rigorous demonstration of a non-Mendelian determinant in yeast came in a series of remarkable papers by Ephrussi and colleagues in 1949, describing the isolation and characterization of the "petite" (or ρ^-) mutant (Ephrussi et al. 1949a,b,c; Slonimski 1949a,b; Slonimski and Ephrussi 1949; Tavlitzki 1949). These initial studies demonstrated that the ρ^- trait was inherited by a non-Mendelian determinant, that normal strains could be converted in mass to ρ^- by acriflavine treatment, that the ρ^- mutation was irreversible, and that ρ^- mutants lacked cytochromes $a \cdot a_3$ and b and were deficient in respiration.

Although Ephrussi's initial work on ρ^- came at the time when *Escherichia coli* was being used to uncover the major principles of molecular biology, the importance of his work on yeast was fully recognized at the onset. Nevertheless, Ephrussi was fond of telling the following sarcastic story during some of his seminar presentations: "I wish to acknowledge advice of my friend and colleague, [name of a prominent French Nobel Laureate]; after showing him the 'petite colonies' induced by acriflavine, he recommended that I throw out the plates and forget the whole thing." Luckily, he did not heed the advice.

During the last 50 years since Ephrussi's initial studies were reported, there have been more than 3000 publications on the function and genetics of yeast mitochondria, including at least one review each year. Obviously, only a few selective papers can be considered in this chapter. Furthermore, we have emphasized only the early contributions in this area. The review by Dujon (1981) is helpful to those interested in more details of some aspects of the early development of the field.

THE LESSON OF "PETITES"

A critical and systematic genetic analysis of ρ^- mutants was reported in one of the initial papers by **Ephrussi et al. (1949b)**, establishing the non-Mendelian nature of the determinant by the results of repeated backcrosses. In this landmark paper, Ephrussi et al. (1949b) relied on the life cycle of *Saccharomyces* and genetic approaches previously reported by Winge and colleagues (Winge 1935; Winge and Lausten 1937) and Lindegren and Lindegren (1943, 1947). The strains used in the study were derived from "Yeast Foam," an American diploid strain of *Saccharomyces cerevisiae*. Two spores from a single ascus were isolated and these gave rise to two haploid strains designated 276/3 D (*MAT+ ADE$^+$ THI$^+$ ρ^+*) (originally denoted + *A T*) and 266/3 B (*MAT– ADE$^+$ thi$^-$ ρ^+*) (originally denoted – *A t*), where *thi$^-$* caused a requirement for thiamine. A mutant 266/3 Br (*MAT– ade$^-$ thi$^-$ ρ^+*) (originally denoted – *a t*) was obtained from 276/3 B on the basis of its red color caused by the *ade$^-$* mutation, which also caused a requirement for adenine. ρ^- mutants were derived from 276/3 Br, either spontaneously (1-sp), or after treatment with acriflavine (1-A), and these were crossed to 276/3 D. Incidentally, this study revealed for the first time that the red color associated with the *ade$^-$* mutation is diminished in ρ^- strains. In addition, they also observed that ρ^- diploids do not sporulate.

The *MAT+ ADE$^+$ THI$^+$ ρ^+* × *MAT– ade$^-$ thi$^-$ ρ^-* crosses gave rise to ρ^+ diploids, indicating that ρ^+ was "dominant." Furthermore, the majority of the meiotic progeny were also ρ^+, establishing that the ρ^- trait was not caused by mutation of a single nuclear gene. In contrast, the nuclear markers exhibited primarily 2:2 segregation. 2:2 segregation was observed in all asci for the mating-type markers, in 129 of the 134 asci for the *ADE$^+$/ade$^-$* markers, and in 120 of the 134 asci for the *THI$^+$/thi$^-$* markers. The lack of 2:2 segregation of these nuclear markers was attributed to imperfection in testing the phenotypes. (Although they do not consider "gene conversion" as a cause of deviation from 2:2 segregation of nuclear genes, they did cite H. Winkler's 1930 treatise *Die Konversion der Gene*.)

The main thrust of the paper was to exclude the possibility that ρ^- was due to at least a limited number multiple recessive alleles of nuclear genes. If ρ^- was indeed due to multiple recessive mutations, the mutant character would reappear only in a fraction of the asci, the proportion of which would depend on the number of genes involved. For example, if ρ^- were due to the simultaneous presence of four recessive genes, the mutant phenotype would reappear only in approximately 12% of the asci. Thus, the lack of recovery of ρ^- in the meiotic progeny could be simply due to the limited number of asci that were analyzed. This question was addressed by backcrossing, i.e., by repeatedly crossing the meiotic segregants formed from hybrids to the initial ρ^- parent. In this way, the genes of the ρ^- parent accumulate in the hybrids of each successive backcross. In their study, the number of asci analyzed in the first cross and in five successive backcrosses were, respectively, 31, 23, 18, 20, 57, and 6, making a total of 620 meiotic segregants. Among these, only five ρ^- mutants were observed, a frequency that was not greater than the frequency of spontaneous mutants. Although their results did not rigorously exclude that ρ^- mutants arose by mutation of nuclear genes, one must assume that at least a dozen unlinked genes must be simultaneously mutated. Their major conclusion was that ρ^- mutants most likely differed by the absence of cytoplasmic units that are endowed with genetic continuity and that are required for the synthesis of certain respiratory enzymes. (There was an obvious wording error in the summary of the paper by Ephrussi et al. [1949b]; the incorrect statement "Il paraît peu probable qu'il s'agisse d'un cas d'hérédité cytoplasmique," stating that there probably was *not* cytoplasmic inheritance, haunted the Ephrussi laboratory for several years.)

As briefly described above, Wright and Lederberg (1957) provided an independent demonstration of the cytoplasmic nature of the ρ determinant, relying on the existence of transient heterokaryons during zygote formation in yeast, as first described by Fowell (1951). After haploid cells mate, there is a short period after cell fusion (but before nuclear fusion) during which haploid buds can appear with nuclei from either one of the parents but with cytoplasm from both. Using double and triple unlinked nuclear markers, and various $MAT\mathbf{a}\ \rho^+ \times MAT\alpha\ \rho^-$ crosses, including those with suppressive ρ^- strains (ρ^- strains that produce significant ρ^- zygotes when crossed to a ρ^+ strain), Wright and Lederberg (1957) observed six cases of $MAT\mathbf{a}\ \rho^-$ and five cases of $MAT\alpha\ \rho^+$ haploid strains from transient heterokaryons, but only the expected uniformity from the $\rho^+ \times \rho^+$ or $\rho^- \times \rho^-$ crosses.

The high rate of induction of ρ^- mutants by numerous agents and by stressful conditions suggested in early studies that the ρ determinant did not have the properties of nuclear genes. Single-cell pedigree analysis revealed that one or two cell divisions in a medium containing 10^{-6} M euflavine, the active ingredient of acriflavine (Ephrussi and Hottinguer 1950), or at the above optimal growth temperature of 40.5°C (Sherman 1959) produced nearly 100% ρ^- buds. Subsequently, an astounding number of diverse chemical agents and conditions were reported by numerous workers to induce ρ^- mutants, including typical mutagens, depletion of essential components, many toxic agents, growth at high and low temperatures, or almost any condition that inhibits growth or kills cells. Notably, ethidium bromide can completely convert a population of normal cells to ρ^- with no lethality under growing or nongrowing conditions (Slonimski et al. 1968). Subsequently, ethidium bromide was shown to induce predominately ρ° mutants, a subclass of ρ^- mutants that completely lack mitochondrial DNA.

The correspondence of the ρ determinant to DNA was suggested by finding mitochondrial DNA, first in vertebrates (Nass and Nass 1963) and later in yeast (Schatz et al. 1964) and numerous other eukaryotic species. The observation that mitochondrial DNA was grossly altered (Mounolou et al. 1966) and was found at reduced amounts (Tewari et al. 1966) or was absent (Corneo et al 1966; Moustacchi and Williamson 1966) in ρ^- mutants firmly established that mitochondrial DNA is the hereditary material corresponding to the ρ determinant.

Numerous studies from many laboratories revealed that mitochondria DNA in different ρ^- mutants can be altered in various ways, from those having complete deficiencies (ρ° mutants) to those containing up to 50% of the normal genome (see Dujon 1981). Typically, ρ^- mutants have extensive deletions with amplification of the retained segment (Bernardi et al. 1970; Faye et al. 1973).

RECOMBINATION OF MITOCHONDRIAL MARKERS

In 1967, Wilkie and co-workers (Wilkie et al. 1967) reported the isolation of mutants capable of growth on glycerol medium in the presence of chloroamphenicol, erythromycin, or tetracyclin, which inhibit mitochondrial protein synthesis or function. Soon after, **Thomas and Wilkie (1968)** and Linnane et al. (1968) demonstrated that the erythromycin resistance was inherited in a non-Mendelian manner and was eliminated by conversion of the strain to ρ^-. The discovery of an antibiotic-resistant marker residing on the ρ determinant ushered in the era of formal mitochondrial genetics, because, in contrast to ρ⁻ mutations, these resistant markers were phenotypically dissimilar and behaved as point mutations, thus allowing studies of recombination.

Recombination of mitochondrial markers was first reported in a brief paper by Thomas and Wilkie (1968) who used spiramycin (spi^R), erythromycin (ery^R), and paromomycin (par^R) markers. In the second landmark paper, Thomas and Wilkie (1968) scored the diploid progeny from two- and three-factor crosses and considered frequencies of parental and recombinant types. The results indicated that the spi^R and ery^R markers were linked to each other but not apparently to the par^R marker. Specifically, cross 2, spi^S ery^S par^R × spi^R ery^R par^S, yielded the following types out of 119 clones tested,

spi^S ery^S par^R	49	spi^R ery^S par^S	3
spi^R ery^R par^S	25	spi^R ery^S par^R	4
spi^S ery^S par^S	14	spi^S ery^R par^S	1
spi^R ery^R par^R	23		

or the following percent recombination between the following pairs of markers: *spi ery*, 7%; *par spi*, 34%; and *par ery*, 35%. These early results are consistent with the current knowledge of the physical location of the mutations; both spi^R and ery^R have alterations in the mitochondrially encoded 21S rRNA, whereas par^R has an alteration in the 15S rRNA, and these small and large rRNAs are encoded by regions nearly opposite on the 74–85-kb circular mitochondrial DNA molecules.

Subsequently, the mechanisms and rules for mitochondrial inheritance were extensively investigated by numerous workers, who also used multifactorial crosses of antibiotic resistance mutants. Mitochondrial recombination was best explained in terms of population genetics and a multicopy genetic system. An elaborate model has been presented by Dujon et al. (1974) and others (see Dujon 1981), who have considered the details of the frequencies of recombination and the distortion of recombination by the presence or absence of an intron in the large rRNA gene.

A critical development in the understanding of the mitochondrial genome was the isolation of mitochondrial *mit*⁻ mutations having defects in the respiratory chain and in oxidative phosphorylation, presumably as a result of single-site alterations. Such *mit*⁻ mutants are expected to have the following properties: the inability to utilize glycerol, a *n*onfermentable carbon *s*ource, defining the Nfs⁻ phenotype; retention of the ability to carry out mitochondrial protein synthesis, in contrast to ρ⁻ mutants, which lack or have grossly altered mitochondrial DNA, and which are deficient in mitochondrial protein synthesis because of the lack of mitochondrially encoded rRNA or tRNA genes; the lack of ability to be complemented by a ρ°; and 4:0 and 0:4 segregation for Nfs⁺/Nfs⁻ of the meiotic progeny from diploids formed from *mit*⁻ mutants crossed to normal ρ⁺ strains.

RESPIRATORY CHAIN AND OXIDATIVE PHOSPHORYLATION MUTANTS

Flury et al. (1974) first reported such *mit*⁻ mutants, which were induced by 2-aminopurine, and which had the desired *mit*⁻ properties, but were highly unstable, converting to Nfs⁺ or ρ⁻. The five examined *mit*⁻ mutants had multiple mitochondrial deficiencies. Concomitantly, **Tzagoloff et al. (1975a,b,c)** systematically isolated *mit*⁻ mutants with single and multiple deficiencies. The success in isolating mit^- mutants after failed attempts for many years in several other laboratories relied on a number of innovative procedures. In the third landmark paper, Tzagoloff et al. (1975c) mutagenized strains with $MnCl_2$, which was previously shown by Putrament et al. (1973, 1975) to induce mitochondrial antibiotic-resistant mutants, under conditions where the induction of ρ^- mutants was minimized. The colonies arising from the mutagenized cells were first tested for the Nfs$^-$ phenotype described above. The second screen was to determine which Nfs$^-$ mutants retained the ability to carry out mitochondrial protein synthesis in vivo (Tzagoloff et al. 1975b). These mit^- candidates were further tested for the lack of ability to be complemented by a ρ° tester strain, a property highly indicative of defects in mitochondrial DNA. In addition, representative mit^- mutants were crossed to a normal ρ^+ strain, the corresponding diploid zygotes were sporulated immediately after mating, and the meiotic progeny were tested. Most of the tetrads exhibited the expected 4:0 and 0:4 segregation for Nfs$^+$/Nfs$^-$, with occasional ones showing 3:1, 1:3, and 2:2 segregation.

Of the 3000 Nfs$^-$ strains examined by Tzagoloff et al. (1975c), 208 had the capacity to carry out mitochondrial protein synthesis, and these could be tentatively assigned as either mit^- mutants (~150) or pet^- mutants (~50), on the basis of the inability or ability, respectively, to be complemented by a ρ° strain. The 150 mit^- mutants could be further assigned as those deficient in cytochrome oxidase, 58 mutants; those deficient in coenzyme QH_2-cytochrome c reductase, 5 mutants; those deficient in rutamycin-sensitive ATPase, 1 mutant; and those with two or three deficiencies, 86 mutants. Furthermore, in a critical paper, Slonimski and Tzagoloff (1976) demonstrated by genetic recombination of mit^- mutants that various regions are dispersed on the mitochondrial genome and, in some instances, regions or clusters of closely linked mutations involved in the same respiratory function, such as cytochrome oxidase, are separated by other regions that code for entirely different functions, such as ribosomal RNA. These and subsequent studies provided a resource for mutants that were invaluable not only for defining the mitochondrial genome, but also for investigating the role of nuclear genes in mitochondrial biogenesis.

These early studies on mitochondrial recombination stimulated subsequent investigations leading to the understanding of homing endonucleases. Homing endonucleases are enzymes that cleave DNA with very high specificity within an extended recognition site and initiate a double-strand-break repair that may lead to the insertion of the sequence coding for the homing endonuclease into an allele that lacks it. For example, a group-I self-splicing intron, encoding a homing endonuclease (I-SceI) in one allele (ω^+) of the mitochondrial 21S rRNA gene, is nearly quantitatively inserted in crosses into a 21S rRNA allele lacking that intron (ω^-) (see Zinn and Butow 1985). These early studies on mitochondrial recombination eventually led to the discovery of maturases. In 1980, mutational analyses suggested that excision of a group-I intron (cob-I3) from the mitochondrial cytochrome b gene required expression of an intron-encoded protein, a maturase (Lazowska et al. 1980), which was later shown to probably act as an allosteric cofactor that stabilizes RNA structure in a reactive conformation for splicing, and which induces or accelerates the formation of secondary and tertiary structures of the self-splicing intron.

The organization of the mitochondrial genome was more or less completely defined about 10 years ago by using a combination of genetic and physical methods, including recombinational mapping with single-site mutants, deletion mapping of single-site mutations with sets of discriminating ρ⁻, restriction mapping, hybridization of tRNA and rRNA to DNA fragments, and DNA sequencing. Mitochondrial DNA maps can be found in Grivell (1995) and numerous genetic textbooks.

Recently, Williamson (2002) has presented a provocative historical review on the nature of mitochondrial DNA of *S. cerevisiae*. On the basis of early experiments with pulsed-field gel electrophoresis (PFGE) (Maleszka et al. 1991), Williamson (2002) proposed that nearly all mitochondrial DNA in vivo is composed of polydisperse linear tandem arrays of the genome, which range in size from approximately 75 kb to 150 kb and that only a small portion is represented by circular forms containing the full genome.

It is indeed a rewarding intellectual experience to review these early papers on mitochondrial genetics in light of the known physical structure. For further insights into the developments in the field, see Schatz (1993), Tzagoloff (1993), and Wilkie (1993), who reminisce about their achievements during the dark ages.

THE NEED FOR NUCLEAR-ENCODED GENES

Early in the 1960s, it was generally believed that mitochondria arose by an increase in mass through the addition of lipids and proteins to preexisting organelles, suggesting that mitochondria serve as a template for their own propagation (Luck 1965). Furthermore, the morphologically intact and partially functional mitochondria in ρ⁻ mutants indicated that the mitochondrial template was presumably determined by nuclear genes. The importance of nuclear genes for mitochondrial function was first reported by Chen et al. (1950), who described a "segregational petite" or *pet* mutation that caused a mitochondrial defect and that segregated in a Mendelian manner. Subsequently, Sherman (1963) and Sherman and Slonimski (1964) reported a more complete analysis of *pet* and related mutants that had a variety of mitochondrial defects, including deficiencies in cytochromes $a \cdot a_3$, b, and c. In the succeeding years, up to the present time, numerous laboratories have systematically identified and characterized nuclear genes affecting a wide range of mitochondrial functions. Tzagoloff and Dieckmann (1990) reviewed the field in 1990 and listed approximately 215 nuclear genes required for functional mitochondria. The availability of the genomic sequence allowed more complete estimates; the number of nuclear genes known to encode mitochondrial proteins is estimated to be 340, where 127 of these 340 are involved in biogenesis and assembly of mitochondria (Grivell et al. 1999). Some of these nuclear genes are required for replication or maintenance of mitochondrial DNA (Berger and Yaffe 2000; Hobbs et al. 2001), and some are required for activating translation of mRNAs encoded by mitochondrial DNA (Fox 1996).

Although mitochondrial DNA encodes components of the respiratory metabolism and translational machinery, it does not encode components essential for maintenance of mitochondria; as mentioned above, ρ° mutants completely lacking mitochondrial DNA are nevertheless viable and have mitochondria. Because mitochondria supply other functions essential for viability, in addition to respiration and oxidative energy production, there are nuclear genes that are essential for mitochondrial inheritance and therefore cell proliferation (Yaffe 1999; Boldogh et al. 2001). Today, numerous laboratories are focused on the role of the numerous nuclear genes that determine the many functions of mitochondria. A complete and detailed understanding of mitochondria is yet to come.

OTHER NON-MENDELIAN SYSTEMS

Another non-Mendelian phenotype, "killer," enables yeast to secrete toxins that kill strains that lack the trait. Like mitochondrial markers, the inheritance of killer is controlled by cytoplasmic nucleic acids, which constitute the family of L-A double-stranded RNA viruses (Table 1). The discovery of killer viruses and their interactions with Mendelian genes represents an important contribution to our understanding of cytoplasmic inheritance (for review, see Bevan et al. 1973; Wickner 1996). Other viral replicons found in the cytoplasm of yeast are the double-stranded RNA L-BC virus, unencapsidated 20S and 23S single-stranded RNAs (Table 1), and the single-stranded RNA Ty retrotransposon particles. None of the yeast viruses are normally able to leave the host cell to infect a new strain. Rather, their transmission depends on mating and cytoplasmic mixing.

ψ, A CYTOPLASMIC SUPPRESSOR OF SUPER-SUPPRESSION

In addition to the success of elucidating the basic principles of mitochondrial inheritance with yeast, the determination of the major properties of prions is also one of the crowning achievements that was accomplished with this tractable microorganism. As described below, the use of yeast for investigating prions began with a fortuitous finding by Cox (1965), who discovered a non-Mendelian determinant, [PSI^+], that affected the level of expression of nonsense suppressors. In addition, another non-Mendelian element, [$URE3$], was uncovered by Lacroute (1971) during his studies of ureidosuccinate metabolism. Attempts to identify the nature of [PSI^+] by genetic, biochemical, and molecular approaches were futile, and the essence of these results, although well known to the yeast community, laid more-or-less dormant until they helped Wickner (1994) to propose that the cytoplasmically inherited elements were prions.

In the fourth landmark paper, Brian Cox (1965) first describes the cytoplasmic element ψ (now called [PSI^+]). At the time he did this work, Mendelian mutations, originally called super-suppressors, that could reverse (i.e., suppress) the phenotypes of some but not all mutations at many different unlinked loci had just been described (Hawthorne and Mortimer 1963). It was also known, as mentioned above, that mutations in the *ad2* Mendelian gene (now called *ade2*) caused not only adenine auxotrophy, but also the accumulation of a red pigment. While working with a red adenine-requiring diploid yeast strain heteroallelic for *ad2,1* and *ad2,c* (now called *ade2-1* and *ade2-c*, respectively), Cox obtained a white revertant that no longer required adenine. In the first part of the paper, he shows that this revertant was due to the presence of an unlinked dominant Mendelian suppressor mutation that he called SUQ_5, which is now also designated *SUP16*.

We now know that the mutations that can be suppressed by super-suppressors contain premature nonsense codons. Translation termination factors normally promote polypeptide chain termination at these nonsense codons, leading to the synthesis of truncated nonfunctional polypeptides. Super-suppressors generally result from mutations in the anticodons of tRNAs that enable them to compete with termination factors and read nonsense codons as sense, thereby allowing synthesis of the complete functional protein (for review, see Hinnebusch and Liebman 1991). For example, SUQ_5 contains a mutation in a serine tRNA that results in the insertion of serine in response to the ochre (UAA) nonsense codon present in the *ad2,1* mutation.

When reading the initial sections of the paper, keep in mind that although SUQ_5 suppresses the adenine requirement and red color associated with the *ad2,1* mutation, it does not suppress the other *ad2* alleles used in their experiments. This is probably because the other *ad2* alleles do not contain UAA mutations, but rather carry a missense or other nonsuppressible mutation. Note, in the initial diploid, that suppression of the *ad2,1* allele is sufficient to cause reversion of the adenine requirement and red color because the nonsuppressible *ad2,c* mutant is recessive. Because of the allele-specific effect of SUQ_5, it is necessary to distinguish the *ad2* alleles in the segregants to score for the presence of the suppressor. This was easily done using allelic complementation (i.e., when the phenotype of a heteroallelic strain is closer to that of the wild type compared to the phenotypes of homozygous mutant strains) because *ad2,1* failed, while *ad2,c* succeeded, in complementing tester alleles of *ad2*. The presence or absence of SUQ_5 can be directly scored in *ad2,1* segregants as white (suppressed) or red (not suppressed), respectively. In contrast, since *ad2,c* is not suppressible, all *ad2,c* segregants are red whether or not SUQ_5 is present. To detect SUQ_5 in *ad2,c* segregants, they were crossed to a red *ad2,1* tester strain. If SUQ_5 were present in the *ad2,c* segregant, suppression of the *ad2,1* allele in the resulting diploid caused it to be white; on the other hand, if SUQ_5 were not present, the diploid was red.

The landmark contribution of this paper is the unraveling of the nature of some sectored meiotic segregants of diploids containing SUQ_5 and *ad2,1*. Surprisingly, even though these spore colonies each arose from a single cell, they were not pure red or pure white, but contained both white and red sectors. It was easily established that the white portions of impure strains had the *ad2,1* mutation and the SUQ_5 suppressor, because red adenine-requiring segregants were recovered in the meiotic progeny from crosses of the white sectors to wild-type strains. When cells from the red portions of impure cultures were crossed to each other, the diploids as well as their progeny and backcrosses of progeny were all red. This suggested that the same genetic event caused the appearance of the red sectors in the different impure segregants.

To explain the appearance of the red sectors in terms of Mendelian mutations, Cox suggests either of the following: (1) the appearance of a new Mendelian mutation in an unlinked gene that hid the suppressor (white) phenotype of SUQ_5; or (2) a reversion of the SUQ_5 suppressor, i.e., loss of the SUQ_5 mutation. He disproved the first hypothesis by finding that the red adenine-requiring phenotype failed to appear in any of the progeny from crosses of the red sectors to suppressed white *ad2,1* strains. He also disproved the second hypothesis because the SUQ_5 suppressor segregated normally in crosses to certain unsuppressed *ad2,1* strains. Surprisingly, however, when other *ad2,1* strains were crossed to the same red sectors, the SUQ_5 suppressor did not emerge in the progeny. It thus appeared that no permanent additional mutation had occurred in the red sectors. Thus, after disposing of both hypotheses involving Mendelian mutations, Cox proposed that in certain SUQ_5 strains, the phenotype of the suppressor remained hidden by a cytoplasmic mutation of what he designated ψ^+ to ψ^-. Like the ρ^- mutation, which had previously been determined to be cytoplasmic (see above), the ψ^- mutation showed a 0:4 non-Mendelian segregation pattern. In addition, it was later shown that ψ^+ was cytoducible (Cox et al. 1988). This first description of a new non-Mendelian element led to years of experiments to uncover its molecular basis (for review, see Cox 1993). Two general hypotheses were considered. The first hypothesis was that a self-replicating cytoplasmic particle corresponded to ψ^+. To test this, much effort was directed toward finding a cytoplasmic nucleic acid responsible for ψ^+, but none were found. The second hypothesis was that ψ^+ was caused by a feedback mechanism. Maintenance of ψ^+ might require mistranslation caused by ψ^+ leading to the indefinite maintenance of ψ^+ by a positive feedback loop.

YEAST PRIONS

Just as scientists became comfortable with the idea that non-Mendelian genetic traits were controlled by cytoplasmic nucleic acid, **Wickner (1994)** in the fifth landmark paper proposed a new type of self-replicating cytoplasmic element, one that is not composed of nucleic acid. This new proposal explained the inheritance of ψ^+ (by then called [*PSI*$^+$]), as well as another cytoplasmic genetic element, [*URE3*], in terms of a self-replicating protein conformation.

When Wickner (1994) proposed his hypothesis, 30 years of studies on the nature of [*PSI*$^+$] were available (for review, see Cox et al. 1988; Cox 1993). Only the most relevant facts are recounted here. Of critical importance was the finding that a mutation of the chromosomal *SUP35* gene caused the loss of [*PSI*$^+$] (Doel et al. 1994; Ter-Avanesyan et al. 1994). In addition, mutations that partially inactivated *SUP35* caused the same phenotype as [*PSI*$^+$], i.e., the misreading of stop codons (Inge-Vechtomov and Andrianova 1970). A variety of treatments "cured" cells of [*PSI*$^+$], the most efficient of which was growth in the presence of low levels of guanidine hydrochloride (Tuite et al. 1981). However, transient overexpression of the *SUP35* gene could cause the de novo appearance of [*PSI*$^+$] after curing (Chernoff et al. 1993).

Another dominant non-Mendelian element, [*URE3*], was also known (Aigle and Lacroute 1975) and was the basis for Wickner's work. The presence of [*URE3*] allowed yeast to take up ureidosuccinate in the presence of ammonia, the same phenotype associated with Mendelian mutations in *URE2* (Lacroute 1971). Like [*PSI*$^+$], [*URE3*] did not segregate 2:2 and was cytoducible. Using segregation analysis and cytoduction experiments, Aigle and Lacroute (1975) had shown that *URE2* was required for the propagation of [*URE3*], even though *ure2* mutations resulted in the opposite phenotype as loss of [*URE3*]. Wickner found this result surprising since by then it was clear that mutations in genes required for the replication of other cytoplasmic elements, e.g., mitochondria, caused the same phenotype, not the opposite, as the loss of these cytoplasmic elements. Nonetheless, when Wickner (1994) disrupted the *URE2* gene in a [*URE3*] strain, he always caused the loss of the [*URE3*] cytoplasmic element.

He also showed that like [*PSI*$^+$], [*URE3*] was efficiently cured by guanidine hydrochloride, whereas guanidine hydrochloride had no effect on a *ure2* deletion. The loss of the [*URE3*] element was not equivalent to ethidium bromide curing of ρ^+, which causes the irrevocable loss of mitochondrial DNA, because [*URE3*] mutants could again be selected in cured strains. This led Wickner to think that the difference between the normal [*ure3*] and abnormal [*URE3*] states was not due to the absence or presence of a cytoplasmic replicon, but rather to different states of something in the cytoplasm.

Different states of what? Since *URE2* was required for the propagation of [*URE3*] and since *ure2* mutants had the same phenotype as [*URE3*], the Ure2 protein seemed to be the logical choice. In further support of this hypothesis, Wickner showed that a high-copy

URE2 plasmid, which caused tenfold overproduction of Ure2, resulted in about a 100-fold increase in the frequency of the de novo appearance of [*URE3*]. Furthermore, the newly induced [*URE3*] elements could be maintained even after the high-copy *URE2* plasmid was lost. Since overexpression of *URE2* encoded on a single-copy plasmid had the same effect, but the overexpression of a mutant *URE2* gene with premature stop codons did not, it appeared that excess protein rather than excess DNA or mRNA was inducing [*URE3*]. This is a critical finding since it suggests that the high level of Ure2 increased the frequency of the chance conversion of Ure2 into [*URE3*].

Wickner also noted that the story concerning [*PSI*+] and Sup35 mirrored the following facts about [*URE3*]: (1) [*URE3*] showed cytoplasmic inheritance, (2) overexpression of Ure2 induced [*URE3*], (3) propagation of [*URE3*] depended on *URE2*, (4) the phenotypes caused by [*URE3*] and mutations in *URE2* were identical, and (5) [*URE3*] could reappear after curing. These facts could not easily be reconciled with inheritance encoded by cytoplasmic nucleic acid, but they were consistent with the concept of protein-based inheritance. The idea of an infectious protein, called a prion, had been previously proposed to explain the transmission of spongiform encephalopathies in mammals (Griffith 1967; Prusiner 1982). The altered, prion, form of a normal cellular protein was proposed to cause the disease and to be infectious because it induced the normal protein to change into the prion form (Gajdusek 1991; Prusiner 1991; Weissmann 1991; Bueler et al. 1993; Prusiner et al. 1993).

Wickner (1994) suggests that prions are not limited to the single mammalian case and that [*URE3*] and [*PSI*+] are, respectively, altered self-replicating forms of Ure2 and Sup35. Considerable evidence now supports this hypothesis (for review, see Liebman and Derkatch 1999; Serio and Lindquist 1999; Masison et al. 2000; Chernoff 2001; Wickner et al. 2001). Notably, both Ure2 and Sup35 have been shown to aggregate specifically in cells with the respective prion phenotype (Masison and Wickner 1995; Patino et al. 1996; Paushkin et al. 1996; Edskes et al. 1999; Speransky et al. 2001). In addition, both proteins can form amyloidlike fibers in vitro, and these fibers, or cell extracts containing prion aggregates, can seed the rapid formation of the corresponding in vitro aggregates (Glover et al. 1997; King et al. 1997; Paushkin et al. 1997; Taylor et al. 1999; Scheibel et al. 2001). Furthermore, the maintenance of both the [*PSI*+] and [*URE3*] prions are influenced by the levels of chaperone proteins whose job is to disaggregate protein aggregates (Chernoff et al. 1995, 1999; Newnam et al. 1999; Moriyama et al. 2000; Ferreira et al. 2001; Jung and Masison 2001; Wegrzyn et al. 2001). In addition, several features characteristic of the putative mammalian prions have been described for [*URE3*] and/or [*PSI*+]: mutations in the prion gene that enhance or inhibit the appearance or propagation of the prion (Aigle and Lacroute 1975; Doel et al. 1994; Wickner 1994; Masison and Wickner 1995; DePace et al. 1998; Kochneva-Pervukhova et al. 1998; Liu and Lindquist 1999; Maddelein and Wickner 1999; Fernandez-Bellot et al. 2000; Borchsenius et al. 2001); the existence of different heritable "strains" of the same prion (Derkatch et al. 1996, 1999; Zhou et al. 1999; Chien and Weissman 2001; King 2001; Kochneva-Pervukhova et al. 2001; Schlumpberger et al. 2001; Uptain et al. 2001); and a barrier that inhibits prion transmission across species lines (Chernoff et al. 2000; Kushnirov et al. 2000; Santoso et al. 2000). There is also evidence for additional prions in yeast (Santoso et al. 2000; Sondheimer and Lindquist 2000; Derkatch et al. 2001; Sondheimer et al. 2001) and other fungi (Coustou et al. 1997).

Although the spontaneous appearances of the [*PSI*⁺] and [*URE3*] prions are enhanced by overproduction of Sup35 and Ure2, respectively, other factors influence this process. [*PSI*⁺] appearance is inhibited by overproduction of a Sup35-binding protein, Sup45 (Derkatch et al. 1998). [*URE3*] appearance requires the presence of Mks1 (Edskes and Wickner 2000). Intriguingly, the presence of heterologous prion aggregates enhances the appearance of other prions (Derkatch et al. 1997, 2000, 2001; Osherovich and Weissman 2001).

Recent important experiments have shown that in-vitro-made "strain"-specific Sup35 fibers have distinct structural properties (Krishnan and Lindquist 2005; Tanaka et al. 2005). Furthermore, when these fibers were inserted into [*psi*⁻] cells they infected the cells with the corresponding specific [*PSI*⁺] prion strain (King and Diaz-Avalos 2004; Tanaka et al. 2004). These experiments prove that [*PSI*⁺] is caused by an infectious form of Sup35 and provides the first definitive proof for the "protein only" prion hypothesis in general.

Who would have thought that Wickner's broad interests in all areas of science, including the rare mammalian diseases believed to be caused by prions, would have enabled him to solve a 30-year-old puzzle in yeast genetics? This finding opened up an entirely new field by suggesting that prions are a general phenomenon and represent a new method by which traits can be inherited and by which the function of gene products can be regulated.

ACKNOWLEDGMENTS

We thank Drs. Reed Wickner, Brian Cox, Thomas Fox, Irina Derkatch, and Alexander Tzagoloff for helpful comments on the manuscript and Patrick Linder for translating numerous articles and helpful suggestions. This work was partially supported by National Institutes of Health grants GM56350 (to S.W.L.) and GM12702 (to F.S.).

STUDY QUESTIONS

1. How would you experimentally test Winge and Lausten's (1940) hypothesis that lower spore germination was due to a diminished number of "chondrisomes," or mitochondria? Would you determine the number of mitochondria or the amount of mitochondrial DNA? (For some current methods, see Jensen et al. 2000.)

2. A $\rho^+ \times \rho^-$ cross can yield both ρ^+ and ρ^- diploids. After sporulation, a ρ^+ diploid gives rise to predominately ρ^+ meiotic segregants. Since ρ^- diploids do not sporulate, predominately ρ^+ meiotic segregants are expected from all diploids of a $\rho^+ \times \rho^-$ cross. However, F. Sherman (1964) reported that more than 90% of the meiotic progeny from strain D-285 were ρ^-. How was this explained?

3. The frequency of ρ^- zygotes from $\rho^+ \times \rho^-$ crosses defines the properties of the ρ^- parental strain, which are designated as follows: neutral, or N ρ^-, no significant ρ^- zygotes; suppressive or S ρ^-, a significant number of ρ^- zygotes; and hypersuppressive or HS ρ^-, a predominate number of ρ^- zygotes. Although $\rho^°$ strains, lacking mitochondrial DNA, are N ρ^-, the extent of suppressiveness is generally a property of the fragment of ρ^+ mitochondrial DNA retained in the ρ^- mutant. Mitochondrial DNAs of all HS ρ^- strains contain one of several short segments from ρ^+ mitochondrial DNA, corresponding to ori sequences implicated in mitochondrial DNA replication, and suggest-

ing that their high density in HS ρ⁻ genomes provides a replicative advantage over ρ⁺ mitochondrial DNA. However, because many ρ⁻ mutants lack an ori sequence and still maintain normal levels of mitochondrial DNA in growing cells, it is evident that ori sequences are not essential for ρ⁻ mitochondrial DNA replication. How can hypersuppressiveness be explained? (For an answer, see MacAlpine 2001.)

4. Genetic codes for mitochondrial DNA differ from the universal genetic code, and the mitochondrial genetic codes vary among different organisms, as first shown by Barrell et al. (1979), Fox (1979), and Macino et al. (1979). For example, the universal chain terminating codon UGA encodes tryptophan in mitochondria of yeast, *Neurospora*, *Drosophila*, and mammals, but is "normal" in mitochondria of plants; the universal isoleucine codon AUA encodes methionine in mitochondria of yeast, *Drosophila*, and mammals, but is "normal" in mitochondria of *Neurospora* and plants. Similarly, the universal leucine codons encode threonine in mitochondria of yeast, but is "normal" in mitochondria of other organisms. Can you suggest the evolutionary advantage, if any, of forming and maintaining a different genetic code in mitochondria? (For one possible answer, see Kurland [1992] and Andersson and Kurland [1995].)

5. It is now known that *SUP35* encodes the translational termination factor eRF3 (Stansfield et al. 1995; Zhouravleva et al. 1995; Frolova et al. 1996). How does this fact explain the [*PSI*⁺] phenotype?

6. Is the proposed ψ⁺ element dominant or recessive to ψ⁻? Which crosses in the Cox (1965) paper demonstrate this? Cytoplasm from ψ⁺ cells is more efficient than cytoplasm from ψ⁻ cells in reading through stop codons in an in vitro assay (see Tuite et al. 1987). What do you think happened when a mixture of cytoplasm from ψ⁺ and ψ⁻ cells was used? Why?

7. How would you determine if ψ⁺, like the ρ⁻ mutation, was controlled by mitochondrial DNA?

8. How would you test the hypothesis that the maintenance of ψ⁺ requires mistranslation caused by ψ⁺, resulting in a positive feedback loop?

9. Given the fact that the presence of a premature nonsense codon makes mRNA a target for degradation, how would you strengthen the evidence presented in Wickner's landmark paper (1994) that induction of [*URE3*] is dependent on Ure2 and not *URE2* mRNA? For a hint, see Derkatch et al. (1996) and Masison et al. (1997).

10. What other genes might be required for the maintenance of ρ⁺, [*PSI*⁺], or [*URE3*]?

11. How would you directly test the idea that the prion form of the protein has an altered shape and that this altered shape can propagate?

12. What would the biological consequences be if, for example, Sup35 could fold in more than one distinct heritable prion shape? What phenomenon in mammalian prions would such a result mirror?

13. What factors might influence the de novo appearance of prions and how could they work?

14. How would you search for other prions?

REFERENCES

Aigle M. and Lacroute F. 1975. Genetic aspects of [URE3], a non-Mendelian, cytoplasmically inherited mutation in yeast. *Mol. Gen. Genet.* **136:** 327–335.

Andersson S.G. and Kurland C.G. 1995. Genomic evolution drives the evolution of the translation system. *Biochem. Cell Biol.* **73:** 775–787.

Barrell B.G., Bankier A.T., and Drouin J. 1979. A different genetic code in human mitochondria. *Nature* **282:** 189–194.

Berger K.H. and Yaffe M.P. 2000. Mitochondrial DNA inheritance in *Saccharomyces cerevisiae*. *Trends Microbiol.* **8:** 508–513.

Bernardi G., Faures M., Piperno G., and Slonimski P.P. 1970. Mitochondrial DNA's from respiratory-sufficient and cytoplasmic respiratory-deficient mutant yeast. *J. Mol. Biol.* **48:** 23–42.

Bevan E.A., Herring A.J., and Mitchell D.J. 1973. Preliminary characterization of two species of dsRNA in yeast and their relationship to the "killer" character. *Nature* **245:** 81–86.

Boldogh I.R., Yang H.C., and Pon L.A. 2001. Mitochondrial inheritance in budding yeast. *Traffic* **2:** 368–374.

Borchsenius A.S., Wegrzyn R.D., Newnam G.P., Inge-Vechtomov S.G., and Chernoff Y.O. 2001. Yeast prion protein derivative defective in aggregate shearing and production of new 'seeds'. *EMBO J.* **20:** 6683–6691.

Bueler H., Aguzzi A., Sailer A., Greiner R.A., Autenried P., Aguet M., and Weissmann C. 1993. Mice devoid of PrP are resistant to scrapie. *Cell* **73:** 1339–1347.

Chen S.Y., Ephrussi B., and Hottinguer H. 1950. Nature génétique des mutants à déficience respiratoire de la souche B-II de la levure de boulangerie. *Heredity* **4:** 337–351.

Chernoff Y.O. 2001. Mutation processes at the protein level: Is Lamarck back? *Mutat. Res.* **488:** 39–64.

Chernoff Y.O., Derkach I.L., and Inge-Vechtomov S.G. 1993. Multicopy *SUP35* gene induces de-novo appearance of *psi*-like factors in the yeast *Saccharomyces cerevisiae*. *Curr. Genet.* **24:** 268–270.

Chernoff Y.O., Lindquist S.L., Ono B., Inge-Vechtomov S.G., and Liebman S.W. 1995. Role of the chaperone protein Hsp104 in propagation of the yeast prion-like factor [*psi*[+]]. *Science* **268:** 880–884.

Chernoff Y.O., Newnam G.P., Kumar J., Allen K., and Zink A.D. 1999. Evidence for a protein mutator in yeast: Role of the Hsp70-related chaperone ssb in formation, stability, and toxicity of the [PSI] prion. *Mol. Cell. Biol.* **19:** 8103–8112.

Chernoff Y.O., Galkin A.P., Lewitin E., Chernova T.A., Newnam G.P., and Belenkiy S.M. 2000. Evolutionary conservation of prion-forming abilities of the yeast Sup35 protein. *Mol. Microbiol.* **35:** 865–876.

Chien P. and Weissman J.S. 2001. Conformational diversity in a yeast prion dictates its seeding specificity. *Nature* **410:** 223–227.

Conde J. and Fink G.R. 1976. A mutant of *Saccharomyces cerevisiae* defective for nuclear fusion. *Proc. Natl. Acad. Sci.* **73:** 3651–3655.

Corneo G., Moore C., Sanadi D.R., Grossman L.I., and Marmur J. 1966. Mitochondrial DNA in yeast and some mammalian species. *Science* **151:** 687–689.

Coustou V., Deleu C., Saupe S., and Begueret J. 1997. The protein product of the *het-s* heterokaryon incompatibility gene of the fungus *Podospora anserina* behaves as a prion analog. *Proc. Natl. Acad. Sci.* **94:** 9773-9778.

Cox B.S. 1965. Ψ, a cytoplasmic suppressor of super-suppressor in yeast. *Heredity* **20:** 505–521.

———. 1993. Psi phenomena in yeast. In *The early days of yeast genetics* (ed. M.N. Hall and P. Linder), pp. 219–239. Cold Spring Harbor Laboratory Press, Cold Spring Harbor, New York.

Cox B.S., Tuite M.F., and McLaughlin C.S. 1988. The *psi* factor of yeast: A problem in inheritance. *Yeast* **4:** 159–178.

DePace A.H., Santoso A., Hillner P., and Weissman J.S. 1998. A critical role for amino-terminal glutamine/asparagine repeats in the formation and propagation of a yeast prion. *Cell* **93:** 1241–1252.

Derkatch I.L., Bradley M.E., and Liebman S.W. 1998. Overexpression of the *SUP45* gene encoding a Sup35p-binding protein inhibits the induction of the de novo appearance of the [*PSI* [+]] prion. *Proc. Natl. Acad. Sci.* **95:** 2400–2405.

Derkatch I.L., Bradley M.E., Hong J.Y., and Liebman S.W. 2001. Prions affect the appearance of other prions: The story of [*PIN* [+]]. *Cell* **106:** 171–182.

Derkatch I.L., Bradley M.E., Zhou P., and Liebman S.W. 1999. The *PNM2* mutation in the prion protein domain of *SUP35* has distinct effects on different variants of the [*PSI*[+]] prion in yeast. *Curr. Genet.* **35:** 59–67.

Derkatch I.L., Bradley M.E., Zhou P., Chernoff Y.O., and Liebman S.W. 1997. Genetic and environmental factors affecting the *de novo* appearance of the [*PSI*⁺] prion in *Saccharomyces cerevisiae*. *Genetics* **147:** 507–519.

Derkatch I.L., Chernoff Y.O., Kushnirov V.V., Inge-Vechtomov S.G., and Liebman S.W. 1996. Genesis and variability of [*PSI*] prion factors in *Saccharomyces cerevisiae*. *Genetics* **144:** 1375–1386.

Derkatch I.L., Bradley M.E., Masse S.V., Zadorsky S.P., Polozkov G.V., Inge-Vechtomov S.G, and Liebman S.W. 2000. Dependence and independence of [*PSI*⁺] and [*PIN*⁺]: A two-prion system in yeast? *EMBO J.* **19:** 1942–1952.

Doel S.M., McCready S.J., Nierras C.R., and Cox B.S. 1994. The dominant *PNM2*[−] mutation which eliminates the ψ factor of *Saccharomyces cerevisiae* is the result of a missense mutation in the *SUP35* gene. *Genetics* **137:** 659–670.

Dujon B. 1981. Mitochondrial genetics and function. In *Molecular biology of the yeast* Saccharomyces: *Life cycle and inheritance* (ed. J.N. Strathern et al.), pp. 505–635. Cold Spring Harbor Laboratory, Cold Spring Harbor, New York.

Dujon B., Slonimski P.P., and Weill L. 1974. Mitochondrial genetics. IX: A model for recombination and segregation of mitochondrial genomes in *Saccharomyces cerevisiae*. *Genetics* **78:** 415–437.

Edskes H.K. and Wickner R.B. 2000. A protein required for prion generation: [URE3] induction requires the Ras-regulated Mks1 protein. *Proc. Natl. Acad. Sci.* **97:** 6625–6629.

Edskes H.K., Gray V.T., and Wickner R.B. 1999. The [URE3] prion is an aggregated form of Ure2p that can be cured by overexpression of Ure2p fragments. *Proc. Natl. Acad. Sci.* **96:** 1498–1503.

Ephrussi B. 1953. *Nucleo-cytoplasmic relations in micro-organisms.* Clarendon Press, Oxford.

Ephrussi B. and Hottinguer H. 1950. Direct demonstration of the mutagenic action of euflavine on baker's yeast. *Nature* **166:** 956.

Ephrussi B., Hottinguer H., and Chimenes A.M. 1949a. Action de l'acriflavine sur les levures. 1. La mutation "petite colonie." *Ann. Inst. Pasteur* **76:** 351–367.

Ephrussi B., Hottinguer H., and Tavlitzki J. 1949b. Action de l'acriflavine sur les levures. II. Étude génétique du mutant "petite colonie." *Ann. Inst. Pasteur* **76:** 419–442.

Ephrussi B., L'Heritier P., and Hottinguer H. 1949c. Action de l'acriflavine sur les levures. VI. Analyse quantitative de la transformation des populations. *Ann. Inst. Pasteur* **77:** 64–83.

Faye G., Fukuhara H., Grandchamp C., Lazowska J., Michel F., Casey J., Getz G.S., Locker J., Rabinowitz M., Bolotin-Fukuhara M., Coen D., Deutsch J., Dujon B., Netter P., and Slonimski P.P. 1973. Mitochondrial nucleic acids in the petite colonie mutants: Deletions and repetition of genes. *Biochimie* **55:** 779–792.

Fernandez-Bellot E., Guillemet E., and Cullin C. 2000. The yeast prion [URE3] can be greatly induced by a functional mutated *URE2* allele. *EMBO J.* **19:** 3215–3222.

Ferreira P.C., Ness F., Edwards S.R., Cox B.S., and Tuite M.F. 2001. The elimination of the yeast [*PSI* ⁺] prion by guanidine hydrochloride is the result of Hsp104 inactivation. *Mol. Microbiol.* **40:** 1357–1369.

Flury U., Mahler H.R., and Feldman F. 1974. A novel respiration-deficient mutant of *Saccharomyces cerevisiae*. I. Preliminary characterization of phenotype and mitochondrial inheritance. *J. Biol. Chem.* **249:** 6130–6137.

Fowell R.R. 1951. Hybridization of yeasts by Lindegren's technique. *J. Inst. Brewing* **57:** 180–195.

Fox T.D. 1979. Five TGA "stop" codons occur within the translated sequence of the yeast mitochondrial gene for cytochrome *c* oxidase subunit II. *Proc. Natl. Acad. Sci.* **76:** 6534–6538.

———. 1996. Translational control of endogenous and recoded nuclear genes in yeast mitochondria: Regulation and membrane targeting. *Experientia* **52:** 1130–1135.

Frolova L., Le Goff X., Zhouravleva G., Davydova E., Philippe M., and Kisselev L. 1996. Eukaryotic polypeptide chain release factor eRF3 is an eRF1- and ribosome-dependent guanosine triphosphatase. *RNA* **2:** 334–341.

Gajdusek D.C. 1991. The transmissible amyloidoses: Genetical control of spontaneous generation of infectious amyloid proteins by nucleation of configurational change in host precursors: kuru-CJD-GSS-scrapie-BSE. *Eur. J. Epidemiol.* **7:** 567–577.

Glover J.R., Kowal A.S., Schirmer E.C., Patino M.M., Liu J.J., and Lindquist S. 1997. Self-seeded fibers formed by Sup35, the protein determinant of [*PSI*⁺], a heritable prion-like factor of *S. cerevisiae*. *Cell* **89:** 811–819.

Griffith J.S. 1967. Self-replication and scrapie. *Nature* **215:** 1043–1044.

Grivell L.A. 1995 Nucleo-mitochondrial interactions in mitochondrial gene expression. *Crit. Rev. Biochem. Mol. Biol.* **30:** 121–164.

Grivell L.A., Artal-Sanz M., Hakkaart G., de Jong L., Nijtmans L.G., van Oosterum K., Siep M., and van der Spek H. 1999. Mitochondrial assembly in yeast. *FEBS Lett.* **452:** 57–60.

Hawthorne D.C. and Mortimer R.K. 1963. Super-suppressors in yeast. *Genetics* **48:** 617–620.

Hinnebusch A. and Liebman S.W. 1991. Protein synthesis and translational control in *Saccharomyces cerevisiae*. In *The molecular and cellular biology of the yeast* Saccharomyces. I. *Genome dynamics, protein synthesis, and energetics* (ed. J.R. Broach et al.), pp. 627–735. Cold Spring Harbor Laboratory Press, Cold Spring Harbor, New York.

Hobbs A.E., Srinivasan M., McCaffery J.M., and Jensen R.E. 2001. Mmm1p, a mitochondrial outer membrane protein, is connected to mitochondrial DNA (mtDNA) nucleoids and required for mtDNA stability. *J. Cell Biol.* **152:** 401–410.

Inge-Vechtomov S.G. and Andrianova V.M. 1970. Recessive super-suppressors in yeast. *Genetica* **6:** 103–115.

Jensen R.E., Hobbs A.E.A., Cerveny K.L., and Sesaki H. 2000. Yeast mitochondrial dynamics: Fusion, division, segregation, and shape. *Microsc. Res. Tech.* **51:** 573–583.

Jung G. and Masison D.C. 2001. Guanidine hydrochloride inhibits Hsp104 activity in vivo: A possible explanation for its effect in curing yeast prions. *Curr. Microbiol.* **43:** 7–10.

King C.Y. 2001. Supporting the structural basis of prion strains: Induction and identification of [*PSI*] variants. *J. Mol. Biol.* **307:** 1247–1260.

King C.Y. and Diaz-Avalos R. 2004. Protein-only transmission of three yeast prion strains. *Nature* **428:** 319–323.

King C.Y., Tittmann P., Gross H., Gebert R., Aebi M., and Wuthrich K. 1997. Prion-inducing domain 2-114 of yeast Sup35 protein transforms in vitro into amyloid-like filaments. *Proc. Natl. Acad. Sci.* **94:** 6618-6622.

Kochneva-Pervukhova N.V., Chechenova M.B., Valouev I.A., Kushnirov V.V., Smirnov V.N., and Ter-Avanesyan M.D. 2001. [*PSI*[+]] prion generation in yeast: Characterization of the 'strain' difference. *Yeast* **18:** 489–497.

Kochneva-Pervukhova N.V., Paushkin S.V., Kushnirov V.V., Cox B.S., Tuite M.F., and Ter-Avanesyan M.D. 1998. Mechanism of inhibition of [*PSI* [+]] prion determinant propagation by a mutation of the N-terminus of the yeast Sup35 protein. *EMBO J.* **17:** 5805–5810.

Krishnan R. and Lindquist S.L. 2005. Structural insights into a yeast prion illuminate nucleation and strain diversity. *Nature* **435:** 765–772.

Kurland C.G. 1992. Evolution of mitochondrial genomes and the genetic code. *Bioessays* **14:** 709–714.

Kushnirov V.V., Kochneva-Pervukhova N.V., Chechenova M.B., Frolova N.S., and Ter-Avanesyan M.D. 2000. Prion properties of the Sup35 protein of yeast *Pichia methanolica*. *EMBO J.* **19:** 324–331.

Lacroute F. 1971. Non-Mendelian mutation allowing ureidosuccinic acid uptake in yeast. *J. Bacteriol.* **106:** 519–522.

Lazowska J., Jacq C., and Slonimski P.P. 1980. Sequence of introns and flanking exons in wild-type and box3 mutants of cytochrome *b* reveals an interlaced splicing protein coded by an intron. *Cell* **22:** 333–348.

Liebman S.W. and Derkatch I.L. 1999. The yeast [*PSI* [+]] prion: Making sense of nonsense. *J. Biol. Chem.* **274:** 1181–1184.

Lindegren C.C. and Lindegren G. 1943. Selecting, inbreeding, recombining, and hybridizing commercial yeasts. *J. Bacteriol.* **46:** 405–419.

———. 1947. Depletion mutation in *Saccharomyces*. *Proc. Natl. Acad. Sci.* **33**: 314–318.

Linnane A.W., Saunders G.W., Gingold E.B., and Lukins H.B. 1968. The biogenesis of mitochondria. V. Cytoplasmic inheritance of erythromycin resistance in *Saccharomyces cerevisiae*. *Proc. Natl. Acad. Sci.* **59:** 903–910.

Liu J.J. and Lindquist S. 1999. Oligopeptide-repeat expansions modulate 'protein-only' inheritance in yeast. *Nature* **400:** 573–576.

Livingston D.M. and Klein H.L. 1977. Deoxyribonucleic acid sequence organization of a yeast plasmid. *J. Bacteriol.* **129:** 472–481.

Luck D.J.L. 1965. Formation of mitochondria in *Neurospora*. A study based on mitochondrial density changes. *J. Cell Biol.* **24:** 461–470.

MacAlpine D.M., Kolesar J., Okamoto K., Butow R.A., and Perlman P.S. 2001. Replication and preferential inheritance of hypersuppressive petite mitochondrial DNA. *EMBO J.* **20:** 1807–1817.

Macino G., Coruzzi G., Nobrega F.G., Li M., and Tzagoloff A. 1979. Use of the UGA terminator as a tryptophan codon in yeast mitochondria. *Proc. Natl. Acad. Sci.* **76:** 3784–3785.

Maddelein M.L. and Wickner R.B. 1999. Two prion-inducing regions of Ure2p are nonoverlapping. *Mol. Cell. Biol.* **19:** 4516–4524.

Maleszka R., Skelly P.J., and Clark-Walker G.D. 1991. Rolling circle replication of DNA in yeast mitochondria. *EMBO J.* **10:** 3923–3929.

Masison D.C. and Wickner R.B. 1995. Prion-inducing domain of yeast Ure2p and protease resistance of Ure2p in prion-containing cells. *Science* **270:** 93–95.

Masison D.C., Maddelein M.L., and Wickner R.B. 1997. The prion model for [URE3] of yeast: Spontaneous generation and requirements for propagation. *Proc. Natl. Acad. Sci.* **94:** 12503–12508.

Masison D.C., Edskes H.K., Maddelein M.L., Taylor K.L., and Wickner R.B. 2000. [URE3] and [PSI] are prions of yeast and evidence for new fungal prions. *Curr. Issues Mol. Biol.* **2:** 51–59.

Moriyama H., Edskes H.K., and Wickner R.B. 2000. [URE3] prion propagation in *Saccharomyces cerevisiae*: Requirement for chaperone Hsp104 and curing by overexpressed chaperone Ydj1p. *Mol. Cell. Biol.* **20:** 8916–8922.

Mortimer R.K. 1993. Øjvind Winge: Founder of yeast genetics. In *The early days of yeast genetics* (ed. M.N. Hall and P. Linder), pp. 3–16. Cold Spring Harbor Laboratory Press, Cold Spring Harbor, New York.

Mounolou J.C., Jakob H., and Slonimski P.P. 1966. Mitochondrial DNA from yeast "petite" mutants: Specific changes of buoyant density corresponding to different cytoplasmic mutations. *Biochem. Biophys. Res. Commun.* **24:** 218–224.

Moustacchi E. and Williamson D.H. 1966. Physiological variations in satellite components of yeast DNA detected by density gradient centrifugation. *Biochem. Biophys. Res. Commun.* **23:** 56–61.

Nass M.M.K. and Nass S. 1963. Intramitochondrial fibers with DNA characteristics II; enzymatic and other hydrolytic treatments. *J. Cell Biol.* **19:** 613–629.

Newnam G.P., Wegrzyn R.D., Lindquist S.L., and Chernoff Y.O. 1999. Antagonistic interactions between yeast chaperones Hsp104 and Hsp70 in prion curing. *Mol. Cell. Biol.* **19:** 1325–1333.

Osherovich L.Z. and Weissman J.S. 2001. Multiple Gln/Asn-rich prion domains confer susceptibility to induction of the yeast [PSI$^+$] prion. *Cell* **106:** 183–194.

Patino M.M., Liu J.J., Glover J.R., and Lindquist S. 1996. Support for the prion hypothesis for inheritance of a phenotypic trait in yeast. *Science* **273:** 622–626.

Paushkin S.V., Kushnirov V.V., Smirnov V.N., and Ter-Avanesyan M.D. 1996. Propagation of the yeast prion-like [*psi*$^+$] determinant is mediated by oligomerization of the *SUP35*-encoded polypeptide chain release factor. *EMBO J.* **15:** 3127–3134.

———. 1997. *In vitro* propagation of the prion-like state of yeast Sup35 protein. *Science* **277:** 381–383.

Pillus L. and Rine J. 1989. Epigenetic inheritance of transcriptional states in *S. cerevisiae*. *Cell* **59:** 637–647.

Prusiner S.B. 1982. Novel proteinaceous infectious particles cause scrapie. *Science* **216:** 136–144.

———. 1991. Molecular biology of prion diseases. *Science* **252:** 1515–1522.

Prusiner S.B., Groth D., Serban A., Koehler R., Foster D., Torchia M., Burton D., Yang S.L., and DeArmond S.J. 1993. Ablation of the prion protein (PrP) gene in mice prevents scrapie and facilitates production of anti-PrP antibodies. *Proc. Natl. Acad. Sci.* **90:** 10608–10612.

Putrament A., Baranowska H., and Prazmo W. 1973. Induction by manganese of mitochondrial antibiotic resistance mutations in yeast. *Mol. Gen. Genet.* **126:** 357–366.

Putrament A., Baranowska H., Ejchart A., and Prazmo W. 1975. Manganese mutagenesis in yeast. IV. The effects of magnesium, protein synthesis inhibitors and hydroxyurea on Ant^R induction in mitochondrial DNA. *Mol. Gen. Genet.* **140:** 339–347.

Santoso A., Chien P., Osherovich L.Z., and Weissman J.S. 2000. Molecular basis of a yeast prion species barrier. *Cell* **100:** 277–288.

Schatz G. 1993. From "granules" to organelles: How yeast mitochondria became respectable. In *The early days of yeast genetics* (ed. M.N. Hall and P. Linder), pp. 241–246. Cold Spring Harbor Laboratory Press, Cold Spring Harbor, New York.

Schatz G., Haslbrunner E., and Tuppy H. 1964. Deoxyribonucleic acid associated with yeast mitochondria. *Biochem. Biophys. Res. Commun.* **15:** 127–132.

Scheibel T., Kowal A.S., Bloom J.D., and Lindquist S.L. 2001. Bidirectional amyloid fiber growth for a yeast prion determinant. *Curr. Biol.* **11:** 366–369.

Schlumpberger M., Prusiner S.B., and Herskowitz I. 2001. Induction of distinct [URE3] yeast prion strains. *Mol. Cell. Biol.* **21:** 7035–7046.

Serio T.R. and Lindquist S.L. 1999. [PSI $^+$]: An epigenetic modulator of translation termination efficiency.

Annu. Rev. Cell. Dev. Biol. **15:** 661–703.

Sherman F. 1959. The effects of elevated temperatures on yeast. II. Induction-respiratory deficient mutants. *J. Cell. Comp. Physiol.* **54:** 37–52.

———. 1963. Respiration-deficient mutants of yeast. I. Genetics. *Genetics* **48:** 375–385.

———. 1964. Mutants of yeast deficient in cytochrome *c*. *Genetics* **49:** 39–48.

———. 2002. Getting started with yeast. *Methods Enzymol.* **350:** 3–41.

Sherman F. and Slonimski P.P. 1964. Respiration-deficient mutants of yeast II. Biochemistry. *Biochim. Biophys. Acta* **90:** 1–15.

Sigurdson D.C., Gaarder M.E., and Livingston D.M. 1981. Characterization of the transmission during cytoductant formation of the 2 micrometers DNA plasmid from *Saccharomyces*. *Mol. Gen. Genet.* **183:** 59–65.

Slonimski P.P. 1949a. Action de l'acriflavine sur les levures. IV. Mode d'utilisation du glucose par les mutants 'petite colonie.' *Ann. Inst. Pasteur* **76:** 510–530.

———. 1949b. Action de l'acriflavine sur les levures. VII. Sur l'activite catalytique du cytochrome *c* des mutants 'petite colonie' de la levure. *Ann. Inst. Pasteur* **77:** 774–776.

Slonimski P.P. and Ephrussi B. 1949. Action de l'acriflavine sur les levures. V. Le systeme, des cytochromes des mutants "petite colonie." *Ann. Inst. Pasteur* **77:** 47–63.

Slonimski P.P. and Tzagoloff A. 1976. Localization in yeast mitochondrial DNA of mutations expressed in a deficiency of cytochrome oxidase and/or coenzyme QH_2-cytochrome *c* reductase. *Eur. J. Biochem.* **61:** 27–41.

Slonimski P.P., Perrodin G., and Croft J.H. 1968. Ethidium bromide induced mutation of yeast mitochondria: Complete transformation of cells into respiratory deficient nonchromosomal "petites." *Biochem. Biophys. Res. Commun.* **30:** 232–239.

Sondheimer N. and Lindquist S. 2000. Rnq1: An epigenetic modifier of protein function in yeast. *Mol. Cell* **5:** 163–172.

Sondheimer N., Lopez N., Craig E.A., and Lindquist S. 2001. The role of Sis1 in the maintenance of the [*RNQ+*] prion. *EMBO J.* **20:** 2435–2442.

Speransky V.V., Taylor K.L., Edskes H.K., Wickner R.B., and Steven A.C. 2001. Prion filament networks in [URE3] cells of *Saccharomyces cerevisiae*. *J. Cell Biol.* **153:** 1327–1336.

Stansfield I., Jones K.M., Kushnirov V.V., Dagkesamanskaya A.R., Poznyakovski A.I., Paushkin S.V., Nierras C.R., Cox B.S., Ter-Avanesyan M.D., and Tuite M.F. 1995. The products of the *SUP45* (eRF1) and *SUP35* genes interact to mediate translation termination in *Saccharomyces cerevisiae*. *EMBO J.* **14:** 4365–4373.

Tanaka M., Chien P., Yonekura K., and Weissman J.S. 2005. Mechanism of cross-species prion transmission: An infectious conformation compatible with two highly divergent yeast prion proteins. *Cell* **121:** 49–62.

Tanaka M., Chien P., Naber N., Cooke R., and Weissman J.S. 2004. Conformational variations in an infectious protein determine prior strain differences. *Nature* **428:** 323–328.

Tavlitzki J. 1949. Action de l'acriflavine sur les levures. III. Etude de la croissance des mutants "petite colonie." *Ann. Inst. Pasteur* **76:** 498–509.

Taylor K.L., Cheng N., Williams R.W., Steven A.C., and Wickner R.B. 1999. Prion domain initiation of amyloid formation in vitro from native Ure2p. *Science* **283:** 1339–1343.

Ter-Avanesyan M.D., Dagkesamanskaya A.R., Kushnirov V.V., and Smirnov V.N. 1994. The *SUP35* omnipotent suppressor gene is involved in the maintenance of the non-Mendelian determinant [*psi*$^+$] in the yeast *Saccharomyces cerevisiae*. *Genetics* **137:** 671–676.

Tewari K.K., Votsch W., Mahler H.R., and Mackler B. 1966. Biochemical correlates of respiratory deficiency. VI. Mitochondrial DNA. *J. Mol. Biol.* **20:** 453–481.

Thomas D.Y. and Wilkie D. 1968. Recombination of mitochondrial drug-resistance factors in *Saccharomyces cerevisiae*. *Biochem. Biophys. Res. Commun.* **30:** 368–372.

Tuite M.F., Cox B.S., and McLaughlin C.S. 1987. A ribosome-associated inhibitor of in vitro nonsense suppression in [*psi*$^-$] strains of yeast. *FEBS Lett.* **225:** 205–208.

Tuite M.F., Mundy C.R., and Cox B.S. 1981. Agents that cause a high frequency of genetic change from [*psi*$^+$] to [*psi*$^-$] in *Saccharomyces cerevisiae*. *Genetics* **98:** 691–711.

Tzagoloff A. 1993. From *MIT* to *PET* genes. In *The early days of yeast genetics* (ed. M.N. Hall and P. Linder), pp. 247–257. Cold Spring Harbor Laboratory Press, Cold Spring Harbor, New York.

Tzagoloff A. and Dieckmann C.L. 1990. *PET* genes of *Saccharomyces cerevisiae*. *Microbiol. Rev.* **54:** 211–225.

Tzagoloff A., Akai A., and Needleman R.B. 1975a. Properties of cytoplasmic mutants of *Saccharomyces cerevisiae* with specific lesions in cytochrome oxidase. *Proc. Natl. Acad. Sci.* **72:** 2054–2057.

———. 1975b. Assembly of the mitochondrial membrane system: Isolation of nuclear and cytoplasmic mutants of *Saccharomyces cerevisiae* with specific defects in mitochondrial functions. *J. Bacteriol.* **122:** 826–831.

Tzagoloff A., Akai A., Needleman R.B., and Zulch G. 1975c. Assembly of the mitochondrial membrane system. Cytoplasmic mutants of *Saccharomyces cerevisiae* with lesions in enzymes of the respiratory chain and in the mitochondrial ATPase. *J. Biol. Chem.* **250:** 8236–8242.

Uptain S.M., Sawicki G.J., Caughey B., and Lindquist S. 2001. Strains of [*PSI* $^+$] are distinguished by their efficiencies of prion-mediated conformational conversion. *EMBO J.* **20:** 6236–6245.

Wegrzyn R.D., Bapat K., Newnam G.P., Zink A.D., and Chernoff Y.O. 2001. Mechanism of prion loss after Hsp104 inactivation in yeast. *Mol. Cell. Biol.* **21:** 4656–4669.

Weissmann C. 1991. A 'unified theory' of prion propagation. *Nature* **352:** 679–683.

Wesolowski M. and Wickner R.B. 1984. Two new double-stranded RNA molecules showing non-Mendelian inheritance and heat inducibility in *Saccharomyces cerevisiae*. *Mol. Cell. Biol.* **4:** 151–157.

Westergaard R. 1963. Øjvind Winge. *C.R. Trav. Lab. Carlsberg Ser. Physiol.* **34:** 1–24.

Wickner R.B. 1994. [URE3] as an altered *URE2* protein: Evidence for a prion analog in *Saccharomyces cerevisiae*. *Science* **264:** 566–569.

———. 1996. Double-stranded RNA viruses of *Saccharomyces cerevisiae*. *Microbiol. Rev.* **60:** 250–265.

Wickner R.B., Taylor K.L., Edskes H.K., Maddelein M.L., Moriyama H., and Roberts B.T. 2001. Yeast prions act as genes composed of self-propagating protein amyloids. *Adv. Protein Chem.* **57:** 313–334.

Wilkie D. 1993. Early recollections of fungal genetics and the cytoplasmic inheritance controversy. In *The early days of yeast genetics* (ed. M.N. Hall and P. Linder), pp. 259–270. Cold Spring Laboratory Press, Cold Spring Harbor, New York.

Wilkie D., Saunders G., and Linnane A.W. 1967. Inhibition of respiratory enzyme synthesis in yeast by chloramphenicol: Relationship between chloramphenicol tolerance and resistance to other antibacterial antibiotics. *Genet. Res.* **10:** 199–203.

Williamson D. 2002. The curious history of yeast mitochondrial DNA. *Nat. Rev. Genet.* **3:** 475–481.

Winge Ø. 1935. On haplophase and diplophase in some *Saccharomycetes*. *C.R. Trav. Lab. Carlsberg Ser. Physiol.* **21:** 77–112.

Winge Ø. and Lausten O. 1937. On two types of spore germination, and on genetic segregation in *Saccharomyces*, demonstrated through single spore cultures. *C.R. Trav. Lab. Carlsberg Ser. Physiol.* **22:** 99–117.

———. 1940. On a cytoplasmic effect on inbreeding in homozygous yeast. *C.R. Trav. Lab. Carlsberg Ser. Physiol.* **23:** 17–39.

Winkler H. 1930. *Die Konversion der Gene*. Fischer, Jena, Germany.

Wright R.E. and Lederberg J. 1957. Extranuclear transmission in yeast heterokaryons. *Proc. Natl. Acad. Sci.* **43:** 919–923.

Yaffe M.P. 1999. The machinery of mitochondrial inheritance and behavior. *Science* **283:** 1493–1497.

Zhou P., Derkatch I.L., Uptain S.M., Patino M.M., Lindquist S., and Liebman S.W. 1999. The yeast non-Mendelian factor [*ETA*$^+$] is a variant of [*PSI*$^+$], a prion-like form of release factor eRF3. *EMBO J.* **18:** 1182–1191.

Zhouravleva G., Frolova L., Le Goff X., Le Guellec R., Inge-Vechtomov S., Kisselev L., and Philippe M. 1995. Termination of translation in eukaryotes is governed by two interacting polypeptide chain release factors, eRF1 and eRF3. *EMBO J.* **14:** 4065–4072.

Zinn A.R. and Butow R.A. 1985. Nonreciprocal exchange between alleles of the yeast mitochondrial 21S rRNA gene: Kinetics and the involvement of a double-strand break. *Cell* **40:** 887–895.

2

Homologous Recombination

Lorraine Symington
Department of Microbiology
Columbia University Medical Center
New York, New York 10032

Early models to explain the mechanism of recombination, 34
Hurst D.D., Fogel S., and Mortimer R.K. 1972. Conversion-associated recombination in yeast (hybrids/meiosis/tetrads/marker loci/models). *Proc. Natl. Acad. Sci.* **69**: 101–105.

Testing the DSBR model: Searching for initiation sites, 39
Nicolas A., Treco D., Schultes N.P., and Szostak J.W. 1989. An initiation site for meiotic gene conversion in the yeast *Saccharomyces cerevisiae*. *Nature* **338**: 35–39.

Testing the DSBR model: Seeing is believing, 40
Cao L., Alani E., and Kleckner N. 1990. A pathway for generation and processing of double-strand breaks during meiotic recombination in *S. cerevisiae*. *Cell* **61**: 1089–1101.

Inconsistencies with the DSBR model, 43
Allers T. and Lichten M. 2001. Differential timing and control of noncrossover and crossover recombination during meiosis. *Cell* **106**: 47–57.

Note: The landmark papers listed above are those discussed in this chapter. Each landmark paper is preceded by the name of the section (with starting page number) where the paper is first discussed in detail.

HOMOLOGOUS RECOMBINATION (HR) IS THE EXCHANGE OR TRANSFER of information between DNA sequences of perfect or near-perfect homology. It has long been appreciated that HR occurs at a high frequency during meiosis to ensure segregation of chromosome homologs and to generate genetic diversity. In mitotic cells, HR occurs at much lower frequencies and is now recognized as a major pathway for the repair of DNA damage.

Much of our understanding of the mechanisms of HR is based on the analysis of meiotic recombination in organisms such as *Saccharomyces cerevisiae*, in which all of the products of an individual meiosis can be recovered in the form of asci containing four haploid spores. Two general classes of recombination events have been identified based on the segregation of heterozygous markers during meiosis: crossing-over and gene conversion. A crossover between linked heterozygous markers results in new linkage arrangements for two spore products, so that parental markers still display Mendelian (2:2) segregation. A single crossover results in a tetratype tetrad instead of the parental ditype (Fig. 1). A nonparental ditype tetrad is diagnostic of two crossovers between two linked heterozygous markers involving all four chromatids. In contrast, gene conversion represents the nonreciprocal transfer of information between two homologous sequences where one allele is duplicated while another is lost. A heterozygous marker, B/b, normally segregates to produce a tetrad with 2B:2b spores, whether or not a crossover has occurred, whereas gene conversion results in a tetrad with 3B:1b or 1B:3b spores (Fig. 1). The convention used for eight-spored fungi (6:2 and 2:6 for gene conversion) is frequently used in *S. cerevisiae*. This can be thought of as referring to the eight DNA strands contained in the four spores. In addition to gene conversion, other types of aberrant (non-Mendelian) segregation are found, the most common being postmeiotic segregation (PMS) of markers, in which 5:3 and 3:5 events are the most common. This type of event arises when one spore within a tetrad contains heteroduplex DNA (a duplex where the two strands contain different parental alleles) at the marker gene. After the first division of the germinating spore, the parental information segregates to the two halves of the spore colony. PMS events can therefore be visualized as sectored spore colonies in *S. cerevisiae* if the appropriate colony color markers are used (e.g., *ADE2*) or by failure of half the colony to grow following replica plating to selective medium.

The ability to follow the segregation of heterozygous markers during fungal meiosis and the discovery of the double-helical structure of DNA led to the development of models to try to explain how recombination occurs at the molecular level. This chapter traces the evolution of these models during the past 40 years, emphasizing some of the key experimental tests that arose from work on yeast. Although we focus to a large extent on the DNA "mechanics" of recombination, it should be kept in mind that there is a rich history of parallel and intertwined studies that address the many proteins involved in recombination from increasingly detailed enzymatic and structural perspectives (Sung et al. 2000; Krogh and Symington 2004).

EARLY MODELS TO EXPLAIN THE MECHANISM OF RECOMBINATION

The most influential early model was that of Holliday (1964) (Fig. 2). The Holliday model suggests that recombination is initiated by nicks on strands of the same polarity of both recombining duplexes (nonsister chromatids). Two single strands exchange, forming a crossed-strand–exchange structure, referred to as the Holliday intermediate or Holliday junction. Branch migration of the Holliday junction by extension of the duplex DNA formed by the invading strands produces heteroduplex DNA (hDNA) on both participating chromatids. Resolution of the intermediate by endonucleolytic cleavage of the crossed strands results in the formation of either crossover or noncrossover products, depending on where the cleavage occurs (see Fig. 2B). The Holliday model proposes that gene conversion events result when hDNA is repaired in such a way that one allele becomes duplicated, and thus a copy of the other is lost, i.e., B/b is repaired to B/B or b/b. The model also suggests that PMS of markers results from failure to repair hDNA and subsequent segregation as a result of postmeiotic DNA synthesis. The Holliday model thus views crossing-over and gene conversion as outcomes of a concerted series of molecular events. One important prediction

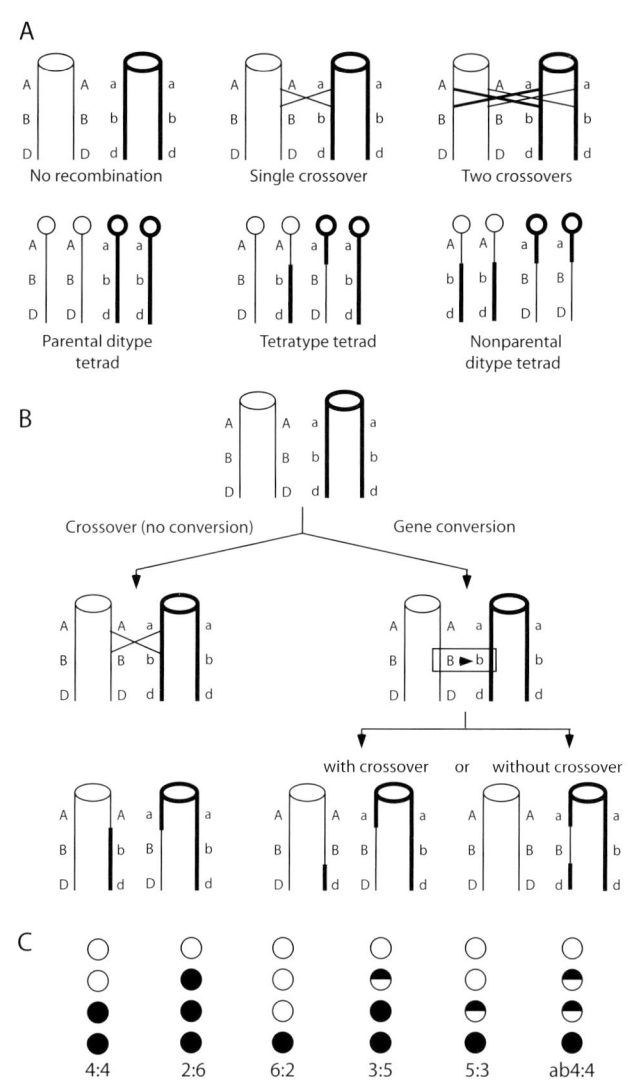

FIGURE 1. Types of homologous recombination events. (*A*) Meiotic crossing-over. The absence of recombination between the indicated linked markers results in two spore products with the same linkage arrangement as one parent and two spores like the other parent (parental ditype). A single crossover produces two spores with new linkage arrangements of the markers and two spores with the pattern of the parents (tetratype). Two crossovers involving all four chromatids result in all four spore products with a recombinant configuration of markers (nonparental ditype). (*B*) Meiotic crossing-over and gene conversion. A crossover between replicated nonsister chromatids generates a new linkage arrangement between the A and B markers. On the right is shown conversion of b to B. Gene conversion is frequently associated with a crossover between markers A and D. (*C*) Types of aberrant segregations observed in yeast. A heterozygous marker normally shows 2:2 segregation (4:4 using nomenclature representative of the eight DNA strands present in meiosis). Departures from this ratio are seen as 6:2 and 2:6 events (gene conversion) and 5:3 and 3:5 events (PMS). Other types of aberrant segregation, e.g., 7:1 and 1:7 events, are rarely observed.

from the Holliday model is that 50% of gene conversions are associated with crossing-over.

At the time the Holliday model was proposed, there were a few studies in fungi suggesting an association between gene conversion and crossing-over. However, it was unclear

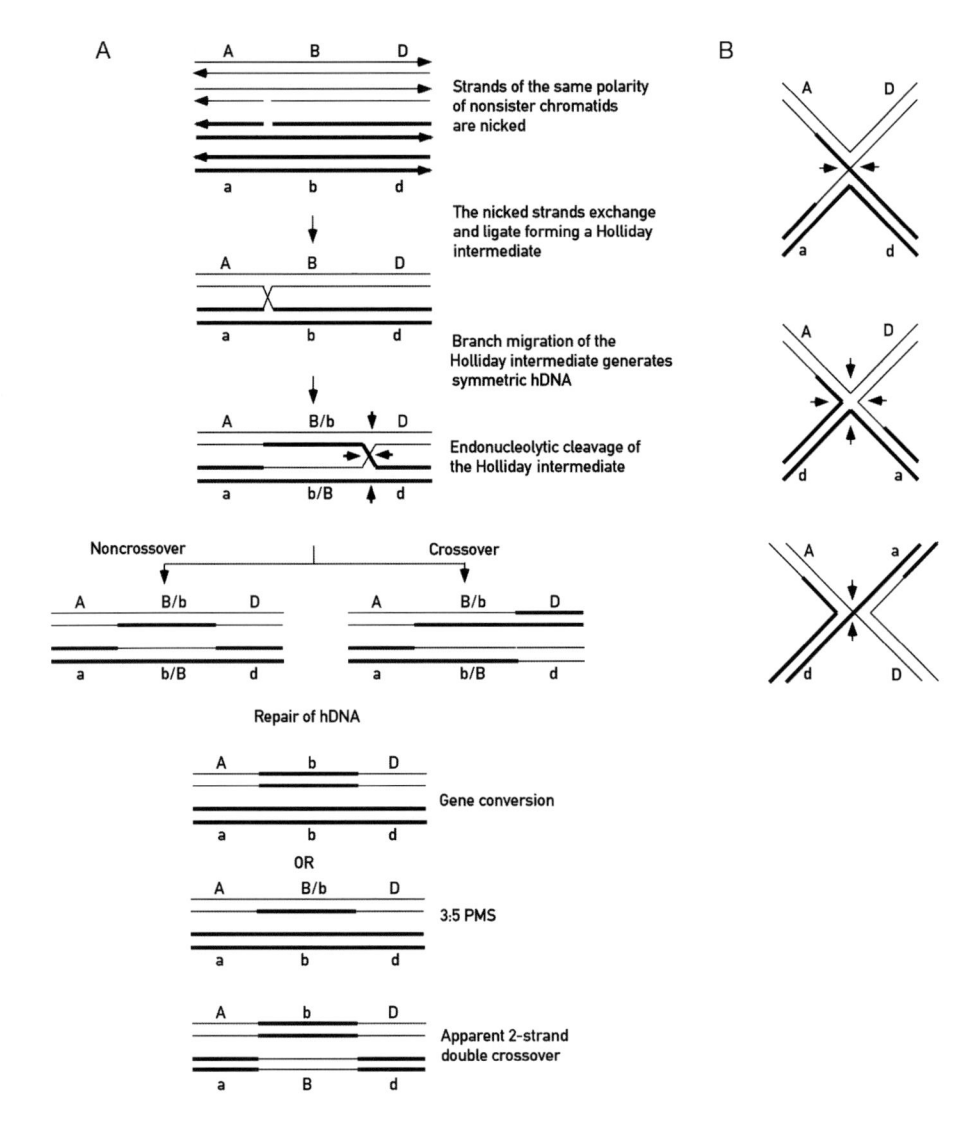

FIGURE 2. (*A*) The Holliday model. Recombination is initiated by nicking strands of the same polarity of two of the four chromatids. For simplicity, only the two chromatids engaged in recombination are shown in the following steps. The strands exchange to form a crossed-strand exchange intermediate referred to as the Holliday junction or intermediate. Migration of the intermediate leads to the formation of symmetric hDNA. The junction is resolved by endonucleolytic cleavage of the two crossed strands to produce noncrossover products that contain hDNA. If left unrepaired, the markers in hDNA will segregate at the first postmeiotic division, producing sectored spore colonies. Full, or partial, mismatch correction of hDNA can yield a variety of products, including gene conversion, restoration, and PMS. Note that only the two recombining duplexes are shown; the other two duplexes present at the time of recombination will be parental and contribute to the segregation patterns observed in tetrads. (*B*) Holliday junction isomerization. Rotation of the strands around the junctions results in an open planar structure. Cleavage in the horizontal plane produces noncrossover products and cleavage in the vertical plane generates crossover products. A second rotation results in isomerization of the junction so that the outer strands are now crossed. Resolution of the crossed strands of the two isomers generates the two types of products.

whether the associated crossovers observed were coincident or mechanistically related to the conversion event, and how general the findings were. As the field of yeast genetics developed, a large number of mutants were identified and genetic maps were generated. This provided the necessary tools to investigate the relationship between gene conversion and crossing-over at several different loci. **Hurst and colleagues (1972)** took up this challenge by first generating diploid strains that were heterozygous for closely linked markers on chromosomes V, VI, and VIII. In each case, the central of the three markers could be scored for gene conversion and the flanking heterozygous markers could then be analyzed to determine whether reciprocal exchange had occurred. To carry out their analysis, Hurst et al. (1972) dissected more than 11,000 tetrads and genotyped each spore colony by the ability of the spore products to grow on selective medium. Of 6871 tetrads dissected from diploids that were heterozygous for *arg4* alleles, 549 exhibited conversion (8% conversion), and of these 48.8% were recombinant for the flanking markers. Analysis of the *thr3*, *his1*, and *SUP6* loci revealed a similar association of crossing-over with conversion. The data of Hurst et al. demonstrated that for a given marker about 50% of conversion events are associated with crossing-over, whereas the other 50% do not show an associated crossover. Subsequent studies showed that the degree of crossover association could vary from locus to locus, with a range of from 18% to 66% (Fogel et al. 1981).

Although the Holliday model provided a plausible explanation for the association between conversion and crossing-over and for the origin of PMS and gene conversion events, as more data were obtained it became clear that several observations were inconsistent with the model. For example, the Holliday model predicts formation of hDNA on two chromatids, but genetic data were more consistent with hDNA being present on only one chromatid (Fogel et al. 1979). These and other observations led Meselson and Radding to propose a variation of the Holliday model in which recombination initiates by strand invasion on only one chromatid, resulting in the formation of asymmetric hDNA (Meselson and Radding 1975). Although the genetic data were more consistent with the Meselson–Radding model, there were still some observations that were not easily reconciled with this model. First, the genetic data were consistent with crossing-over occurring on either side of the conversion event and, second, the initiating chromatid was usually the recipient of information in the conversion event. The Meselson–Radding model predicts that the initiating chromatid is the donor of information and crossing-over occurs downstream from the initiating event.

THE DOUBLE-STRAND-BREAK REPAIR MODEL

A key theoretical paper by Szostak et al. (1983) discussed the limitations of the early models and presented a new model, the double-strand-break repair (DSBR) model, that was more consistent with the existing tetrad data. This model had been originally proposed to explain the integration of linearized yeast plasmids into the genome during transformation of mitotic cells (Orr-Weaver et al. 1981). Transformation of yeast with a nonreplicating plasmid containing a region of DNA homologous to chromosomal DNA occurs by integration of the plasmid at the homologous chromosomal site. These events occur with low efficiency, but Orr-Weaver et al. found that linearization of the plasmid within the region homologous to chromosomal sequences dramatically stimulated recombination and therefore the frequency of transformation. Most of the transformants were found to contain integrated plasmid sequences flanked by a duplication of the fragment of yeast DNA contained on the plasmid. When plasmid DNA was digested with restriction endonucleases to delete an internal fragment from the yeast DNA, it was still found to transform yeast at high frequency. The resulting transformants were indistinguishable in structure from those obtained using uncut DNA, indicating that repair of the gapped region had occurred during integration. This repair of a double-strand gap, a so-called "gap repair" event, is equivalent to gene conversion without mismatch repair.

Interestingly, Orr-Weaver et al. noted that linear plasmid integration is dependent on the *RAD52* gene. *RAD52* was first identified because it is required for resistance to ionizing radiation, such as X-rays and γ-rays, which were known to cause double-strand DNA breaks (DSBs). Subsequently, *RAD52* was shown to be required for most types of mitotic and meiotic recombination. Therefore, the *RAD52* dependence of linear plasmid integration indicated that plasmid DNA DSBs are repaired by the normal pathway of homologous recombination. Because nonreplicating plasmids were used in the original studies of Orr-Weaver et al., transformants could only arise by integration of the plasmid into the genome (a crossover event). In subsequent studies using autonomously replicating plasmids containing deletions within the yeast target sequence, gap repair was found to be associated with integration of the plasmid 50% of the time (Orr-Weaver and Szostak 1983). Furthermore, the DNA molecule with the initiating lesion is the recipient of information during repair. These results suggested that plasmid gap repair is mechanistically similar to meiotic recombination.

Further evidence in support of DSBs as initiators of recombination came from studies of mating-type switching in yeast (for a more detailed description of the mating-type system in yeast, see Chapter 8). Mating-type switching is a gene conversion event in which sequences at the mating-type locus (*MAT*) are replaced by sequences from one of the two donor cassettes, *HML* or *HMR*. This conversion event is unusual because the *MAT* locus is always the recipient of information and because it is rarely associated with crossing-over. Mating-type switching requires the *HO* gene, which encodes a site-specific endonuclease that makes a DSB at the *MAT* locus, but not at the homologous sites at *HML* or *HMR* (Strathern et al. 1982; Kostriken et al. 1983).

In the original conception of the DSBR model, the initiating lesion was proposed to be a DSB that was enlarged to a gap with 3′ single-stranded DNA tails (see Fig. 3). One of the 3′ ends was proposed to invade a homologous duplex, priming DNA synthesis from the donor duplex. Continued DNA synthesis extends the displaced loop (D-loop) from the donor, which is able to pair with the other 3′ end. This end can then prime DNA synthesis using the displaced donor strand as a template, resulting in repair of the gap. The double Holliday junction intermediate can be resolved to generate crossover or noncrossover products. The model was modified in 1991 (Fig. 3) to accommodate data from several labs indicating that most gene conversion is the result of hDNA repair rather than gap repair, and the 3′ ends at break sites are not degraded (Nag et al. 1989; Sun et al. 1991; Liu et al. 1995). Although the Holliday and DSBR models differ in the mode of initiation, both models propose that crossovers and noncrossover products arise by alternate resolution of the Holliday intermediate.

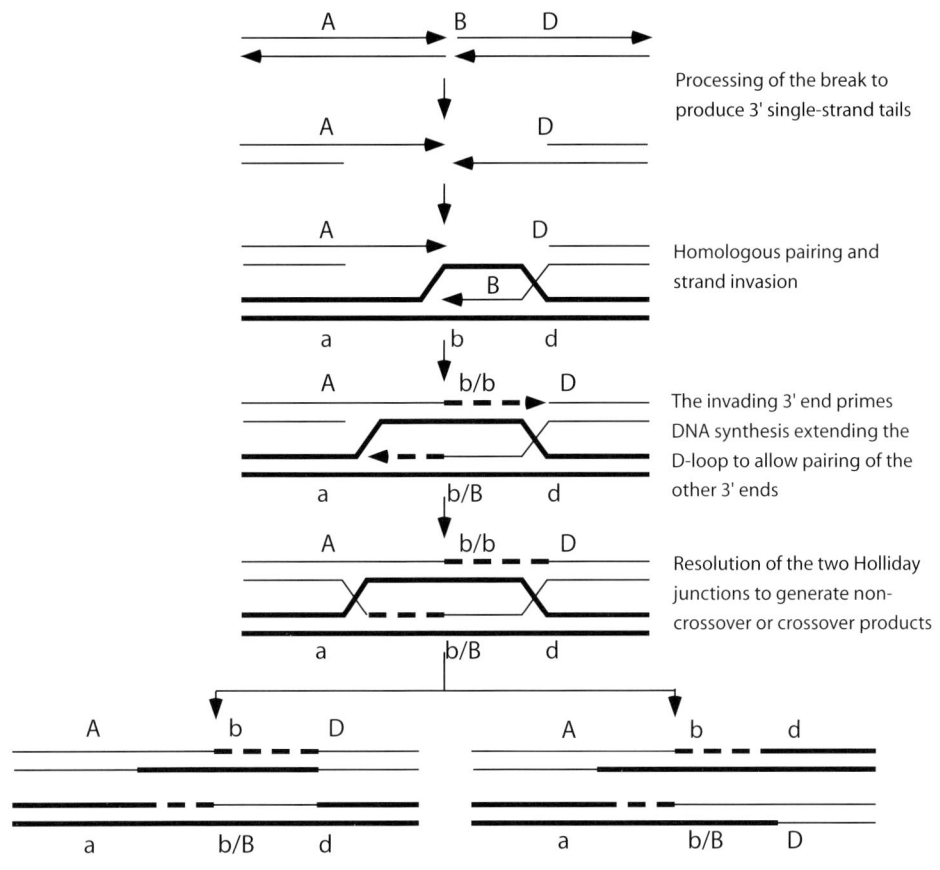

FIGURE 3. The double-strand-break repair (DSBR) model. Recombination is initiated by a DSB on one duplex. The ends are processed to generate 3′ single-stranded tails. One of the 3′ ends invades the homologous duplex displacing a strand. DNA synthesis primed from the invading 3′ ends extends the displacement loop (D-loop) until the other side of the break can pair with the D-loop. The other 3′ end is then extended by DNA synthesis and ligated resulting in repair of the break. The two Holliday junctions can be resolved to form either crossover or noncrossover products.

Testing the DSBR Model: Searching for Initiation Sites

Just as the Holliday and Meselson–Radding models were rigorously tested in the 1960s and 1970s, research into the mechanisms of meiotic recombination since 1983 has been driven by the predictions of the DSBR model. It was hypothesized that recombination initiates at specific sites, but no such site had been identified or characterized. **Nicolas et al. (1989)** set out to characterize an initiation site for meiotic gene conversion at the DNA level. The *ARG4* locus was chosen for analysis—first, because extensive studies by Fogel and colleagues had demonstrated an unusually high frequency of conversion of markers at one end of the gene and, second, because the DNA sequence of the locus had been determined. As in many areas of yeast genetics, recombinant DNA technology revolutionized the study of meiotic recombination. In this case, Nicolas et al. were able to generate *arg4* alleles with defined nucleotide alterations in vitro, replace these at the chromosomal locus, and then apply molecular tools of the day (Southern analysis of restriction fragment polymorphisms) to classical tetrad analysis. Consistent with the earlier work of Fogel and Mortimer, the *arg4* markers generated by Nicolas et al. showed a steep gradient of conversion with markers close to the 5´ end of the gene showing the highest frequency of conversion. The initiation site was mapped using overlapping deletions to within a 142-bp region in the promoter of the *ARG4* gene and, consistent with the prediction of the DSBR model, was the recipient of information during gene conversion.

Testing the DSBR Model: Seeing Is Believing

The key prediction of the DSBR model is initiation of meiotic recombination by DSBs. Given the estimate of 150 conversion events per meiosis (Hurst et al. 1972), this would correspond to 150 DSBs in a single nucleus! The idea of intentionally creating 150 DSBs in the yeast genome was heretical to some in the field, and during the late 1980s, there were intensive efforts by proponents of the DSBR model to demonstrate that DSBs were in fact associated with recombination initiation sites. As described above, the first step was to define initiation sites for gene conversion (see Nicolas et al. 1989), which was followed by the use of physical methods to detect breaks in DNA at the initiation sites. This ushered in a new approach to study the mechanisms of recombination with the emphasis on direct physical analysis (by gel electrophoresis) of DNA isolated from sporulating cultures of yeast. At hot spots (initiation sites), recombination can occur in up to 30% of meioses, a level high enough to allow detection not only of the products of recombination reactions, but also of possible DNA intermediates, by Southern blot analysis. In this case, a kinetic analysis can be used to order intermediates in the recombination pathway. Furthermore, physical methods can be used to identify possible recombination defects of mutant strains that fail to sporulate.

In 1989, two papers were published showing formation of meiosis-specific DSBs. Sun et al. (1989) used a low-copy-number plasmid containing a 15-kb insert including the *ARG4* locus to monitor changes in DNA conformation during meiosis by DNA extraction and Southern blot analysis. A linear form of the plasmid was detected during meiosis at the time of recombination. More importantly, these breaks were dependent on the 142-bp sequence identified by Nicolas et al. as the initiation site for gene conversion at *ARG4*. DSBs were also transiently formed at the chromosomal *ARG4* locus coincident with the time of meiotic recombination. Game et al. (1989) used a circular derivative of chromosome III that is unable to enter pulsed-field gels, except when linearized, to show the transient appearance of DSBs coincident with the time at which meiotic recombination occurs.

These two studies were quickly followed by experiments by **Cao et al. (1990)**, who combined classical genetic analysis with the new physical methods to characterize a chromosomal hot spot for meiotic crossing-over. Insertion of a 2.8-kb *LEU2*-containing fragment adjacent to the *HIS4* locus in strain SK1 (this strain shows rapid, synchronous meiosis) fortuitously created a strong hot spot for meiotic crossovers, as determined by tetrad analysis. In strains homozygous for this insertion and heterozygous for flanking *HIS4* and *URA3* markers, the frequency of recombination was increased sixfold, compared with a strain lacking the *LEU2* insertion. The strains created by Cao et al. contained restriction site polymorphisms within the *HIS4* locus and at the site of the *URA3* insertion. In addition to providing genetic markers, these were used to monitor the formation of crossover products by generation of novel restriction fragments during meiosis. This innovative method had been used previously to study formation of recombinants in sporulation-defective mutants (Borts et al. 1984). In agreement with the genetic results, physical analysis revealed a sixfold increase in the percent of recombinant fragments in the strain with the *LEU2* insertion, compared with the strain lacking the insertion. During the course of the physical analysis, two novel DNA species were identified corresponding to DSB fragments. No DSBs were detected in the strain lacking the *LEU2* insertion.

Cao et al. went on to exploit their experimental system in order to gain insight into the role of *trans*-acting factors required for recombination. The *RAD50* and *SPO11* genes were known from genetic studies to be required for initiation of meiotic recombination. DSBs were not detected in strains homozygous for *rad50Δ* or *spo11Δ* mutations. However, DSBs were observed in strains containing certain non-null alleles of *RAD50*, called collectively *rad50S* alleles. The *rad50S* alleles were recovered in a screen for mutations that separate the meiotic and mitotic functions of *RAD50*, hence the designation "S" (Alani et al. 1990). The *rad50S* alleles confer only weak sensitivity to the DNA-damaging agent methyl methane sulfonate (MMS), but cause a complete defect in sporulation. In contrast, the *rad50* null mutants are very sensitive to MMS as well as sporulation defective. Analysis of DNA extracted from a *rad50S* diploid revealed the presence of DSBs at the same positions and with the same kinetics of appearance as found in the *RAD50* parental strain. However, the DSBs observed in the *rad50S* strain differed from those in wild type in two ways. First, although the breaks appeared at the same time as in the wild-type strain, they were stable rather than transient. Second, the DSB fragments in *rad50S* mutants were discrete and of higher molecular weight than the breaks produced in the wild-type strain. Mapping of the breaks using restriction enzymes that cut on either side of the break sites revealed that in the *rad50S* strain, the two ends mapped to the same site. In contrast, the ends of the breaks were about 200 bp shorter in the wild-type strain. Thus, the sporulation defect of the *rad50S* strain is due to the inability to process DSBs.

Characterization of Meiotic DSBs

The studies of Sun et al. (1989) and Cao et al (1990) established that DSBs coincide with hot spots for gene conversion and crossing-over, a key prediction of the DSBR model. The focus of the field was then to characterize the breaks, map break sites at the nucleotide level, and identify the meiotic endonuclease responsible for generating the breaks. Further characterization of the breaks produced at the *ARG4* locus revealed that the ends terminated in 3´ single-stranded tails (Sun et al. 1991). In the *rad50S* mutant, the ends were not processed to yield 3´ single-stranded tails, indicating that 3´ tails are an important intermediate in meiotic recombination. The heterogeneity of DSB fragments suggested that the length of the single-stranded tails was variable.

The stability of DSBs and the discrete nature of the fragments in *rad50S* mutants proved to be invaluable for mapping DSB sites. The sites of DSBs at several recombination hot spots were resolved at the nucleotide level and found to be distributed over a region of about 100 bp, rather than at specific sequences (de Massy et al. 1995; Keeney and Kleckner 1995; Liu et al. 1995; Xu and Petes 1996). In a more global approach, Baudat and Nicolas mapped meiotic DSBs on chromosome III by a combination of pulsed-field and standard gel electrophoresis followed by Southern blotting with specific probes (Baudat and Nicolas 1997). They found that a common feature of meiotic DSBs is their location in intergenic regions and, more particularly, in promoters of transcribed genes.

All hot spots analyzed to date are coincident with regions of open chromatin, as measured by DNase I and micrococcal nuclease hypersensitive sites (Wu and Lichten 1994). However, not all DNase I hypersensitive sites are sites for meiotic DSBs, and some transcription factors, for example, Rap1, are known to influence meiotic DSB formation (White et al. 1993). Nevertheless, transcription per se is not required for meiotic DSB formation. These results suggest that meiotic DSBs are not made by a site-specific nuclease and that their formation is linked with some ill-defined aspect of chromatin structure.

Meiosis-specific DSBs Are Catalyzed by Spo11

Efforts to fine-map DSBs in *rad50S* mutants led to the surprising finding of a protein covalently bound to the 5′ ends (de Massy et al. 1995; Keeney and Kleckner 1995; Liu et al. 1995). Keeney et al. (1997) purified protein-DNA complexes from *rad50S* cells and, by peptide analysis, identified Spo11 as the end-binding protein. Epitope-tagged Spo11 immunoprecipitated from *rad50S* cells was found covalently bound to DNA fragments derived from the *HIS4-LEU2* hot spot, confirming the identity of Spo11 as the protein attached to 5′ ends at break sites (Keeney et al. 1997). At around the same time, Forterre's lab discovered a novel type II topoisomerase (a class of enzymes that make DNA DSBs), topoisomerase VI, in archaebacteria (Bergerat et al. 1997). Topoisomerase VI is composed of two subunits, one of which is homologous to Spo11. Mutation of a conserved tyrosine residue of Spo11 postulated to be required for the transesterification reaction resulted in failure to induce meiotic DSBs (Bergerat et al. 1997). Typically, type II topoisomerases make a transient double-strand cleavage with the catalytic subunit attached to the 5′ ends, followed by strand passage, and then rejoin the broken ends. In yeast, there is no clear homolog of the B subunit of topoisomerase VI, suggesting that Spo11 carries out only the first step of the reaction and the resulting breaks are repaired by homologous recombination instead of rejoining. At this point, it is unknown how Spo11 is removed from 5′ ends or why it remains attached in certain mutant backgrounds. Gerton and collaborators capitalized on the covalent attachment of Spo11 to DNA at hot spots to generate hybridization probes for DNA microarrays and were able to measure the global distribution of DSBs at single-gene resolution (Gerton et al. 2000). Their analysis was consistent with the transcriptional profile noted by Baudat and Nicolas for chromosome III and also indicated nonrandom association of DSBs to DNA of high G + C content.

Identification of Branched DNA Intermediates during Meiosis

Another key prediction of the DSBR model is the formation of an intermediate containing two Holliday junctions. The pioneering work of Bell and Byers established gel electrophoresis conditions in which rare branched DNA molecules could be separated from bulk genomic DNA (Bell and Byers 1983; see also Chapter 3). Bell and Byers used electron microscopy to examine DNA molecules with retarded migration on two-dimensional gels and presented evidence for several types of branched molecules during meiosis. Among the

X-shaped forms, they observed molecules with eye loops, which could be interpreted as double Holliday junction intermediates. Using the same gel method and Southern blotting to identify specific DNA sequences, two groups reported identification of X-form molecules during meiosis (Collins and Newlon 1994; Schwacha and Kleckner 1994). The X forms appeared at recombination hot spots and were dependent on *SPO11*. Analysis of the constituent strands of the X forms was most consistent with a double Holliday junction structure (Schwacha and Kleckner 1994). Kinetic analysis was consistent with the double Holliday junction intermediates appearing after the formation of DSBs and preceding the formation of crossover fragments and the meiosis I division. The purified X forms could be cleaved in vitro with an *Escherichia coli* Holliday junction resolvase to generate both crossover and noncrossover products consistent with their identity as Holliday intermediates (Schwacha and Kleckner 1995).

INCONSISTENCIES WITH THE DSBR MODEL

Although there is no doubt that DSBs initiate meiotic recombination, some observations are inconsistent with the pathway envisioned by the DSBR model being the sole mechanism for repair of DSBs. In mitotic recombination, only about 20% of gene conversion events are associated with exchange of flanking markers, and mating-type switching, a DSB-induced gene conversion event, is associated with crossing-over less than 1% of the time. The synthesis-dependent strand-annealing (SDSA) model was proposed as a mechanism for mitotic DSBR without formation of crossovers (Nassif et al. 1994; Fig. 1 in **Allers and Lichten 2001**). The SDSA model also predicts that the donor duplex is unaltered during repair, which is consistent with studies of mitotic DSBR (Paques et al. 1998).

In 1990, Engebrecht et al. (1990) suggested that meiotic gene conversion could be uncoupled from crossing-over. This conclusion was based on the *mer1* mutant, which exhibits a tenfold decrease in both gene conversion and crossing-over. *MER2* was isolated as a high-copy suppressor of the meiotic gene conversion defect of the *mer1* mutant, but did not suppress the crossover defect. Many other mutants have since been found that exhibit normal gene conversion, but show decreased intergenic recombination (see Allers and Lichten 2001). One explanation for these observations is that several factors influence resolution of the junction. Alternatively, these gene products could convert one type of intermediate, which is only resolved in the noncrossover configuration, to another that gives rise to crossovers. These observations have triggered a reevaluation of the association between conversion and crossing-over and have led to the suggestion of more than one pathway for the repair of meiosis-induced DSBs (Allers and Lichten 2001; Hunter and Kleckner 2001).

If crossovers and noncrossovers are derived from the same intermediate, then the timing of their appearance is expected to be the same. Testing this hypothesis by physical methods is technically challenging because noncrossover products are the same size as parental molecules. Allers and Lichten devised a clever experimental system to distinguish between parental, noncrossover, and crossover products by restriction digestion and Southern blot analysis (Allers and Lichten 2001). The recombination system consists of *ARG4* alleles inserted into the closely linked *HIS4* and *LEU2* loci on chromosome III. Because *ARG4* is inserted at nonhomologous positions, this is referred to as an ectopic recombination assay. The flanking heterologies present physical markers to unambiguously distinguish between parental and recombinant molecules by restriction digestion and strand-specific probes. To facilitate detection of hDNA, which should be present in both noncrossover and crossover products, Allers and Lichten inserted a short palindrome within the coding region of the *ARG4* gene, about 100 bp from a strong DSB site. When palindromes are incorporated into hDNA, they form a stem-loop structure that is refractory to mismatch repair; hence, the palindrome provides a physical marker for hDNA formation (Nag et al. 1989).

Genetic analysis confirmed that the *ARG4* recombination system exhibited normal properties of meiotic recombination: 15% of tetrads showed ectopic crossovers, 6.4% of tetrads displayed full conversion, and 6.0% of tetrads displayed PMS of the *ARG4* alleles. DSBs were made preferentially on the *arg4* allele inserted at the *HIS4* locus, and this gave rise to the expected parity for gene conversion and PMS. Finally, 37% of the conversion tetrads had an associated crossover.

Physical assays were used to show the expected kinetics of appearance and disappearance of DSB fragments and joint molecules, followed by the appearance of crossover fragments. Using this elegant system, Allers and Lichten were able to demonstrate formation of hDNA about 1 hour after DSBs and at around the same time as the appearance of joint molecules. However, the hDNA in the noncrossover products was detected at the same time as the appearance of joint molecules, and this preceded the appearance of hDNA in crossover products by 1 hour. The clear separation in time between the appearance of noncrossover recombinants and crossover recombinants suggests that these two products do not result from alternate resolution of the same intermediate. Further evidence in support of a distinction between crossover and noncrossover products came from analysis of an *ndt80* strain. Together, the kinetic analysis of wild-type strains and *ndt80* mutants suggests two pathways for meiotic recombination. The pathway that gives rise to noncrossover recombinants could occur as envisioned by the SDSA model and the pathway that gives rise to crossovers by the canonical DSBR model. The pathway choice at a particular locus may result in the varying levels of crossing-over associated with gene conversion observed at different loci (18 to 66%).

CONCLUDING REMARKS

Our understanding of the mechanisms of meiotic recombination has been advanced by the combined use of genetic and physical methods in yeast. Studies in yeast were the first to define an initiation site for recombination, DSBs as instigators of recombination, and the protein that makes DSBs. Spo11 is highly conserved in eukaryotes, suggesting that the pathway identified and characterized in yeast is common to all eukaryotes. How DSBs are repaired to generate crossover and noncrossover products remains an open question. This chapter began with a discussion of the Hurst et al. (1972) paper and their demonstration of crossing-over associated with gene conversion. Their data were interpreted as resolution of a single intermediate to generate two types of products: gene conversion unassociated with crossing-over and gene conversion associated with crossing-over. The work of Allers and Lichten (2001) has challenged this view and suggests two pathways for generating the two classes of recombination products.

This chapter has focused on genetic and physical assays to monitor the transfer of genetic information between two DNA duplexes. Our understanding of the proteins that catalyze these events has advanced considerably, but it is far from complete. Genes in the

RAD52 epistasis group were originally identified by their requirement for repair of ionizing radiation–induced DNA damage and subsequently shown to be important for meiotic recombination (Paques and Haber 1999). In addition, many meiosis-specific genes are required for DSB formation and processing. Meiotic DSBs are formed in *rad51*, *rad52*, *rad55*, and *rad57* mutants, but the level of recombinant molecules detected by physical methods is greatly reduced. *RAD51* encodes a homolog of the *E. coli* RecA protein and, like RecA, promotes strand transfer in vitro to generate hDNA, the central intermediate in recombination (Shinohara et al. 1992; Sung 1994). *DMC1* encodes a meiosis-specific RecA homolog that functions with Rad51 to promote strand invasion during meiotic recombination (Bishop et al. 1992). In 1964, Holliday proposed the existence of a repair system to recognize and remove mismatched bases from DNA (Holliday 1964). His hypothesis was correct, and the mismatch repair pathway, as it is now known, is important for the repair of hDNA during recombination resulting in gene conversion, as well as removal of DNA mismatches during DNA synthesis (Harfe and Jinks-Robertson 2000). Although our knowledge of the pairing and strand invasion steps of the recombination reaction, and how DNA mismatches are repaired, has significantly advanced during the last decade, we still know little about selection of DSB sites, DSB processing, and how Holliday junctions are resolved. These issues are likely to keep the field busy for many years to come.

ACKNOWLEDGMENTS

The author would like to thank members of the Symington lab, W.K. Holloman and M. Lichten, for critical comments on the manuscript. Research in the author's lab was supported by grants from the National Institutes of Health (GM41784 and GM54099).

STUDY QUESTIONS

1. What is meant by coconversion and what does coconversion indicate about the mechanism of recombination?

2. What was the evidence that crossing-over in intervals flanking *ARG4* was due to conversion-associated crossovers rather than incidental exchanges (see Table 2 in the paper by Hurst et al. 1972)?

3. What is the interpretation of gene conversion polarity gradients?

4. Does the nature of the mutation affect the type of non-Mendelian segregation or the frequency?

5. Explain the experimental strategy to identify the *ARG4* initiation site for gene conversion.

6. How did Nicolas et al. (1989) test whether the initiation site defined for *ARG4* is the recipient of information during gene conversion? Is disparity a general feature of crosses including heterozygous insertions?

7. What was the evidence that recombination events stimulated by the *LEU2* insertion are characteristic of normal meiotic recombination in yeast?

8. What is the evidence that novel bands correspond to DSB fragments and what was the unusual feature of the DSB fragments revealed by mapping?

9. What is the physical and genetic evidence that DSBs initiate recombination?

10. Describe the marker used by Allers and Lichten (2001) to detect hDNA. How did they propose to distinguish between hDNA in crossover and noncrossover DNA molecules?

11. The conclusions of Allers and Lichten (2001) are critically dependent on the kinetic analysis. Is this justified? What is the genetic evidence for separation of crossover and noncrossover recombinants?

REFERENCES

Alani E., Padmore R., and Kleckner N. 1990. Analysis of wild-type and *rad50* mutants of yeast suggests an intimate relationship between meiotic chromosome synapsis and recombination. *Cell* **61:** 419–436.

Allers T. and Lichten M. 2001. Differential timing and control of noncrossover and crossover recombination during meiosis. *Cell* **106:** 47–57.

Baudat F. and Nicolas A. 1997. Clustering of meiotic double-strand breaks on yeast chromosome III. *Proc. Natl. Acad. Sci.* **94:** 5213–5218.

Bell L.R. and Byers B. 1983. Homologous association of chromosomal DNA during yeast meiosis. *Cold Spring Harbor Symp. Quant. Biol.* **47:** 829–840.

Bergerat A., de Massy B., Gadelle D., Varoutas P.C., Nicolas A., and Forterre P. 1997. An atypical topoisomerase II from Archaea with implications for meiotic recombination. *Nature* **386:** 414–417.

Bishop D.K., Park D., Xu L., and Kleckner N. 1992. DMC1: A meiosis-specific yeast homolog of *E. coli* recA required for recombination, synaptonemal complex formation, and cell cycle progression. *Cell* **69:** 439–456.

Borts R.H., Lichten M., Hearn M., Davidow L.S., and Haber J.E. 1984. Physical monitoring of meiotic recombination in *Saccharomyces cerevisiae*. *Cold Spring Harbor Symp. Quant. Biol.* **49:** 67–76.

Cao L., Alani E., and Kleckner N. 1990. A pathway for generation and processing of double-strand breaks during meiotic recombination in *S. cerevisiae*. *Cell* **61:** 1089–1101.

Collins I. and Newlon C.S. 1994. Meiosis-specific formation of joint DNA molecules containing sequences from homologous chromosomes. *Cell* **76:** 65–75.

de Massy B., Rocco V., and Nicolas A. 1995. The nucleotide mapping of DNA double-strand breaks at the *CYS3* initiation site of meiotic recombination in *Saccharomyces cerevisiae*. *EMBO J.* **14:** 4589–4598.

Engebrecht J., Hirsch J., and Roeder G.S. 1990. Meiotic gene conversion and crossing over: Their relationship to each other and to chromosome synapsis and segregation. *Cell* **62:** 927–937.

Fogel S., Mortimer R., and Lusnak K. 1981. Mechanisms of meiotic gene conversion, or "wanderings on a foreign strand." In *The molecular biology of the yeast* Saccharomyces: *Life cycle and inheritance* (ed. J.N. Strathern et al.), pp. 289–339. Cold Spring Harbor Laboratory, Cold Spring Harbor, New York.

Fogel S., Mortimer R., Lusnak K., and Tavares F. 1979. Meiotic gene conversion: A signal of the basic recombination event in yeast. *Cold Spring Harbor Symp. Quant. Biol.* **43:** 1325–1341.

Game J.C., Sitney K.C., Cook V.E., and Mortimer R.K. 1989. Use of a ring chromosome and pulsed-field gels to study interhomolog recombination, double-strand DNA breaks and sister-chromatid exchange in yeast. *Genetics* **123:** 695–713.

Gerton J.L., DeRisi J., Shroff R., Lichten M., Brown P.O., and Petes T.D. 2000. Inaugural article: Global mapping of meiotic recombination hotspots and coldspots in the yeast *Saccharomyces cerevisiae*. *Proc. Natl. Acad. Sci.* **97:** 11383–11390.

Harfe B.D. and Jinks-Robertson S. 2000. DNA mismatch repair and genetic instability. *Annu. Rev. Genet.* **34:** 359–399.

Holliday R. 1964. A mechanism for gene conversion in fungi. *Genet. Res.* **5:** 282–304.

Hunter N. and Kleckner N. 2001. The single-end invasion: An asymmetric intermediate at the double-strand break to double-holliday junction transition of meiotic recombination. *Cell* **106:** 59–70.

Hurst D.D., Fogel S., and Mortimer R.K. 1972. Conversion-associated recombination in yeast (hybrids/meiosis/tetrads/marker loci/models). *Proc. Natl. Acad. Sci.* **69:** 101–105.

Keeney S. and Kleckner N. 1995. Covalent protein-DNA complexes at the 5′ strand termini of meiosis-specific double-strand breaks in yeast. *Proc. Natl. Acad. Sci.* **92:** 11274–11278.

Keeney S., Giroux C.N., and Kleckner N. 1997. Meiosis-specific DNA double-strand breaks are catalyzed by Spo11, a member of a widely conserved protein family. *Cell* **88:** 375–384.

Kostriken R., Strathern J.N., Klar A.J., Hicks J.B., and Heffron F. 1983. A site-specific endonuclease essential for mating-type switching in *Saccharomyces cerevisiae*. *Cell* **35:** 167–174.

Krogh B.O. and Symington L.S. 2004. Recombination proteins in yeast. *Annu. Rev. Genet.* **38:** 233–271.

Liu J., Wu T.C., and Lichten M. 1995. The location and structure of double-strand DNA breaks induced during yeast meiosis: Evidence for a covalently linked DNA-protein intermediate. *EMBO J.* **14:** 4599–4608.

Meselson M.S. and Radding C.M. 1975. A general model for genetic recombination. *Proc. Natl. Acad. Sci.* **72:** 358–361.

Nag D.K., White M.A., and Petes T.D. 1989. Palindromic sequences in heteroduplex DNA inhibit mismatch repair in yeast. *Nature* **340:** 318–320.

Nassif N., Penney J., Pal S., Engels W.R., and Gloor G.B. 1994. Efficient copying of nonhomologous sequences from ectopic sites via P-element-induced gap repair. *Mol. Cell. Biol.* **14:** 1613–1625.

Nicolas A., Treco D., Schultes N.P., and Szostak J.W. 1989. An initiation site for meiotic gene conversion in the yeast *Saccharomyces cerevisiae*. *Nature* **338:** 35–39.

Orr-Weaver T.L. and Szostak J.W. 1983. Yeast recombination: The association between double-strand gap repair and crossing-over. *Proc. Natl. Acad. Sci.* **80:** 4417–4421.

Orr-Weaver T.L., Szostak J.W., and Rothstein R.J. 1981. Yeast transformation: A model system for the study of recombination. *Proc. Natl. Acad. Sci.* **78:** 6354–6358.

Paques F. and Haber J.E. 1999. Multiple pathways of recombination induced by double-strand breaks in *Saccharomyces cerevisiae*. *Microbiol. Mol. Biol. Rev.* **63:** 349–404.

Paques F., Leung W.Y., and Haber J.E. 1998. Expansions and contractions in a tandem repeat induced by double-strand break repair. *Mol. Cell. Biol.* **18:** 2045–2054.

Schwacha A. and Kleckner N. 1994. Identification of joint molecules that form frequently between homologs but rarely between sister chromatids during yeast meiosis. *Cell* **76:** 51–63.

———. 1995. Identification of double Holliday junctions as intermediates in meiotic recombination. *Cell* **83:** 783–791.

Shinohara A., Ogawa H., and Ogawa T. 1992. Rad51 protein involved in repair and recombination in *S. cerevisiae* is a RecA-like protein (erratum in *Cell* [1992] **71:** following 180). *Cell* **69:** 457–470.

Strathern J.N., Klar A.J., Hicks J.B., Abraham J.A., Ivy J.M., Nasmyth K.A., and McGill C. 1982. Homothallic switching of yeast mating type cassettes is initiated by a double-stranded cut in the MAT locus. *Cell* **31:** 183–192.

Sun H., Treco D., and Szostak J.W. 1991. Extensive 3′-overhanging, single-stranded DNA associated with the meiosis-specific double-strand breaks at the *ARG4* recombination initiation site. *Cell* **64:** 1155–1161.

Sun H., Treco D., Schultes N.P., and Szostak J.W. 1989. Double-strand breaks at an initiation site for meiotic gene conversion. *Nature* **338:** 87–90.

Sung P. 1994. Catalysis of ATP-dependent homologous DNA pairing and strand exchange by yeast RAD51 protein. *Science* **265:** 1241–1243.

Sung P., Trujillo K.M., and Van Komen S. 2000. Recombination factors of *Saccharomyces cerevisiae*. *Mutat. Res.* **451:** 257–275.

Szostak J.W., Orr-Weaver T.L., Rothstein R.J., and Stahl F.W. 1983. The double-strand-break repair model for recombination. *Cell* **33:** 25–35.

White M.A., Dominska M., and Petes T.D. 1993. Transcription factors are required for the meiotic recombination hotspot at the HIS4 locus in *Saccharomyces cerevisiae*. *Proc. Natl. Acad. Sci.* **90:** 6621–6625.

Wu T.C. and Lichten M. 1994. Meiosis-induced double-strand break sites determined by yeast chromatin structure. *Science* **263:** 515–518.

Xu F. and Petes T.D. 1996. Fine-structure mapping of meiosis-specific double-strand DNA breaks at a recombination hotspot associated with an insertion of telomeric sequences upstream of the *HIS4* locus in yeast. *Genetics* **143:** 1115–1125.

3

Chromosome Replication and Segregation

Carol S. Newlon
Department of Microbiology and Molecular Genetics
UMDNJ–New Jersey Medical School
Newark, New Jersey 07101-1709

Replicators: Defining the first eukaryotic replicon, 50
Stinchcomb D.T., Struhl K., and Davis R.W. 1979. Isolation and characterisation of a yeast chromosomal replicator. *Nature* **282:** 39–43.

Replicators: Mapping origins in vivo, 51
Brewer B.J. and Fangman W.L. 1987. The localization of replication origins on ARS plasmids in *S. cerevisiae*. *Cell* **51:** 463–471.

Replicators: Identifying initiator proteins, 52
Bell S.P. and Stillman B. 1992. ATP-dependent recognition of eukaryotic origins of DNA replication by a multiprotein complex. *Nature* **357:** 128–134.

Telomeres, 54
Lundblad V. and Szostak J.W. 1989. A mutant with a defect in telomere elongation leads to senescence in yeast. *Cell* **57:** 633–643.

Centromeres, 57
Clarke L. and Carbon J. 1980a. Isolation of a yeast centromere and construction of functional small circular chromosomes. *Nature* **287:** 504–509.

Note: The landmark papers listed above are those discussed in this chapter. Each landmark paper is preceded by the name of the section (with starting page number) where the paper is first discussed in detail.

THE STABLE TRANSMISSION OF EUKARYOTIC CHROMOSOMES DEPENDS on three *cis*-acting elements: replicators, telomeres, and centromeres. Replicators are the DNA sequences required to direct replication initiation. Telomeres are the physical ends of the linear chromosomal DNA molecules, and they function both to allow complete replication of chromosomal DNA molecules and to "cap" the ends of chromosomes, protecting them from being joined together by the recombinational or end-joining reactions that repair internal double-stranded breaks. Centromeres direct the assembly of kinetochores, which direct chromosome segregation in mitosis and meiosis by attaching chromosomes to the spindle.

Studies of these elements in *Saccharomyces cerevisiae* were instrumental in developing an understanding of how they work. Replicators and centromeres were first cloned and char-

Note: Boldfaced references in the text denote landmark papers that are on the accompanying CD.

acterized in yeast, and studies of telomeres have benefited immensely from the combination of genetic and molecular approaches so readily available in yeast. The analysis of replicators and centromeres also benefited from the serendipity that these elements are much smaller in yeast than in most other organisms, including the fission yeast, *Schizosaccharomyces pombe*. Although it seemed at first that because of their small size the *S. cerevisiae* replicators and centromeres might function in a manner fundamentally different from the corresponding elements in other organisms, most of the proteins that interact with these elements are highly conserved throughout eukaryotes. The papers selected for this chapter focus on the three *cis*-acting chromosomal stability elements.

REPLICATORS: DEFINING THE FIRST EUKARYOTIC REPLICON

By the early 1970s, it was clear that DNA replication in *Escherichia coli* was regulated at the level of initiation, with forks initiated at a single origin of replication replicating the entire chromosome (Bird et al. 1972; Hiraga 1976). Thus, the control of replication in *E. coli* was consistent with the replicon hypothesis of Jacob, Brenner, and Cuzin (Jacob et al. 1964), which postulated a genetically defined "replicator" element controlled by a positive, *trans*-acting regulator. The replicator controlled the replication of the DNA attached to it. The replicator was likely to coincide with the site at which replication actually initiated, the origin of replication, but, in principle, the origin of replication could be some distance away from the replicator.

How replication was regulated in eukaryotic chromosomes was much less clear. In contrast to prokaryotic chromosomes and plasmids, replication initiated at multiple sites along the lengths of eukaryotic chromosomal DNAs (Fig. 1). The larger genome sizes and slower replication fork rates found in eukaryotes required the presence of multiple replicons. However, the observed size of replicons was much smaller than was necessary, based on the measured rates of fork movement and the length of S phase. Even *S. cerevisiae*, with a genome only four times the size of the *E. coli* genome divided among 16 chromosomes, had multiple replication origins per chromosome (Newlon et al. 1974). Furthermore, the spacing between replication origins varied with developmental stage in several organisms, with embryos showing more closely spaced origins than somatic tissues. Therefore, it was questioned whether the replicon model applied to eukaryotes. A competing hypothesis was that replication initiated at random sites determined by the organization of DNA in chromosomes (for review, see DePamphilis 1996).

One approach to this problem was to map the positions of replication bubbles on a chromosome by electron microscopy, an approach that had been used to map origins of replication on several bacteriophage chromosomes. However, the larger size of yeast chromosomes and the problem of uniquely identifying DNA from a particular chromosome complicated this approach. It was the development of a method for transforming yeast (Beggs 1978; Hinnen et al. 1978) and the ability to identify differences in the properties of cells transformed with plasmids carrying different selectable markers that provided the first evidence of replicators in eukaryotic chromosomes (Hsiao and Carbon 1979; Struhl et al. 1979). Each group found that plasmids carrying one selectable marker (*TRP1* or *ARG4*) transformed yeast two to three orders of magnitude more efficiently than plasmids carrying another marker (e.g., *LEU2*) and that the *TRP1* and *ARG4* plasmids were maintained extra-

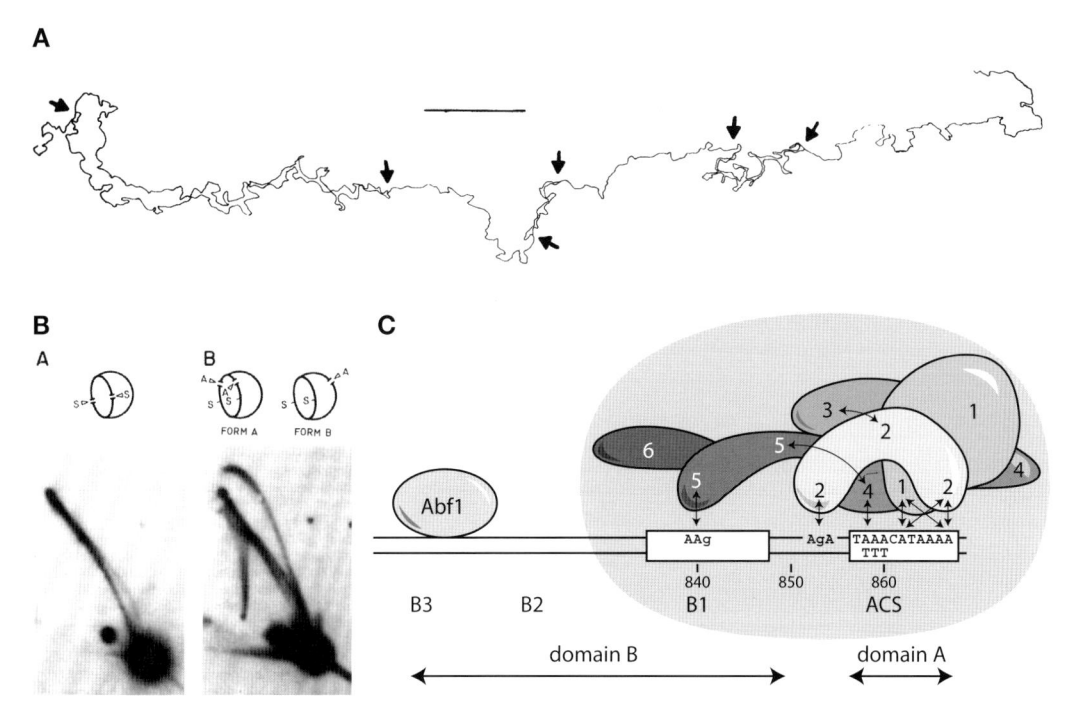

FIGURE 1. Chromosomes contain multiple origins that are recognized by large protein complexes. (*A*) Line drawing from an electron micrograph showing several replication bubbles on one DNA molecule (Newlon et al. 1974). (*B*) Two-dimensional gel electrophoresis of replication intermediates was a crucial element in showing that ARS elements function as chromosomal replication origins. The first dimension (*left to right*) separates the molecules in proportion to their mass at low agarose concentration and low voltage. Electrophoresis in the second dimension (*top to bottom*) at higher agarose concentration and higher voltage exaggerates the contribution of the three-dimensional shape of molecules (Brewer and Fangman 1987). (*C*) Schematic presentation of a yeast origin with the ARS consensus sequence (ACS), the elements of the B domain of the origin and Abf1, and the ORC complex binding to the origin. The ORC complex and its intracomplex interactions are taken from Lee and Bell (1997). (*A*, Reprinted, with permission, from Newlon et al. 1974 [©Nature Publishing Group]; *B*, reprinted, with permission, from Brewer and Fangman 1987 [©Cell Press]; *C*, redrawn, with permission, from Lee and Bell 1997 [©ASM].)

chromosomally. These plasmids must be capable of replicating in yeast, and the sequences that permit replication in yeast were termed ARSs (autonomous replication sequences). These plasmids were, together with the 2-μm plasmids constructed by Jean Beggs, the first "shuttle vectors," capable of replicating in both yeast and *E. coli* and thus able to shuttle DNA sequences between the two organisms. Stinchcomb, Struhl, and Davis (**Stinchcomb et al. 1979**) provided the first characterization of *ARS1*, the replicator associated with *TRP1*.

REPLICATORS: MAPPING ORIGINS IN VIVO

Settling the issue of whether ARS elements represent bona fide replicators required developments of a way to map replication initiation sites relative to the positions of known ARS elements. This question simmered in the background for nearly 10 years as characterization of ARS elements proceeded in several labs. The breakthrough came with the application of two-dimensional agarose gel electrophoresis to replication intermediates (RIs) achieved independently by **Brewer and Fangman (1987)** and Huberman et al. (1987). The Huberman approach was to examine the size of nascent daughter strands that hybridize to small regions of a restriction fragment to deduce the direction of replication fork movement through the restriction fragment. The Brewer and Fangman approach was based on the earlier demonstration by Bell and Byers (1983) that branched DNA molecules do not migrate through gels at the same rate as linear DNA molecules of the same mass (Fig. 1). The key insight that led to the application of the technique to RIs was the realization that RIs of a given restriction fragment could be detected by hybridization, avoiding the need for purifying RIs prior to analysis (B. Brewer, pers. comm.). The first use of both techniques was to map the position of the replication origin of the endogenous yeast 2-μm plasmid, which had been shown by electron microscopy to have Θ-form RIs. Although both techniques have been used for the analysis of chromosomal RIs in a variety of organisms, the Brewer and Fangman technique allows a wide variety of branched molecules, including replication and recombination intermediates, to be directly recognized and analyzed.

Comparisons of the sequences of several ARS elements revealed that they were more A+T-rich than the bulk of the genome and that each one contained a match to an 11-bp sequence, the ARS consensus sequence (ACS) (Broach et al. 1983). Mutagenesis of the ACS in *ARS307* confirmed the importance of this sequence for ARS activity, with a large number of single-base-pair changes abolishing ARS activity (Van Houten and Newlon 1990). These mutations provided the basis for a direct test of whether ARS elements are indeed yeast chromosomal replicators (Deshpande and Newlon 1992). The excellent correlation between the effect of mutations in the ACS on plasmid ARS activity and the effect of the same mutations on chromosomal replication origin activity, measured by two-dimensional gel analysis, provided conclusive evidence that ARS elements are the *cis*-acting replicator sequences that direct replication initiation in yeast chromosomes.

Although a match to the ACS is essential, it is not sufficient for replicator activity. Other required elements are located 3′ to the T-rich strand of the ACS in a region designated domain B (Fig. 1). Any single B element can be mutated or deleted from *ARS1* with only a modest loss of activity, but deletion of any two of the three inactivates the ARS (Marahrens and Stillman 1992).

REPLICATORS: IDENTIFYING INITIATOR PROTEINS

The ACS was likely to represent the binding site for the yeast replicator initiator protein, and from the time that ARS elements were first identified, researchers had been searching for proteins that bound to them. The first ARS-binding protein found was Abf1 (ARS-binding factor 1) (Diffley and Stillman 1988; Sweder et al. 1988), an abundant DNA-binding protein with roles in transcriptional activation and transcriptional silencing as well as DNA replication and repair (for review, see Campbell and Newlon 1991). It did not bind the ACS of *ARS1*, but it did bind about 100 bp away in domain B. Disappointingly, deletion of the Abf1-binding site from *ARS1* had only a modest effect on ARS activity (Diffley and Stillman 1988; Marahrens and Stillman 1992).

Thirteen years after ARS elements were first identified, the six-subunit protein complex that binds the ACS, the origin-recognition complex (ORC), was finally identified (**Bell and Stillman 1992**). ORC is unusual in that it requires ATP for binding to DNA, a property it shares with the *E. coli* DNA replication initiator protein, DnaA. Adding ATP to the DNA-binding reactions was the trick that allowed the identification of ORC. Two other approaches identified subunits of ORC. A genetic approach based on the known role of the ACS in transcriptional silencing at the silent mating-type locus *HMR* identified silencing-defective alleles of *orc2* and *orc5*, which were also shown to have defects in DNA replication initiation (Foss et al. 1993; Micklem et al. 1993). *ORC6* was identified in a pioneering one-hybrid screen that used a fragment of *ARS1* containing the ACS as bait (Li and Herskowitz 1993). A second paper provided at the same time important support for the interaction of ORC with replicators in vivo (Diffley and Cocker 1992). Using high-resolution in vivo footprinting to examine the structure of replicators, Diffley and Cocker found patterns of DNase I protection that strongly resembled the pattern produced by ORC binding in vitro. Further analysis of these in vivo footprints, which revealed a change consistent with the binding of additional components to replicators during G_1, was key to understanding how DNA replication is limited to one and only one round per S phase (see Chapter 6). The proteins recruited to these prereplicative complexes (pre-RCs) include Cdc6 and six minichromosome maintenance (MCM) proteins, which had been identified in genetic screens for mutants defective in the initiation of replication (Aparicio et al. 1997; Donovan et al. 1997).

Early on, it was appreciated that DNA replication in eukaryotes is temporally controlled, with particular DNA sequences replicating at reproducible times in every S phase. Experiments in *S. cerevisiae*, which made use of density transfer methods similar to those of Meselson and Stahl (1959), demonstrated that centromeres tend to replicate early and telomeres late in S phase (McCarroll and Fangman 1988) and that replicators on chromosome III initiate at reproducible times during S phase (Reynolds et al. 1989). The observation that the time of initiation of the *ARS1* replicator could be changed from early to late in S phase by inserting it near a telomere suggested that telomeres establish a late-replicating domain on chromosomes (Ferguson and Fangman 1992). In an elegant set of experiments that made use of a replicator flanked by site-specific recombination sites that could be excised from a late-replicating region of a chromosome, Raghuraman and colleagues showed that initiation timing is determined at the same time during G_1 that replicators are assembled into pre-RCs competent for initiating replication (Raghuraman et al. 1997). The molecular basis of timing determinants is not well understood, although recent work suggests that chromatin structure has an important role (Vogelauer et al. 2002).

The past few years have seen tremendous progress in genomic approaches to studying chromosome replication. Systematic mapping of ARS elements and chromosomal replicators on chromosome III (Newlon et al. 1991, 1993; Poloumienko et al. 2001), chromosome VI (Shirahige et al. 1993; Friedman et al. 1997; Yamashita et al. 1997), and part of chromosome XIV (Friedman et al. 1996) provided a collection of ARS elements for structural analysis as well as an important means to validate genome-wide approaches to the identification of ARS elements and chromosomal replicators. Three independent approaches have yielded a comfortingly consistent view of genome replication. The first approach combined classical density transfer experiments with microarray analysis to determine the time of replication of thousands of sequences spanning the entire genome (Raghuraman et al. 2001). The replication profiles of each chromosome could be used to deduce the positions of active replicators, which replicate earlier than flanking sequences, and the rates of replication fork movement. In the second approach, chromatin immunoprecipitation analysis was used to map the positions of binding of ORC and MCM proteins throughout the genome (Wyrick et al 2001). Finally, using the sequences of 26 ARS elements whose essential ACS had been identified by mutational analysis to develop an algorithm for predicting the locations of ARS elements, Breier et al. (2004) succeeded in identifying a large fraction of replicators in the yeast genome.

TELOMERES

Since the experiments of H.J. Muller (1938) and B. McClintock (1939, 1941) on X-ray-induced breakage of chromosomes in the 1930s and 1940s, it has been recognized that the natural ends of chromosomes behave very differently from ends created by chromosomal breakage. Broken ends are highly recombinogenic, fusing with other chromosomes to produce chromosomal rearrangements and dicentric chromosomes. In contrast, the natural ends of chromosomes are stable, as if they are "capped" in some way. Muller coined the word telomere, meaning end part, in recognition of the special function required for sealing the ends of chromosomes (Gall 1995). When it became clear in the 1960s and early 1970s that eukaryotic chromosomes contain single linear DNA molecules, it also became clear that telomeres must solve another problem, crystallized by Watson (1972) as the "end-replication" problem. DNA polymerases cannot initiate DNA synthesis, but instead extend a preexisting primer (usually RNA). Because DNA polymerases synthesize DNA only in the 5´ to 3´ direction, only one of the two daughter strands can be replicated continuously in the 5´ to 3´ direction, the "leading strand." The other strand, the "lagging strand," is replicated as a series of short fragments, each of which is initiated with an RNA primer, that are processed and ligated together. As a fork approaches the end of a linear DNA, the lagging strand cannot be replicated completely because no DNA polymerase could fill in the gap left at the very end after removal of the primer.

Whereas circular bacterial genomes avoid the problem, studies of the replication of linear bacteriophage and mammalian viral genomes had revealed a number of different solutions. Bacteriophage DNAs often form circles or concatamers by the interaction of terminally redundant sequences, effectively eliminating the ends and thus the end-replication problem. Adenoviruses and some bacteriophages use a nucleotide covalently linked to a terminal protein to initiate replication at the very ends of the linear genome. Vaccinia virus has hairpin termini on its DNA, with replication producing a dimer circle, which is thought to be resolved by endonucleolytic nicking and reformation of monomer linear molecules with hairpin ends. Other viruses have palindromic sequences at the ends of linear molecules, so that the 3´ end can fold back on itself and prime synthesis to fill in the gap on the other strand. All of these models can explain the replication of telomeres. However, the first telomeres that were characterized were isolated from *Tetrahymena* ribosomal DNA, which is present in the macronucleus as linear DNA molecules that could be separated from other genomic DNA on the basis of its higher density (Blackburn and Gall 1978). Although the replication mechanism was unclear, these ciliate telomeres were composed of tandem repeats of a simple 6-bp sequence, C_4A_2/T_2G_4.

Szostak and Blackburn (1982) collaborated on a bold and technically difficult set of experiments to clone yeast telomeres. Their approach to the problem was to determine whether *Tetrahymena* telomeres would function in yeast. They created an "artificial chromosome" made from a linearized yeast autonomously replicating plasmid flanked by *Tetrahymena* telomeres and asked whether or not the linear plasmid would be maintained in yeast. At the time, ciliates such as *Tetrahymena* appeared to be very different from other eukaryotes, and the idea that telomeres would be conserved over the evolutionary distance between ciliates and yeast was met with skepticism. Since linear plasmids are not maintained by *E. coli*, circular plasmids had to be isolated and converted to linear molecules in sufficient quantity to directly transform yeast. They proceeded by ligating purified fragments carrying *Tetrahymena* telomeres to a linearized shuttle vector carrying *ARS1* and *LEU2* and used purified DNA of the expected length to transform yeast. Amazingly, the experiment worked! The linear plasmids, still carrying *Tetrahymena* telomeric sequences, were maintained in yeast, which seemed to add a few hundred base pairs of DNA onto the ends. By cutting off one of the *Tetrahymena* telomeres and ligating restriction-digested yeast DNA to the cut plasmid, they were able to clone several yeast telomeres. All of these cloned telomeres were very similar, suggesting that all yeast chromosomes carry common telomere sequences.

The next few years saw rapid advances in understanding the structure of telomeres (Fig. 2). Unlike the regular hexanucleotide repeats found in *Tetrahymena*, the yeast terminal repeat is somewhat degenerate, with the sequence of the G-rich strand, which forms the 3´ end of the telomere, represented by a $T(G)_{1-3}$ motif (Shampay et al. 1984; Walmsley et al. 1984). On the centromere-facing side of the approximately 300 bp of telomeric repeats are sequences known as telomere-associated DNAs, the Y´ and X elements (Chan and Tye 1983). About two thirds of telomeres carry one to four copies of the 6.7-kb Y´ element. The X element, which is composed of several small repeats (Louis et al. 1994), is located interior to Y´, or adjacent to the telomeric repeats on ends that lack Y´. Chromosomes lack-

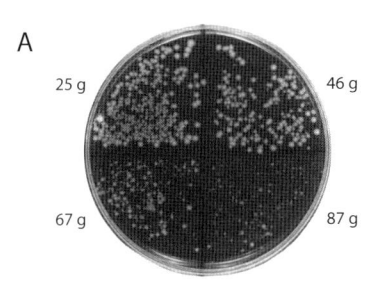

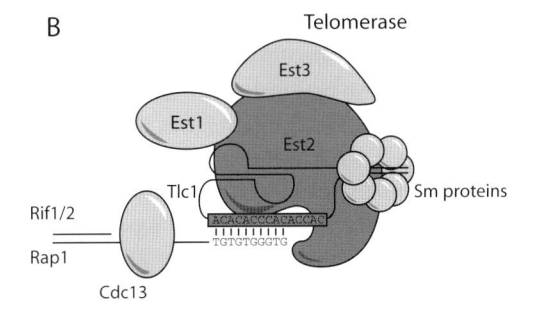

FIGURE 2. Telomeres prevent shortening of chromosomes and end-to-end joining. (*A*) The *est1-1* mutant was streaked after 25, 46, 67, and 87 generations of growth of a freshly germinated spore. The reduced viability is clearly seen on this plate (Lundblad and Szostak 1989). (*B*) Schematic representation of the telomerase as discussed in the text. (*A*, Reprinted, with permission, from Lundblad and Szostak 1989 [©Cell Press].)

The small size of yeast chromosomes made it difficult to observe them by light microscopy. However, centromeres can be recognized genetically by their ability to direct the first meiotic division segregation of genes adjacent to them because recombination is limited in short intervals. Centromere-linked markers had been mapped on yeast chromosomes by the 1960s.

With the advent of gene cloning, John Carbon and colleagues set out to isolate and characterize yeast centromeric DNA. Their strategy was to identify overlapping fragments of chromosomal DNA that covered the interval between a pair of genes, *LEU2* and *PGK1*, that mapped on opposite sides of the centromere of yeast chromosome III. They cloned *LEU2* by virtue of its ability to complement an *E. coli leuB6* mutation (Ratzkin and Carbon 1977) and attempted to "walk" in both directions from the cloned gene by overlap hybridization (Chinault and Carbon 1979). Using this approach, they were able to assemble a contiguous segment, now called a "contig," of approximately 30 kb of DNA in one direction, but they were stymied by the presence of a repetitive element (which turned out to be the first Ty2 element) on the other side of *LEU2* (Kingsman et al. 1981). It was not clear whether their contig extended toward the centromere or toward the telomere from *LEU2*. They developed an immunologic screening approach to identify clones containing the *PGK1* gene (Hitzeman et al. 1980). However, before an extensive chromosome walk was undertaken from *PGK1*, the development of yeast transformation (Beggs 1978; Hinnen et al. 1978) provided the opportunity to directly complement the temperature-sensitive *cdc10* mutation, which was very tightly linked to *CEN3* (Clarke and Carbon 1980b). The *CDC10* clone overlapped the *LEU2* contig, demonstrating that the chromosome walk from *LEU2* was toward the centromere, and additional hybridizations with the other end of the *CDC10* clone were used to extend the "contig." Assays of the stability and meiotic segregation of plasmids carrying segments of this contig ultimately provided evidence of the presence of a centromere (*CEN3*) adjacent to *CDC10* (**Clarke and Carbon 1980a**).

Other centromeres were identified quickly, using the mitotic stabilization of plasmids as an assay. The development of colony color assays, based on the red pigment that accumulates in certain adenine auxotrophs, that allow chromosome or plasmid loss to be detected by simple visual examination of colonies facilitated this effort (Hieter et al. 1985a; Koshland et al. 1985; Koshland and Hieter 1987). Although ARS plasmids are greatly stabilized by the addition of a centromere, CEN-ARS plasmids are still about three orders of magnitude less stable than natural chromosomes (nondisjunction rates of 10^{-2} vs. 10^{-5}). The lower stability of plasmids is largely the result of their small size (Murray and Szostak 1983). Therefore, the detection of subtle defects in centromere function required the use of more sensitive assay systems, which were provided by the introduction of mutant centromeres into nonessential chromosome fragments (Hegemann et al. 1988).

DNA sequence comparisons of the centromeres revealed a region of about 125 bp that consisted of two short, highly conserved elements (8-bp CDEI and 25-bp CDEIII) separated by an approximately 80-bp region, CDEII, that was highly A+T-rich, but whose sequence varied from centromere to centromere (Fig. 3) (Fitzgerald-Hayes et al. 1982; Hieter et al. 1985b). Mutational analysis examining mitotic segregation revealed that mutations in CDEIII had the largest effect on centromere function, with some point mutations in the center of the 25-bp imperfect palindrome completely abrogating activity (McGrew et al. 1986; Hegemann et al. 1988). While removal of CDEII inactivated the centromere,

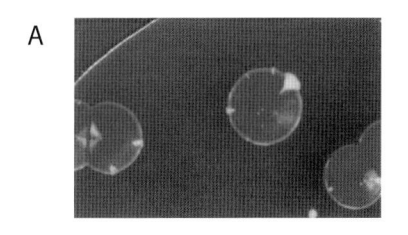

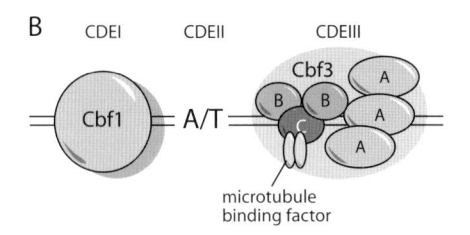

FIGURE 3. Centromeres are important for chromosome stability. (*A*) Colonies from a strain disomic for chromosome VII. One copy of chromosome VII harbors the *ADE3* allele, whereas the other harbors a *ade3* mutant allele. The segregation rate is 2×10^{-5} (Hartwell et al. 1982). The photograph, in its original publication, presents red colonies with white sectors (Koshland et al. 1985). (*B*) Schematic drawing of a *S. cerevisiae* centromere with Cbf1 binding as a dimer to CDEI and the Cbf3 complex binding to CDEIII. At least 70 proteins are now known to interact with centromeres. (*A*, Reprinted, with permission, from Koshland et al. 1985 [©Cell Press].)

small deletions within CDEII had only modest effects (Gaudet and Fitzgerald-Hayes 1987). Mutations in CDEI, including a complete deletion, caused only a modest 10–30-fold increase in loss rate (Niedenthal et al. 1991).

Both biochemical and genetic methods were used to identify kinetochore proteins. Biochemical approaches made use of wild-type and mutant centromeres as substrates for gel shift and footprinting assays, and for DNA-affinity purifications. The primary genetic approach, taken by a large number of labs, was to identify mutations that affected chromosome stability. These mutations affect genes required for a variety of processes, including DNA replication, spindle assembly, sister chromatid cohesion, and kinetochore assembly. The identification of the kinetochore-specific class of mutations required the development of secondary screens, exemplified by the work of Doheny et al. (1993), who looked for the subset of mutants that increased the stability of a dicentric plasmid and also relieved a transcriptional block caused by inserting a centromere between a reporter gene and its promoter.

The two approaches converged in the identification of the protein complex that interacts with CDEIII, CBF3. This complex was first purified by affinity chromatography, a technically demanding purification because of the low abundance of the protein (Ng and Carbon 1987; Lechner and Carbon 1991). The purified complex contained three major subunits of 110, 64, and 58 kD and several minor bands. As the three subunits of the complex were purified and sequenced to isolate the genes encoding them, the genes were identified in genetic screens for mutations that might affect kinetochore function. Thus, the purified 110-kD subunit was used to identify the *CBF2* gene (Jiang et al. 1993); the same gene was identified as a mutation that prevents chromosome segregation, *ndc10-1* (Goh and Kilmartin 1993), and as a chromosome transmission fidelity mutation that impaired kinetochore activity in two secondary screens for kinetochore function, *ctf14-2* (Doheny et al. 1993). Similarly, the second mutation with impaired kinetochore activity, *ctf13-30*, was shown by immunological methods to be part of CBF3 (Doheny et al. 1993) and was later shown by protein sequencing to encode the 58-kD subunit (Lechner 1994). The 64-kD subunit was also identified in genetic screens and by protein sequencing (Lechner 1994;

Strunnikov et al. 1995). Finally, a fourth subunit encoded by *SKP1*, which was not identified in the purified CBF3 complex, was found as a high-copy suppressor of *ctf13-30* (Connelly and Hieter 1996).

These same approaches have been used to identify proteins interacting directly with CDEI and CDEII, as well as further complexes that form the central layer of the kinetochore and are recruited to the kinetochore by the proteins that interact directly with centromere DNA. Connections are being made between the central layer of the kinetochore and outer kinetochore proteins, which are thought to mediate the interaction of the kinetochore with spindle microtubules. The observation that many of these kinetochore proteins have homologs in other eukaryotes strongly suggests that yeast kinetochores work in fundamentally the same way as the much larger kinetochores of fission yeast and metazoans (for review, see McAinsh et al. 2003).

SUMMARY

The isolation and genetic characterization of the *cis*-acting chromosomal stability elements of *Saccharomyces cerevisiae* answered fundamental questions about the replication and segregation of eukaryotic chromosomes, providing the first direct evidence that replicators are at defined places, defining the structure of telomeres, and providing the first molecular view of a centromere. These elements have been used to identify proteins in replication initiation, telomere replication and capping, and chromosome segregation. They have also been combined to produce valuable tools for chromosome manipulations, gene expression, and cloning DNA from other organisms in yeast.

ACKNOWLEDGMENTS

I thank Drs. Michael Newlon and James Theis for helpful comments on the manuscript. Work in the author's lab was supported by National Institutes of Health grant GM-35679.

STUDY QUESTIONS

1. Why do ARS-containing plasmids transform yeast at high frequency?

2. What is the logic behind the *cis–trans* test of ARS function?

3. How can the instability of ARS plasmids be explained?

4. Is it surprising that ARS1 failed to cross-hybridize to other chromosomal sequences in the *S. cerevisiae* genome? What are the limits on the length of DNA sequence homology shared between ARS1 and other ARS elements in the genome?

5. What is a "Cairns" structure? What is the evidence that the arc of Cairns structures represents replication intermediates?

6. Is the resolution of the two-dimensional gel technique sufficient to conclude that the replication origin is precisely coincident with the ARS element?

7. What properties of ORC made it so hard to detect?

8. What is the evidence that ORC is a six-subunit complex? What further experiments would address this issue?

9. How might ATP be involved in DNA binding by ORC? Did the use of ATP analogs significantly limit the possible functions?

10. What was the basis of the *EST* mutant screen and what were the underlying assumptions?

11. What might have been the defects in the mutants that showed an increased frequency of conversion of circular to linear plasmids?

12. What was the challenge in cloning the wild-type *EST1* gene?

13. What explanations other than the presence of a centromere might account for the increased mitotic stability of a subset of the plasmids in the *LEU2* contig?

14. Explain how the meiotic segregation of pYe(*CDC10*)1 supports the idea that the plasmid carries a centromere. How were sister spores identified?

15. What is the likely explanation for tetrads showing a 4+:0– segregation of *TRP1* for those showing a 0+:4– segregation of *TRP1*? Why would you not expect 1+:3– and 3+:1– segregations of a centromere-containing plasmid?

16. What is the genetic evidence that the small centromere-containing plasmid does not pair with chromosomes?

REFERENCES

Aparicio O.M., Weinstein D.M., and Bell S.P. 1997. Components and dynamics of DNA replication complexes in *S. cerevisiae*: Redistribution of MCM proteins and Cdc45p during S phase. *Cell* **91:** 59–69.

Beggs J. 1978. Transformation of yeast by a replicating hybrid plasmid. *Nature* **275:** 104–109.

Bell L. and Byers B. 1983. Separation of branched from linear DNA by two-dimensional gel electrophoresis. *Anal. Biochem.* **130:** 527–535.

Bell S.P. and Stillman B. 1992. ATP-dependent recognition of eukaryotic origins of DNA replication by a multiprotein complex. *Nature* **357:** 128–134.

Bird R.E., Louarn J., Martuscelli J., and Caro L. 1972. Origin and sequence of chromosome replication in *Escherichia coli*. *J. Mol. Biol.* **70:** 549–566.

Blackburn E.H. and Gall J.G. 1978. A tandemly repeated sequence at the termini of the extrachromosomal ribosomal RNA genes in *Tetrahymena*. *J. Mol. Biol.* **120:** 33–53.

Breier A., Chatterji S., and Cozzarelli N. 2004. Prediction of *Saccharomyces cerevisiae* replication origins. *Genome Biol.* **5:** R22.

Brewer B.J. and Fangman W.L. 1987. The localization of replication origins on ARS plasmids in *S. cerevisiae*. *Cell* **51:** 463–471.

Broach J.R., Li Y.Y., Feldman J., Jayaram M., Abraham J., Nasmyth K.A., and Hicks J.B. 1983. Localization and sequence analysis of yeast origins of DNA replication. *Cold Spring Harbor Symp. Quant. Biol.* **47:** 1165–1173.

Campbell J.L. and Newlon C.S. 1991. Chromosomal DNA replication. In *The Molecular and cellular biology of the yeast* Saccharomyces: *Genome dynamics, protein synthesis, and energetics* (ed. J.R. Broach et al.), pp. 41–146. Cold Spring Harbor Laboratory Press, Cold Spring Harbor, New York.

Chan C.S. and Tye B.K. 1983. Organization of DNA sequences and replication origins at yeast telomeres. *Cell* **33:** 563–573.

Chinault A.C. and Carbon J. 1979. Overlap hybridization screening: Isolation and characterization of overlap-

ping DNA fragments surrounding the *leu2* gene on yeast chromosome III. *Gene* **5**: 111–126.
Clarke L. and Carbon J. 1980a. Isolation of a yeast centromere and construction of functional small circular chromosomes. *Nature* **287**: 504–509.
———. 1980b. Isolation of the centromere-linked *CDC10* gene by complementation in yeast. *Proc. Natl. Acad. Sci.* **77**: 2173–2177.
Connelly C. and Hieter P. 1996. Budding yeast SKP1 encodes an evolutionarily conserved kinetochore protein required for cell cycle progression. *Cell* **86**: 275–285.
DePamphilis M.L. 1996. Origins of DNA replication. In *DNA replication in eukaryotic cells* (ed. M.L. DePamphilis), pp. 45–86. Cold Spring Harbor Laboratory Press, Cold Spring Harbor, New York.
Deshpande A.M. and Newlon C.S. 1992. The ARS consensus sequence is required for chromosomal origin function in *Saccharomyces cerevisiae*. *Mol. Cell. Biol.* **12**: 4305–4513.
Diffley J.F. and Cocker J.H. 1992. Protein-DNA interactions at a yeast replication origin. *Nature* **357**: 169–172.
Diffley J.F. and Stillman B. 1988. Purification of a yeast protein that binds to origins of DNA replication and a transcriptional silencer. *Proc. Natl. Acad. Sci.* **85**: 2120–2124.
Doheny K.F., Sorger P.K., Hyman A.A., Tugendreich S., Spencer F., and Hieter P. 1993. Identification of essential components of the *S. cerevisiae* kinetochore. *Cell* **73**: 761–774.
Donovan S., Harwood J., Drury L.S., and Diffley J.F. 1997. Cdc6p-dependent loading of Mcm proteins onto pre-replicative chromatin in budding yeast. *Proc. Natl. Acad. Sci.* **94**: 5611–5616.
Dunn B., Szauter P., Pardue M.L., and Szostak J.W. 1984. Transfer of yeast telomeres to linear plasmids by recombination. *Cell* **39**: 191–201.
Evans S.K. and Lundblad V. 1999. Est1 and Cdc13 as comediators of telomerase access. *Science* **286**: 117–120.
Ferguson B.M. and Fangman W.L. 1992. A position effect on the time of replication origin activation in yeast. *Cell* **68**: 333–339.
Fitzgerald-Hayes M., Clarke L., and Carbon J. 1982. Nucleotide sequence comparisons and functional analysis of yeast centromere DNAs. *Cell* **29**: 235–244.
Foss M., McNally F.J., Laurenson P., and Rine J. 1993. Origin recognition complex (ORC) in transcriptional silencing and DNA replication in *S. cerevisiae*. *Science* **262**: 1838–1844.
Friedman K.L., Brewer B.J., and Fangman W.L. 1997. Replication profile of *Saccharomyces cerevisiae* chromosome VI. *Genes Cells* **2**: 667–678.
Friedman K.L., Diller J.D., Ferguson B.M., Nyland S.V., Brewer B.J., and Fangman W.L. 1996. Multiple determinants controlling activation of yeast replication origins late in S phase. *Genes Dev.* **10**: 1595–1607.
Gall J.G. 1995. History of telomeres. In *Telomeres* (ed. E.H. Blackburn and C.W. Greider), pp. 1–10. Cold Spring Harbor Laboratory Press, Cold Spring Harbor, New York.
Gaudet A. and Fitzgerald-Hayes M. 1987. Alterations in the adenine-plus-thymine-rich region of CEN3 affect centromere function in *Saccharomyces cerevisiae*. *Mol. Cell. Biol.* **7**: 68–75.
Goh P.Y. and Kilmartin J.V. 1993. *NDC10*: A gene involved in chromosome segregation in *Saccharomyces cerevisiae*. *J. Cell Biol.* **121**: 503–512.
Gottschling D.E., Aparicio O.M., Billington B.L., and Zakian V.A. 1990. Position effect at *S. cerevisiae* telomeres: Reversible repression of Pol II transcription. *Cell* **63**: 751–762.
Greider C.W. and Blackburn E.H. 1985. Identification of a specific telomere terminal transferase activity in *Tetrahymena* extracts. *Cell* **43**: 405–413.
———. 1989. A telomeric sequence in the RNA of *Tetrahymena* telomerase required for telomere repeat synthesis. *Nature* **337**: 331–337.
Hartwell L.H., Dutcher S.K., Wood J.S., and Garvik B. 1982. The fidelity of mitotic chromosome reproduction in *S. cerevisiae*. *Recent Adv. Yeast Mol. Biol.* **1**: 28–38.
Hegemann J.H., Shero J.H., Cottarel G., Philippsen P., and Hieter P. 1988. Mutational analysis of centromere DNA from chromosome VI of *Saccharomyces cerevisiae*. *Mol. Cell. Biol.* **8**: 2523–2535.
Henson J.D., Neumann A.A., Yeager T.R., and Reddel R.R. 2002. Alternative lengthening of telomeres in mammalian cells. *Oncogene* **21**: 598–610.
Hieter P., Mann C., Snyder M., and Davis R.W. 1985a. Mitotic stability of yeast chromosomes: A colony color assay that measures nondisjunction and chromosome loss. *Cell* **40**: 381–392.
Hieter P., Pridmore D., Hegemann J.H., Thomas M., Davis R.W., and Philippsen P. 1985b. Functional selection and analysis of yeast centromeric DNA. *Cell* **42**: 913–921.

Hinnen A., Hicks J.B., and Fink G.R. 1978. Transformation of yeast. *Proc. Natl. Acad. Sci.* **75:** 1929–1933.

Hiraga S. 1976. Novel F prime factors able to replicate in *Escherichia coli* Hfr strains. *Proc. Natl. Acad. Sci.* **73:** 198–202.

Hitzeman R.A., Clarke L., and Carbon J. 1980. Isolation and characterization of the yeast 3-phosphoglycerokinase gene (*PGK*) by an immunological screening technique. *J. Biol. Chem.* **255:** 12073–12080.

Hsiao C.L. and Carbon J. 1979. High-frequency transformation of yeast by plasmids containing the cloned yeast *ARG4* gene. *Proc. Natl. Acad. Sci.* **76:** 3829–3833.

Huberman J.A., Spotila L.D., Nawotka K.A., el-Assouli S.M., and Davis L.R. 1987. The in vivo replication origin of the yeast 2 micron plasmid. *Cell* **51:** 473–481.

Hughes T.R., Evans S.K., Weilbaecher R.G., and Lundblad V. 2000. The Est3 protein is a subunit of yeast telomerase. *Curr. Biol.* **10:** 809–812.

Jacob F., Brenner S., and Cuzin F. 1964. On the regulation of DNA replication in bacteria. *Cold Spring Harbor Symp. Quant. Biol.* **28:** 329–348.

Jiang W., Lechner J., and Carbon J. 1993. Isolation and characterization of a gene (*CBF2*) specifying a protein component of the budding yeast kinetochore. *J. Cell Biol.* **121:** 513–519.

Kingsman A.J., Gimlich R.L., Clarke L., Chinault A.C., and Carbon J. 1981. Sequence variation in dispersed repetitive sequences in *Saccharomyces cerevisiae*. *J. Mol. Biol.* **145:** 619–632.

Koshland D. and Hieter P. 1987. Visual assay for chromosome ploidy. *Methods Enzymol.* **155:** 351–372.

Koshland D., Kent J.C., and Hartwell L.H. 1985. Genetic analysis of the mitotic transmission of minichromosomes. *Cell* **40:** 393–403.

Lechner J. 1994. A zinc finger protein, essential for chromosome segregation, constitutes a putative DNA binding subunit of the *Saccharomyces cerevisiae* kinetochore complex, Cbf3. *EMBO J.* **13:** 5203–5211.

Lechner J. and Carbon J. 1991. A 240 kd multisubunit protein complex, CBF3, is a major component of the budding yeast centromere. *Cell* **64:** 717–725.

Lee D.G. and Bell S.P. 1997. Architecture of the yeast origin recognition complex bound to origins of DNA replication. *Mol. Cell. Biol.* **17:** 7159–7168.

Lendvay T.S., Morris D.K., Sah J., Balasubramanian B., and Lundblad V. 1996. Senescence mutants of *Saccharomyces cerevisiae* with a defect in telomere replication identify three additional EST genes. *Genetics* **144:** 1399–1412.

Li J.J. and Herskowitz I. 1993. Isolation of *ORC6*, a component of the yeast origin recognition complex by a one-hybrid system. *Science* **262:** 1870–1874.

Lingner J., Hughes T.R., Shevchenko A., Mann M., Lundblad V., and Cech T.R. 1997. Reverse transcriptase motifs in the catalytic subunit of telomerase. *Science* **276:** 561–567.

Louis E.J., Naumova E.S., Lee A., Naumov G., and Haber J.E. 1994. The chromosome end in yeast: Its mosaic nature and influence on recombinational dynamics. *Genetics* **136:** 789–802.

Lundblad V. and Blackburn E.H. 1993. An alternative pathway for yeast telomere maintenance rescues *est1*-senescence. *Cell* **73:** 347–360.

Lundblad V. and Szostak J.W. 1989. A mutant with a defect in telomere elongation leads to senescence in yeast. *Cell* **57:** 633–643.

Marahrens Y. and Stillman B. 1992. A yeast chromosomal origin of DNA replication defined by multiple functional elements. *Science* **255:** 817–823.

Marcand S., Gilson E., and Shore D. 1997. A protein-counting mechanism for telomere length regulation in yeast. *Science* **275:** 986–990.

McAinsh A.D., Tytell J.D., and Sorger P.K. 2003. Structure, function, and regulation of budding yeast kinetochores. *Annu. Rev. Cell Dev. Biol.* **19:** 519–539.

McCarroll R.M. and Fangman W.L. 1988. Time of replication of yeast centromeres and telomeres. *Cell* **54:** 505–513.

McClintock B. 1939. The behavior in successive nuclear divisions of a chromosome broken at meiosis. *Proc. Natl. Acad. Sci.* **25:** 405–416.

―――. 1941. The stability of broken ends of chromosomes in *Zea mays*. *Genetics* **26:** 234–282.

McGrew J., Diehl B., and Fitzgerald-Hayes M. 1986. Single base-pair mutations in centromere element III cause aberrant chromosome segregation in *Saccharomyces cerevisiae*. *Mol. Cell. Biol.* **6:** 530–538.

Meselson M. and Stahl F.W. 1959. The replication of DNA. *Cold Spring Harbor Symp. Quant. Biol.* **23:** 9–11.

Micklem G., Rowley A., Harwood J., Nasmyth K., and Diffley J.F. 1993. Yeast origin recognition complex is involved in DNA replication and transcriptional silencing. *Nature* **366:** 87–89.

Moazed D. 2001. Common themes in mechanisms of gene silencing. *Mol. Cell* **8:** 489–498.

Muller H.J. 1938. The remaking of chromosomes. *The Collecting Net* **13:** 181–195, 198.

Murray A.W. and Szostak J.W. 1983. Construction of artificial chromosomes in yeast. *Nature* **305:** 189–193.

———. 1986. Construction and behavior of circularly permuted and telocentric chromosomes in *Saccharomyces cerevisiae*. *Mol. Cell. Biol.* **6:** 3166–3172.

Murray A.W., Claus T.E., and Szostak J.W. 1988. Characterization of two telomeric DNA processing reactions in *Saccharomyces cerevisiae*. *Mol. Cell. Biol.* **8:** 4642–4650.

Newlon C.S., Petes T.D., Hereford L.M., and Fangman W.L. 1974. Replication of yeast chromosomal DNA. *Nature* **247:** 32–35.

Newlon C.S., Collins I., Dershowitz A., Deshpande A.M., Greenfeder S.A., Ong L.Y., and Theis J.F. 1993. Analysis of replication origin function on chromosome III of *Saccharomyces cerevisiae*. *Cold Spring Harbor Symp. Quant. Biol.* **58:** 415–423.

Newlon C.S., Lipchitz L.R., Collins I., Deshpande A., Devenish R.J., Green R.P., Klein H.L., Palzkill T.G., Ren R.B., Synn S., and Woody S.T. 1991. Analysis of a circular derivative of *Saccharomyces cerevisiae* chromosome III: A physical map and identification and location of ARS elements. *Genetics* **129:** 343–357.

Ng R. and Carbon J. 1987. Mutational and in vitro protein-binding studies on centromere DNA from *Saccharomyces cerevisiae*. *Mol. Cell. Biol.* **7:** 4522–4534.

Niedenthal R., Stoll R., and Hegemann J.H. 1991. In vivo characterization of the *Saccharomyces cerevisiae* centromere DNA element I, a binding site for the helix-loop-helix protein CPF1. *Mol. Cell. Biol.* **11:** 3545–3553.

Pennock E., Buckley K., and Lundblad V. 2001. Cdc13 delivers separate complexes to the telomere for end protection and replication. *Cell* **104:** 387–396.

Pluta A.F. and Zakian V.A. 1989. Recombination occurs during telomere formation in yeast. *Nature* **337:** 429–433.

Poloumienko A., Dershowitz A., De J., and Newlon C. 2001. Completion of replication map of *Saccharomyces cerevisiae* chromosome III. *Mol. Biol. Cell* **12:** 3317–3327.

Qi H. and Zakian V.A. 2000. The *Saccharomyces* telomere-binding protein Cdc13p interacts with both the catalytic subunit of DNA polymerase alpha and the telomerase-associated est1 protein. *Genes Dev.* **14:** 1777–1788.

Raghuraman M.K., Brewer B.J., and Fangman W.L. 1997. Cell cycle-dependent establishment of a late replication program. *Science* **276:** 806–809.

Raghuraman M.K., Winzeler E.A., Collingwood D., Hunt S., Wodicka L., Conway A., Lockhart D.J., Davis R.W., Brewer B.J., and Fangman W.L. 2001. Replication dynamics of the yeast genome. *Science* **294:** 115–121.

Ratzkin B. and Carbon J. 1977. Functional expression of cloned yeast DNA in *Escherichia coli*. *Proc. Natl. Acad. Sci.* **74:** 487–491.

Reynolds A.E., McCarroll R.M., Newlon C.S., and Fangman W.L. 1989. Time of replication of ARS elements along yeast chromosome III. *Mol. Cell. Biol.* **9:** 4488–4494.

Shampay J., Szostak J.W., and Blackburn E.H. 1984. DNA sequences of telomeres maintained in yeast. *Nature* **310:** 154–157.

Shirahige K., Iwasaki T., Rashid M.B., Ogasawara N., and Yoshikawa H. 1993. Location and characterization of autonomously replicating sequences from chromosome VI of *Saccharomyces cerevisiae*. *Mol. Cell. Biol.* **13:** 5043–5056.

Singer M.S. and Gottschling D.E. 1994. TLC1: Template RNA component of *Saccharomyces cerevisiae* telomerase. *Science* **266:** 404–409.

Stinchcomb D.T., Struhl K., and Davis R.W. 1979. Isolation and characterisation of a yeast chromosomal replicator. *Nature* **282:** 39–43.

Struhl K., Stinchcomb D.T., Scherer S., and Davis R.W. 1979. High-frequency transformation of yeast: Autonomous replication of hybrid DNA molecules. *Proc. Natl. Acad. Sci.* **76:** 1035–1039.

Strunnikov A.V., Kingsbury J., and Koshland D. 1995. CEP3 encodes a centromere protein of *Saccharomyces cerevisiae*. *J. Cell Biol.* **128:** 749–760.

Sweder K.S., Rhode P.R., and Campbell J.L. 1988. Purification and characterization of proteins that bind to yeast ARSs. *J. Biol. Chem.* **263:** 17270–17277.

Szostak J.W. and Blackburn E.H. 1982. Cloning yeast telomeres on linear plasmid vectors. *Cell* **29:** 245–255.

Van Houten J.V. and Newlon C.S. 1990. Mutational analysis of the consensus sequence of a replication origin from yeast chromosome III. *Mol. Cell. Biol.* **10:** 3917–3925.

Vogelauer M., Rubbi L., Lucas I., Brewer B.J., and Grunstein M. 2002. Histone acetylation regulates the time of replication origin firing. *Mol. Cell* **10:** 1223–1233.

Walmsley R.W., Chan C.S., Tye B.K., and Petes T.D. 1984. Unusual DNA sequences associated with the ends of yeast chromosomes. *Nature* **310:** 157–160.

Watson J.D. 1972. Origin of concatemeric T7 DNA. *Nat. New Biol.* **239:** 197–201.

Wyrick J.J., Aparicio J.G., Chen T., Barnett J.D., Jennings E.G., Young R.A., Bell S.P., and Aparicio O.M. 2001. Genome-wide distribution of ORC and MCM proteins in *S. cerevisiae*: High-resolution mapping of replication origins. *Science* **294:** 2357–2360.

Yamashita M., Hori Y., Shinomiya T., Obuse C., Tsurimoto T., Yoshikawa H., and Shirahige K. 1997. The efficiency and timing of initiation of replication of multiple replicons of *Saccharomyces cerevisiae* chromosome VI. *Genes Cells* **2:** 655–665.

4

Transcription

Fred Winston
Department of Genetics
Harvard Medical School
Boston, Massachusetts 02115

The yeast galactose system: A new model for gene regulation, 68

Matsumoto K., Toh-e A., and Oshima Y. 1978. Genetic control of galactokinase synthesis in *Saccharomyces cerevisiae:* Evidence for constitutive expression of the positive regulatory gene *gal4*. *J. Bacteriol.* **134:** 446–457.

Anatomy of a simple eukaryotic promoter, 71

Guarente L., Yocum R.R., and Gifford P. 1982. A *GAL10-CYC1* hybrid yeast promoter identifies the *GAL4* regulatory region as an upstream site. *Proc. Natl. Acad. Sci.* **79:** 7410–7414.

Modularity and mechanism of transcriptional activators, 72

Brent R. and Ptashne M. 1985. A eukaryotic transcriptional activator bearing the DNA specificity of a prokaryotic repressor. *Cell* **43:** 729–736.

Chromatin structure and gene regulation, 73

Hirschhorn J.N., Brown S.A., Clark C.D., and Winston F. 1992. Evidence that SNF2/SWI2 and SNF5 activate transcription in yeast by altering chromatin structure. *Genes Dev.* **6:** 2288–2298.

Discovery of the RNA polymerase II mediator complex, 76

Thompson C.M., Koleske A.J., Chao D.M., and Young R.A. 1993. A multisubunit complex associated with the RNA polymerase II CTD and TATA-binding protein in yeast. *Cell* **73:** 1361–1375.

Note: The landmark papers listed above are those discussed in this chapter. Each landmark paper is preceded by the name of the section (with starting page number) where the paper is first discussed in detail.

STUDIES IN *SACCHAROMYCES CEREVISIAE* HAVE BEEN EXTRAORDINARILY IMPORTANT in elucidating fundamental mechanisms in the control of eukaryotic transcription. Historically, these studies can be viewed as occurring in three phases. The earliest phase, from the 1960s until the late 1970s, began with the belief, based on little data, that *S. cerevisiae* was a larger version of *Escherichia coli*, similar not only in being a microorganism, but also in its mechanism of transcriptional control. During this period, however, it came to be realized that gene organization is fundamentally different in *S. cerevisiae* than in *E. coli*. In *S.*

cerevisiae, the genes of a biosynthetic pathway are generally unlinked and are not cotranscribed as operons, the usual organizational mode found in *E. coli*. The second phase, from the late 1970s until the mid 1980s, was an era when investigators began to realize that *S. cerevisiae* is also significantly different from *E. coli* in terms of mechanisms of transcriptional regulation. During this time, results began to hint at the strong similarity between *S. cerevisiae* and larger eukaryotes. The third phase, beginning in the mid 1980s and still in progress, is a time when investigators understand that there exists a remarkably high degree of conservation for transcriptional mechanisms among all eukaryotes. During this time, there has been a rich cross-feeding of information between studies of *S. cerevisiae*, primarily but not exclusively genetic, and studies of larger eukaryotes, primarily biochemical, that have resulted in rapid and large advances in understanding transcriptional control.

In this chapter, we discuss, in chronological order, seven historically significant papers, each of which constitutes an important breakthrough, that occurred during the latter two phases of *S. cerevisiae* studies. (Of these seven key papers, the five listed at the beginning of this chapter are included on the accompanying disk.) In each case, other papers were also central in elucidating the new concept being discussed, and these papers are also cited and discussed. Most of these studies came to be of clear relevance to other eukaryotes studied in the laboratory, and even in humans.

THE YEAST GALACTOSE SYSTEM: A NEW MODEL FOR GENE REGULATION

The *GAL1, GAL10*, and *GAL7* genes, required for galactose utilization, were among the first group of genes whose regulation was extensively studied in *S. cerevisiae*. At the time that studies of the *S. cerevisiae GAL* genes were being performed in the 1960s, studies of the *E. coli* lactose operon by Jacob and Monod (1962) formed the prevailing concepts of gene regulation. The *S. cerevisiae GAL* genes had properties that suggested that they might fit the *lac* paradigm. First, they were known to be subjected to strong transcriptional control, strongly repressed when cells were grown on glucose, and induced to a high level when cells were grown on galactose. Second, the three genes were tightly linked. This coregulation and strong genetic linkage suggested that, similar to *E. coli lac* (Fig. 1), the *GAL1, GAL10*, and *GAL7* genes might be cotranscribed as a polycistronic mRNA whose synthesis is regulated by a repressor-operator relationship. The finding that the operon model also seemed to apply to the *E. coli* galactose operon (Buttin 1963a,b) strengthened the suspicion that the *S. cerevisiae GAL* genes were regulated in a similar fashion. However, later studies showed that the *S. cerevisiae GAL* genes, although coregulated, are three separate transcription units (Hopper et al. 1978; Broach 1979).

On the basis of studies of the *E. coli lac* operon, specific classes of regulatory mutants were predicted and sought for the *S. cerevisiae GAL* genes. One expected class, constitutive mutants, analogous to mutations in the *E. coli lacI* gene (encoding *lac* repressor) was identified (Douglas and Pelroy 1963), and the gene, today named *GAL80*, was originally named *i*, after its believed *E. coli lacI* counterpart. However, other classes of mutations, based on analogy to mutations in the *lac* repressor-binding site, were not identified. Instead, different types of regulatory mutations were identified, and these began to reshape the models for *GAL* gene regulation, although these models continued to include important elements of the *lac* model. First, uninducible *S. cerevisiae* mutations, today known to be in the *GAL4*

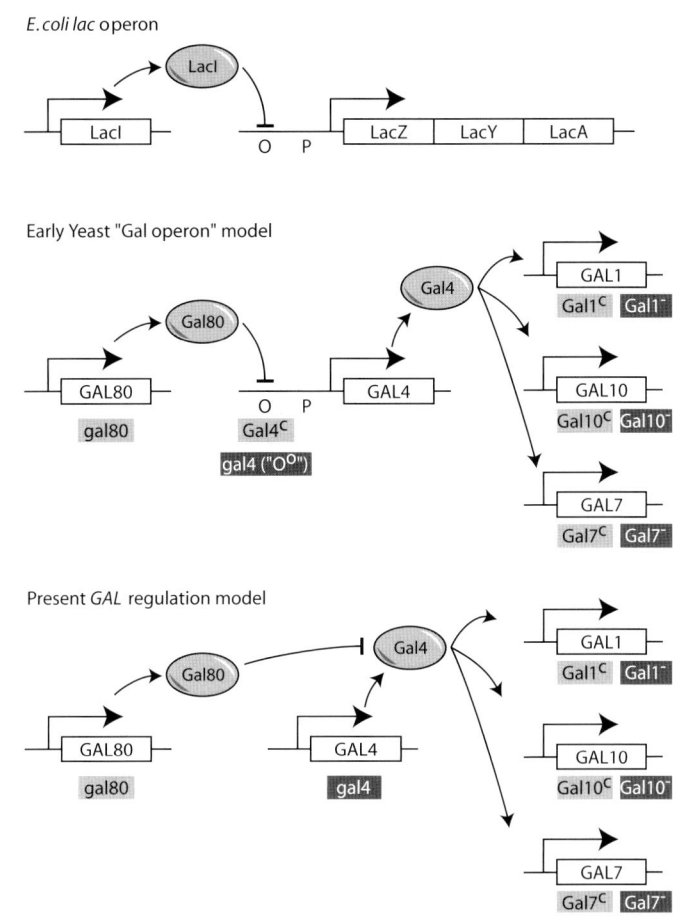

FIGURE 1. From bacterial to eukaryotic transcription: Evolution of gene regulation models. Models of transcriptional regulation in *E. coli*, particularly the lactose operon model (*top*) expounded by Jacob and Monod (1962), had an overwhelming influence on early models for galactose regulation in yeast (*middle*). The isolation and physiological characterization of thermolabile *gal4* mutants by Oshima and colleagues (Matsumoto et al. 1978; see text) led to a very different model of *GAL* regulation that still stands today.

gene, were identified. (Note that the *S. cerevisiae* nomenclature has changed over time; here, we use the current nomenclature.) These mutations failed to express *GAL1*, *GAL10*, and *GAL7* and were initially thought to be similar to *lac* O° mutations, although the original bacterial *lac* O° mutations are now known to be polar mutations in *lacZ* and not within the bacterial operator (Beckwith 1963). However, unlike *lac* O° mutations or any other class of *E. coli lac* mutations, the *gal4* mutations were unlinked to the *GAL1*, *GAL10*, and *GAL7* genes and they were recessive. This finding led Douglas and Hawthorne (1964) to conclude that these mutations identified a "cytoplasmic product that is required in some unspecified way" for the functions of *GAL1*, *GAL10*, and *GAL7*. Second, dominant-constitutive mutations were isolated by Douglas and Hawthorne (1966), who were seeking mutations analogous to *E. coli lac* O^c mutations, *cis*-dominant mutations that fail to bind the repressor.

However, linkage studies demonstrated that these mutations, rather than being linked to *GAL1, GAL10,* and *GAL7,* were tightly linked to *GAL4.* These two classes of mutations in *GAL4* shifted the repressor-operator model from direct regulation of the *GAL1, GAL10,* and *GAL7* genes to regulation of the *GAL4* gene. In this model of Douglas and Hawthorne, later shown to be incorrect, the Gal80 (i) gene product binds to a site, adjacent to *GAL4,* to repress *GAL4* expression. Thus, Gal4 synthesis would occur upon induction by galactose and would, in turn, activate expression of *GAL1, GAL10,* and *GAL7* (Fig. 1).

This model for regulation of the *S. cerevisiae GAL* genes stood until the late 1970s, when new genetic experiments provided strong evidence for a very different model for *GAL* gene regulation, a model that turned out to be correct (**Matsumoto et al. 1978**). These experiments were designed to test an important component of the Douglas-Hawthorne model: that Gal80 regulates synthesis of Gal4. The key result of Matsumoto et al. (1978), described in more detail below, provided evidence that Gal4 is preexisting, prior to induction by galactose. These data were not consistent with the repressor-operator hypothesis of Douglas and Hawthorne (1966). Importantly, these studies also demonstrated that not all gene regulation is controlled by a repressor-operator relationship.

The approach taken by Matsumoto et al. (1978) allowed measurement of the time required for Gal4 to become active in vivo, comparing a wild-type strain to a *gal4* mutant in which Gal4 had to be synthesized de novo. These studies were based on earlier studies of the *E. coli lac* operon that used temperature-sensitive *lacI* mutations to study the kinetics of *lac* gene induction (Horiuchi et al. 1961; Horiuchi and Novick 1965; Sadler and Novick 1965). Matsumoto et al. (1978) isolated temperature-sensitive mutations in the *GAL4* and *GAL80* genes that were shown to be of two classes: (1) thermolabile (TL) mutants, encoding proteins that are heat-sensitive at elevated temperature, and (2) temperature-sensitive synthesis (TSS) mutants in which synthesis is heat-sensitive (i.e., a protein synthesized at permissive temperature is not heat-sensitive). The identification of these two classes of Ts$^-$ mutants was the key to providing compelling evidence for a new model for *GAL* regulation.

The central experiment in this paper is the measurement of induction kinetics that compares a wild-type strain to a *gal4-4* TL mutant (Fig. 6 in Matsumoto et al. 1978). This experiment measures the time required to induce the *GAL* genes under conditions where new Gal4 must be synthesized due to the *gal4-4* TL mutation. If de novo synthesis of Gal4 is normally required, then the time of induction should be the same for the wild-type and the *gal4-4* mutant. If, however, Gal4 is normally preexisting, then the wild-type strain should induce more rapidly than the mutant. The results of Matsumoto et al. (1978) clearly met the second possibility. These results led them to suggest that the Douglas-Hawthorne model was incorrect.

At the time these experiments were performed, many methods commonly used today to measure RNA and protein levels, including northern and western analyses, had not yet been developed. The most commonly used method to analyze gene expression at that time was enzyme assays, and Matsumoto et al. (1978) measured expression of *GAL1* by galactokinase assays. There was no direct way to assay for Gal4 levels in vivo; therefore, the use of the *gal4* TL mutant was critical in this analysis. Interestingly, Gal4 is present at extremely low levels and even today it cannot be detected easily by western analysis.

The results of Matsumoto et al. (1978) were strongly supported by another study (Perlman and Hopper 1979) that took a different approach to the question of Gal4 and whether it needed to be newly synthesized upon induction by galactose. The Perlman and Hopper (1979) study provided two lines of evidence for the existence of Gal4 prior to induction by galactose. First, they showed that the *GAL* mRNAs are induced by galactose even in the presence of cycloheximide, a protein synthesis inhibitor. (At this time, they isolated the RNA, used it to direct translation in vitro in a wheat germ system, and then measured the level of Gal gene products by immunoprecipitation of radiolabeled material.) Therefore, new protein synthesis (including Gal4 synthesis) is not required for Gal4-dependent activation. Second, they showed that in newly formed zygotes, Gal4 is able to induce *GAL* gene synthesis, also suggesting its constitutive expression.

ANATOMY OF A SIMPLE EUKARYOTIC PROMOTER

During the late 1970s, as studies began to show that *S. cerevisiae* gene regulation was significantly different from that of *E. coli*, critical new tools were developed to study transcription in yeast. First, recombinant DNA technologies greatly expanded the methods available to study gene regulation. In addition, another methodology was developed that is key to all molecular genetic analysis in *S. cerevisiae*—DNA-mediated transformation and "shuttle" vectors (plasmids) that can propagate in both yeast and *E. coli* (Beggs 1978; Hinnen et al. 1978). These major advances allowed the cloning of *S. cerevisiae* genes, their manipulation in vitro, and their subsequent analysis in vivo. These were quickly recognized as extremely powerful approaches to learning about regulation.

An early use of recombinant DNA was motivated by an important method widely used to study gene expression in *E. coli*: the use of gene fusions to study regulation (Beckwith et al. 1967). Such studies fuse the promoter of one gene to the coding region of another gene, often *lacZ*, in order to facilitate regulatory studies. Thus, by simple assay of β-galactosidase levels, gene expression can be accurately measured. Recombinant DNA made the construction of such fusions possible in *S. cerevisiae*. An important first step was the demonstration that fusions to *lacZ* could provide the same type of information in yeast as they did in *E. coli* (Guarente and Ptashne 1981; Rose et al. 1981).

Two gene systems were obvious candidates to be studied by recombinant DNA methods. Both of these, the *GAL* system, described in the previous section, and the *CYC1* gene (see also Chapter 5), encoding iso-1-cytochrome *c*, had been extensively studied for several years. Therefore, the nutritional and environmental conditions that controlled their expression were well understood. Both the *GAL1-GAL10-GAL7* locus and the *CYC1* gene were among the first yeast genes cloned (Smith et al. 1979; St. John and Davis 1981), and their regulation became the subject of intense investigation. Prior to this time, the *S. cerevisiae HIS3* gene was cloned (Struhl and Davis 1977), and its regulation also became a major focus of study. These three regulatory systems provided some of the most significant advances in our understanding of transcriptional control in yeast during the next several years.

A major question in the field at the time was the nature of an *S. cerevisiae* promoter. Studies of promoters in larger eukaryotes were revealing in that they were significantly different from those in *E. coli*. Studies of mammalian viruses, primarily SV40, as well as studies in *Drosophila*, identified a consensus sequence, the TATA box, as an important promoter element (Saltzman and Weinmann 1989). Other studies identified upstream promoter elements in mammalian viruses, called enhancers, that were remarkable in their ability to activate transcription over a range of distances in either orientation, two characteristics not previously observed in prokaryotes (Guarente 1988). Given the dramatic differences between promoters in *E. coli* and in larger eukaryotes, there was great interest in *S. cerevisiae* promoters, particularly whether or not they contained enhancers.

The second landmark paper was one of the first demonstrations of a key *S. cerevisiae* promoter element called an upstream activation site (UAS) (**Guarente et al. 1982**). The work is a clear demonstration that the *S. cerevisiae GAL* UAS confers regulation in a gene fusion to the *S. cerevisiae CYC1* gene and thus provided the first demonstration of the autonomous function of an *S. cerevisiae* UAS. The plasmids described in this paper had a large impact on *S. cerevisiae* studies: Those that contained the *GAL* UAS became widely used to induce high-level expression of any gene by growth on galactose, and those lacking any UAS became widely used to assay promoter sequences for UAS activity. In two key papers that appeared shortly after (Guarente et al. 1982), the *GAL* and *CYC1* UAS elements were tested for enhancer-like properties with respect to orientation and position (Guarente and Hoar 1984; Struhl 1984). Their functions were orientation and distance independent (over a modest range), but unlike mammalian enhancers, they were unable to function when placed 3′ of the transcription initiation site. Nevertheless, the yeast system can also be successfully transposed into higher eukaryotes (Phelps and Brand 1998). Taken together, these studies demonstrated that UAS elements possess many, but not all, properties of mammalian enhancers.

The elucidation of the UAS as a key promoter element in yeast raised questions regarding the mechanism by which a transcriptional activator, bound to a UAS, stimulated transcription initiation by RNA polymerase II. In *E. coli*, where activators and repressors generally bind directly adjacent to promoter sequences, experiments showed that this proximity was essential for their functions. In particular, for activation by λ repressor (cI), studies had elegantly demonstrated that contact between cI and RNA polymerase stimulated transcriptional activation (Ptashne 1992). However, in *S. cerevisiae*, the ability of activators to function over distances greater than in *E. coli* raised the possibility that they acted by a fundamentally different mechanism, likely reflecting the roles of multiple classes of transcription factors found in *S. cerevisiae* and other eukaryotes that do not exist in prokaryotes, including histones and complexes that control chromatin structure.

MODULARITY AND MECHANISM OF TRANSCRIPTIONAL ACTIVATORS

During the early to mid 1980s, as mammalian enhancers and *S. cerevisiae* UASs were identified and studied, several different types of models were considered for activation at a distance. The paper by **Brent and Ptashne (1985)** considers two such models for the activator Gal4. The first model stated that binding of Gal4 transmits a structural change along the DNA, from the UAS to the TATA region, allowing the binding of the general factors and RNA polymerase. The second model, closer to what had been observed in *E. coli*, stipulated that activation occurred by Gal4 directly contacting other proteins required for initiation. The experiments in this paper provide the first strong evidence in favor of the second model.

In their studies, Brent and Ptashne (1985) constructed a hybrid protein, containing the DNA-binding domain of an *E. coli* repressor, LexA, and the activation domain of Gal4. They then demonstrated that this protein could activate transcription in *S. cerevisiae*, in a manner similar to that of Gal4, if LexA-binding sites were present in place of the normal UAS. Thus, the Gal4 activation domain was sufficient for normal galactose-dependent activation of transcription, strongly supporting the second model described above.

In addition to the important evidence that this paper provides with respect to the mechanism of transcriptional activation, it had a very significant impact on the methods used by the transcription field. The construction and analysis of the LexA-Gal4 hybrid protein provided the framework for a general method to assay for a transcription activation domain in any protein in vivo. The analysis of suspected transcriptional activation domains by their fusion to a well-characterized DNA-binding domain, generally either that of LexA or Gal4, followed by their analysis in vivo, quickly became a standard method that has had an enormous impact on the transcription field. Significantly, the impact of this method has also been felt in other areas, including developmental studies in *Drosophila* (Brand and Perrimon 1993). Most notably, this approach led directly to the development of another method of huge impact, the two-hybrid screen for in vivo protein-protein interactions (Fields and Song 1989).

Consideration of the experiments in Brent and Ptashne (1985) is a good place to think about the basis for significant scientific advances. In many cases, including this work, the results are considered a striking and surprising leap forward, taking a step that few had imagined. In many cases, however, a full awareness of preceding work reveals that the "leap" was an insightful, but also logical, next step in a progression of studies. The foundation for the advances by Brent and Ptashne (1985) came from studies, in the same laboratory, of λ cI. Those studies defined two domains of cI, one for DNA binding and one for transcriptional activation (for review, see Ptashne 1992). Furthermore, it was shown that these domains were separable both physically and genetically. Finally, it was shown that other DNA-binding regulatory proteins, such as LexA of *E. coli* and Gal4 of *S. cerevisiae*, also fit this two-domain model (Brent and Ptashne 1981; Keegan et al. 1986). With this knowledge as background, the significant insight that led to the advance of Brent and Ptashne (1985) was to mix organisms: i.e., to use an *E. coli* DNA-binding domain to test a hypothesis for the role of an activation domain in *S. cerevisiae*. Similarly, the development of the two-hybrid system can be viewed as another very insightful, yet logical step, in this case to put the two domains into two separate molecules. The development of this method can be traced from the elucidation of the separate DNA-binding and activation domains (Ptashne 1992) to the construction of hybrid activators (Brent and Ptashne 1985), to the demonstration of activation by protein-protein interaction (Ma and Ptashne 1988), and, finally, to the clear articulation of the generality of the two-hybrid approach to detect protein–protein interactions (Fields and Song 1989).

CHROMATIN STRUCTURE AND GENE REGULATION

One of the most important areas of gene expression in which *S. cerevisiae* studies have made original contributions is chromatin structure and transcription. Although chromatin structure had been studied for decades, studies were generally limited to examining the nuclease sensitivity of chromatin. These studies suggested a correlation between transcriptional activity and particular chromatin structures. The field changed dramatically with the cloning of the *S. cerevisiae* genes encoding the four core histones, H2A, H2B, H3, and H4 (Hereford et al. 1979; Smith and Andresson 1983; Smith and Murray 1983). The identification of these genes opened up the field of chromatin structure to genetic studies in *S. cerevisiae*. An important advantage for these genetic studies is that each histone-encoding gene is present in only two copies per cell, as opposed to significantly greater copy numbers in larger eukaryotes. Early genetic studies in yeast demonstrated that histones were, as expected, essential for viability in vivo and examined the requirement for the histone amino-terminal tails (Rykowski et al. 1981; Wallis et al. 1983; Kayne et al. 1988).

The initial demonstration that histones control transcription in vivo came from two different sets of genetic studies, both showing that altered levels of histones can compensate for promoter defects. First, Clark-Adams et al. (1988) showed that particular histone genes, when either mutant or overexpressed, altered transcription of genes adjacent to transposable element insertion mutations. The overexpression phenotypes were similar to those previously observed in studies of chromosome stability (Meeks-Wagner and Hartwell 1986). In the second set of studies, transcription was examined under conditions where histone H4 was depleted (Han and Grunstein 1988; Han et al. 1988). In these studies, *S. cerevisiae* strains were constructed in which there is only one histone H4 gene and its transcription was dependent on the carbon source, being expressed in galactose and repressed in glucose. When cells were shifted to glucose and became depleted for H4, the *PHO5* promoter became activated under normally repressing conditions (high phosphate levels), and a number of mutant promoters were activated even though they were lacking their UASs. The studies of *PHO5* were in strong support of earlier elegant studies of *PHO5* chromatin structure that had demonstrated the removal of specifically positioned nucleosomes over the *PHO5* promoter upon induction of this gene (Almer and Horz 1986; Almer et al. 1986). This group of studies then showed that reduced levels of histones activate transcription from inactive promoters, suggesting that nucleosomes normally repress transcription in vivo.

Shortly after these initial genetic studies, the analysis of particular histone mutants provided strong evidence that chromatin structure plays a key role in transcriptional silencing in *S. cerevisiae* (Kayne et al. 1988). Transcriptional silencing is analogous in many respects to the control of heterochromatin in larger eukaryotes as it works over large domains rather than by promoter-specific repression. Early genetic studies demonstrated the existence of two silent copies of *S. cerevisiae* mating-type information, called *HML* and *HMR*. Since then, other silenced regions, at telomeres and in the ribosomal RNA genes, have been studied (for reviews on silencing, see Laurenson and Rine 1992; Moazed 2001). Three early classic studies in this field helped to define both the *trans*-acting proteins and *cis*-acting sequences required for silencing (Brand et al. 1987; Rine and Herskowitz 1987; Shore and Nasmyth 1987). Silencing in *S. cerevisiae* was the first system in which repression at a distance could be studied in detail.

The role of chromatin structure in transcriptional silencing became clear with the historically significant studies of Kayne et al. (1988). These studies demonstrated that histone H4 mutants lacking their amino-terminal tails fail to silence *HML* and *HMR*. These demonstrations were accomplished by both mating assays and northern hybridization analysis. An important subsequent paper showed that single amino acid changes in the amino-terminal tail of H4 caused loss of silencing (Johnson et al. 1990). Significantly, this silencing defect could be suppressed by a mutation that altered the Sir3 protein, thus providing the initial connection between histone tails and Sir proteins. Subsequent studies have led the way in elucidating how histone modifications, particularly acetylation, are critical modifications in the control of transcription by chromatin structure. One early example of this type of analysis (Braunstein et al. 1993) was one of the first uses of chromatin immunoprecipitation (ChIP) in *S. cerevisiae* studies. Later studies popularized polymerase chain reaction (PCR)-based ChIP to demonstrate the association of Sir proteins with silenced chromatin in vivo (Hecht et al. 1996). This field is an outstanding example of how a combination of genetic, molecular, and biochemical approaches has elucidated a complex network of interactions.

Another area of chromatin and transcription that has become widely studied throughout all eukaryotes is the mechanism by which nucleosomal repression is overcome in vivo to allow transcriptional activation. During the 1980s, multiple studies had identified two overlapping sets of genes required for transcriptional activation. One set, called *SNF* genes (sucrose *n*onfermentor), was identified by mutations that impair the expression of the *S. cerevisiae SUC2* gene (Carlson et al. 1981; Neigeborn and Carlson 1984). The second set, called *SWI* genes (*sw*itching-defective), was identified by mutations that block mating-type switching (Stern et al. 1984; Breeden and Nasmyth 1987; Sternberg et al. 1987). Several studies suggested that *SNF* and *SWI* genes have common roles in transcriptional control (for review, see Winston and Carlson 1992). Later, it was shown that the Snf and Swi proteins form a multisubunit complex, the Swi/Snf complex (Cairns et al. 1994; Peterson et al. 1994).

The results of **Hirschhorn et al. (1992)** provided the initial evidence that the Snf and Swi gene products are required to overcome transcriptional repression by nucleosomes in vivo. The critical observation leading to this model was a genetic one—a deletion that reduces the levels of histones H2A and H2B (one of the same mutations studied by Clark-Adams et al. [1988]) suppresses the transcriptional defects caused by *snf* and *swi* mutants. (Later studies showed similar suppression by mutations in H3- and H4-encoding genes [Kruger et al. 1995].) Hirschhorn et al. (1992) went on to provide genetic evidence and MNase (micrococcal nuclease) studies that strongly suggested that the Snf and Swi proteins cause an alteration in chromatin structure that is required for active transcription. Thus, these proteins were envisioned to "function by removing or otherwise modifying nucleosomes" to allow transcription factors such as TATA-binding protein accessibility to their sites (Fig. 2).

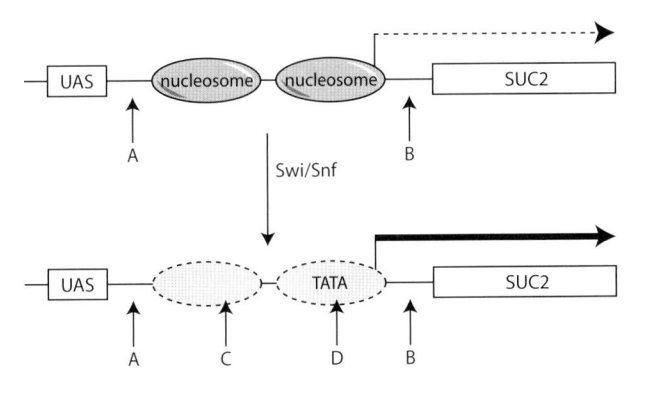

FIGURE 2. A schematic presentation of the Swi/Snf-dependent activation of *SUC2* transcription. The possible chromatin structures of the *SUC2* promoter region are depicted either before or after the action of Swi/Snf factors. Arrows below each line represent cleavage sites for MNase. Solid and open ovals represent present and absent nucleosomes, respectively.

CLOSING COMMENTS

Many studies of *S. cerevisiae*, including the seven papers discussed in detail in this section, have provided new insights into transcriptional control in *S. cerevisiae* and throughout eukaryotes in general. Space constraints restrict this chapter to covering only a subset of the significant findings that have made *S. cerevisiae* studies a vital contributor to our current understanding of transcription. For example, studies of TATA-binding protein (TBP) in *S. cerevisiae* made critical advances in the study of this key factor. Although TBP was originally identified biochemically in HeLa nuclear extracts (Matsui et al. 1980), its purification and analysis remained elusive for several years (see, e.g., Nakajima et al. 1988). The purification of TBP, the identification of the TBP-encoding gene, and the first demonstration of a role for TBP in vivo all came from studies in *S. cerevisiae* (Cavallini et al. 1989; Eisenmann et al. 1989; Hahn et al. 1989; Horikoshi et al. 1989; Schmidt et al. 1989). These studies not only served to gain a greater understanding of this central factor in RNA polymerase II-dependent transcription, but also highlighted the remarkable degree of conservation between *S. cerevisiae* and larger eukaryotes that has fueled this field ever since. In addition to TBP studies, studies in other areas, including global repressors (Keleher et al. 1992) and coactivators (Eisenmann et al. 1992), have made critical contributions. The papers included in this chapter, then, can serve as the starting point for further reading to gain familiarity with other classic papers in *S. cerevisiae* transcription.

Finally, it is important to emphasize the crucial contributions of mutant hunts in this field. Many studies, only some of which are mentioned in this chapter, have identified genes known as *SRB* (Nonet and Young 1989; Thompson et al. 1993; Hengartner et al. 1995), *SPT* (Winston et al. 1984, 1987; Fassler and Winston 1988), *SNF* (Carlson et al. 1981; Neigeborn and Carlson 1984), *SWI* (Stern et al. 1984; Breeden and Nasmyth 1987; Sternberg et al. 1987), *SSN* (Carlson et al. 1984; Neigeborn et al. 1986), *ADA* (Berger et al. 1992), and others, each of which has contributed importantly to this field. The identification of these mutations was usually performed in the absence of clear hypotheses, but they have been the driving force in the formation of many of today's accepted concepts for mechanisms of transcriptional control.

ACKNOWLEDGMENTS

I would like to thank David Shore and Mark Johnston for valuable suggestions. Work in my laboratory is supported by grants from the National Institutes of Health.

STUDY QUESTIONS

1. How did Matsumoto et al. (1978) distinguish between TL and TSS mutations in *GAL80* and in *GAL4*?

2. Why was it important for Matsumoto et al. to show that they could incubate cells in the minimal salt vitamin medium?

3. Although the results of Matsumoto et al. (1978) were incompatible with the Douglas-Hawthorne model, their results do not rule out several other possible models. What models are compatible with their data?

4. What features made the *GAL10* promoter region a good choice for the experiments described by Guarente et al. (1982)?

5. What was the basis for choosing the exact *GAL* DNA fragment to be used for these experiments?

6. Although expression is measured by assays of β-galactosidase expression, the authors are confident they are measuring transcription. Why?

7. In a wild-type strain, glucose repression of Gal4-mediated activation is extremely tight. Yet, in the case of the Gal4-LexA fusions (Brent and Ptashne 1985), it is much weaker. Why?

8. Do you think that DNA binding is completely eliminated as an important aspect of transcriptional activation by Gal4?

9. The MNase experiments in Figures 3, 4, and 5 of Hirschhorn et al. (1992) suggest that Swi/Snf controls chromatin structure over both the UAS and TATA regions of the *SUC2* promoter. Do these experiments prove that Swi/Snf is acting directly at the *SUC2* promoter? Propose a model in which it acts indirectly.

10. Why was Mediator not identified as one of the original fractions required for initiation by RNA polymerase II in vitro?

11. Some Mediator components are essential for viability while others are not. What are possible essential and nonessential roles for different Mediator components?

12. How might dominant mutations in *SRB* genes suppress the defects of CTD deletion mutations?

REFERENCES

Almer A. and Horz W. 1986. Nuclease hypersensitive regions with adjacent positioned nucleosomes mark the gene boundaries of the PHO5/PHO3 locus in yeast. *EMBO J.* **5:** 2681–2687.

Almer A., Rudolph H., Hinnen A., and Horz W. 1986. Removal of positioned nucleosomes from the yeast PHO5 promoter upon PHO5 induction releases additional upstream activating DNA elements. *EMBO J.* **5:** 2689–2696.

Beckwith J. 1963. Restoration of operon activity by suppressors. *Biochim. Biophys. Acta* **76:** 162–164.

Beckwith J.R., Signer E.R., and Epstein W. 1967. Transposition of the Lac region of *E. coli*. *Cold Spring Harbor Symp. Quant. Biol.* **31:** 393–401.

Beggs J. 1978. Transformation of yeast by a replicating hybrid plasmid. *Nature* **275:** 104–109.

Berger S.L., Pina B., Silverman N., Marcus G.A., Agapite J., Regier J.L., Triezenberg S.J., and Guarente L. 1992. Genetic isolation of ADA2: A potential transcriptional adaptor required for function of certain acidic activation domains. *Cell* **70:** 251–265.

Bhoite L.T., Yu Y., and Stillman D.J. 2001. The Swi5 activator recruits the Mediator complex to the HO promoter without RNA polymerase II. *Genes Dev.* **15:** 2457–2469.

Brand A.H. and Perrimon N. 1993. Targeted gene expression as a means of altering cell fates and generating dominant phenotypes. *Development* **118:** 401–415.

Brand A.H., Micklem G., and Nasmyth K. 1987. A yeast silencer contains sequences that can promote autonomous plasmid replication and transcriptional activation. *Cell* **51:** 709–719.

Braunstein M., Rose A.B., Holmes S.G., Allis C.D., and Broach J.R. 1993. Transcriptional silencing in yeast is associated with reduced nucleosome acetylation. *Genes Dev.* **7:** 592–604.

Breeden L. and Nasmyth K. 1987. Cell cycle control of the yeast HO gene: *cis*- and *trans*-acting regulators. *Cell* **48:** 389–397.

Brent R. and Ptashne M. 1981. Mechanism of action of the lexA gene product. *Proc. Natl. Acad. Sci.* **78:** 4204–4208.
———. 1985. A eukaryotic transcriptional activator bearing the DNA specificity of a prokaryotic repressor. *Cell* **43:** 729–736.
Broach J.R. 1979. Galactose regulation in *Saccharomyces cerevisiae*. The enzymes encoded by the GAL7, 10, 1 cluster are co-ordinately controlled and separately translated. *J. Mol. Biol.* **131:** 41–53.
Brownell J.E., Zhou J., Ranalli T., Kobayashi R., Edmondson D.G., Roth S.Y., and Allis C.D. 1996. Tetrahymena histone acetyltransferase A: A homolog to yeast Gcn5p linking histone acetylation to gene activation. *Cell* **84:** 843–851.
Buratowski S., Hahn S., Guarente L., and Sharp P.A. 1989. Five intermediate complexes in transcription initiation by RNA polymerase II. *Cell* **56:** 549–561.
Buttin G. 1963a. Mécanismes régulateurs dans la biosynthèse des enzymes du métabolisme du galactose chez *Escherichia coli* K12. I. La biosynthèse induite de la galactokinase et l'induction simultanée de la séquence enzymatique. *J. Mol. Biol.* **7:** 164–182.
———. 1963b. Mécanismes régulateurs dans la biosynthèse des enzymes du métabolisme du galactose chez *Escherichia coli* K12. II. Le déterminisme génétique de la regulation. *J. Mol. Biol.* **7:** 183–205.
Cairns B.R., Kim Y.J., Sayre M.H., Laurent B.C., and Kornberg R.D. 1994. A multisubunit complex containing the SWI1/ADR6, SWI2/SNF2, SWI3, SNF5, and SNF6 gene products isolated from yeast. *Proc. Natl. Acad. Sci.* **91:** 1950–1954.
Carlson M., Osmond B.C., and Botstein D. 1981. Mutants of yeast defective in sucrose utilization. *Genetics* **98:** 25–40.
Carlson M., Osmond B.C., Neigeborn L., and Botstein D. 1984. A suppressor of SNF1 mutations causes constitutive high-level invertase synthesis in yeast. *Genetics* **107:** 19–32.
Cavallini B., Faus I., Matthes H., Chipoulet J.M., Winsor B., Egly J.M., and Chambon P. 1989. Cloning of the gene encoding the yeast protein BTF1Y, which can substitute for the human TATA box-binding factor. *Proc. Natl. Acad. Sci.* **86:** 9803–9807.
Clark-Adams C.D., Norris D., Osley M.A., Fassler J.S., and Winston F. 1988. Changes in histone gene dosage alter transcription in yeast. *Genes Dev.* **2:** 150–159.
Cosma M.P., Panizza S., and Nasmyth K. 2001. Cdk1 triggers association of RNA polymerase to cell cycle promoters only after recruitment of the mediator by SBF. *Mol. Cell* **7:** 1213–1220.
Cote J., Quinn J., Workman J.L., and Peterson C.L. 1994. Stimulation of GAL4 derivative binding to nucleosomal DNA by the yeast SWI/SNF complex. *Science* **265:** 53–60.
Douglas H.C. and Hawthorne D.C. 1964. Enzymatic expression and genetic linkage of genes controlling galactose utilization in *Saccharomyces*. *Genetics* **49:** 837–844.
———. 1966. Regulation of genes controlling synthesis of the galactose pathway enzymes in yeast. *Genetics* **54:** 911–916.
Douglas H.C. and Pelroy G. 1963. A gene controlling inducibility of the galactose pathway enzymes in *Saccharomyces*. *Biochim. Biophys. Acta* **68:** 155–156.
Eisenmann D.M., Dollard C., and Winston F. 1989. SPT15, the gene encoding the yeast TATA binding factor TFIID, is required for normal transcription initiation in vivo. *Cell* **58:** 1183–1191.
Eisenmann D.M., Arndt K.M., Ricupero S.L., Rooney J.W., and Winston F. 1992. SPT3 interacts with TFIID to allow normal transcription in *Saccharomyces cerevisiae*. *Genes Dev.* **6:** 1319–1331.
Fassler J.S. and Winston F. 1988. Isolation and analysis of a novel class of suppressor of Ty insertion mutations in *Saccharomyces cerevisiae*. *Genetics* **118:** 203–212.
Fields S. and Song O. 1989. A novel genetic system to detect protein-protein interactions. *Nature* **340:** 245–246.
Flanagan P.M., Kelleher R.J., III, Sayre M.H., Tschochner H., and Kornberg R.D. 1991. A mediator required for activation of RNA polymerase II transcription in vitro. *Nature* **350:** 436–438.
Georgakopoulos T. and Thireos G. 1992. Two distinct yeast transcriptional activators require the function of the GCN5 protein to promote normal levels of transcription. *EMBO J.* **11:** 4145–4152.
Guarente L. 1988. UASs and enhancers: Common mechanism of transcriptional activation in yeast and mammals. *Cell* **52:** 303–305.
Guarente L. and Hoar E. 1984. Upstream activation sites of the CYC1 gene of *Saccharomyces cerevisiae* are active when inverted but not when placed downstream of the "TATA box." *Proc. Natl. Acad. Sci.* **81:** 7860–7864.

Guarente L. and Ptashne M. 1981. Fusion of *Escherichia coli lacZ* to the cytochrome *c* gene of *Saccharomyces cerevisiae*. *Proc. Natl. Acad. Sci.* **78:** 2199–2203.

Guarente L., Yocum R.R., and Gifford P. 1982. A *GAL10-CYC1* hybrid yeast promoter identifies the *GAL4* regulatory region as an upstream site. *Proc. Natl. Acad. Sci.* **79:** 7410–7414.

Hahn S., Buratowski S., Sharp P.A., and Guarente L. 1989. Isolation of the gene encoding the yeast TATA binding protein TFIID: A gene identical to the SPT15 suppressor of Ty element insertions. *Cell* **58:** 1173–1181.

Han M. and Grunstein M. 1988. Nucleosome loss activates yeast downstream promoters in vivo. *Cell* **55:** 1137–1145.

Han M., Kim U.J., Kayne P., and Grunstein M. 1988. Depletion of histone H4 and nucleosomes activates the PHO5 gene in *Saccharomyces cerevisiae*. *EMBO J.* **7:** 2221–2228.

Hecht A., Strahl-Bolsinger S., and Grunstein M. 1996. Spreading of transcriptional repressor SIR3 from telomeric heterochromatin. *Nature* **383:** 92–96.

Hengartner C.J., Thompson C.M., Zhang J., Chao D.M., Liao S.M., Koleske A.J., Okamura S., and Young R.A. 1995. Association of an activator with an RNA polymerase II holoenzyme. *Genes Dev.* **9:** 897–910.

Hereford L., Fahrner K., Woolford J., Jr., Rosbash M., and Kaback D.B. 1979. Isolation of yeast histone genes H2A and H2B. *Cell* **18:** 1261–1271.

Hinnen A., Hicks J.B., and Fink G.R. 1978. Transformation of yeast. *Proc. Natl. Acad. Sci.* **75:** 1929–1933.

Hirschhorn J.N., Brown S.A., Clark C.D., and Winston F. 1992. Evidence that SNF2/SWI2 and SNF5 activate transcription in yeast by altering chromatin structure. *Genes Dev.* **6:** 2288–2298.

Hopper J.E., Broach J.R., and Rowe L.B. 1978. Regulation of the galactose pathway in *Saccharomyces cerevisiae*: Induction of uridyl transferase mRNA and dependency on GAL4 gene function. *Proc. Natl. Acad. Sci.* **75:** 2878–2882.

Horikoshi M., Wang C.K., Fujii H., Cromlish J.A., Weil P.A., and Roeder R.G. 1989. Cloning and structure of a yeast gene encoding a general transcription initiation factor TFIID that binds to the TATA box. *Nature* **341:** 299–303.

Horiuchi T. and Novick A. 1965. Studies of a thermolabile repressor. *Biochim. Biophys. Acta* **108:** 687–696.

Horiuchi T., Horiuchi S., and Novick A. 1961. A temperature-sensitive regulatory system. *J. Mol. Biol.* **3:** 703–704.

Imbalzano A.N., Kwon H., Green M.R., and Kingston R.E. 1994. Facilitated binding of TATA-binding protein to nucleosomal DNA. *Nature* **370:** 481–485.

Jacob F. and Monod J. 1962. On the regulation of gene activity. *Cold Spring Harbor Symp. Quant. Biol.* **26:** 193–211.

Johnson L.M., Kayne P.S., Kahn E.S., and Grunstein M. 1990. Genetic evidence for an interaction between SIR3 and histone H4 in the repression of the silent mating loci in *Saccharomyces cerevisiae*. *Proc. Natl. Acad. Sci.* **87:** 6286–6290.

Kayne P.S., Kim U.J., Han M., Mullen J.R., Yoshizaki F., and Grunstein M. 1988. Extremely conserved histone H4 N terminus is dispensable for growth but essential for repressing the silent mating loci in yeast. *Cell* **55:** 27–39.

Keegan L., Gill G., and Ptashne M. 1986. Separation of DNA binding from the transcription-activating function of a eukaryotic regulatory protein. *Science* **231:** 699–704.

Keleher C.A., Redd M.J., Schultz J., Carlson M., and Johnson A.D. 1992. Ssn6-Tup1 is a general repressor of transcription in yeast. *Cell* **68:** 709–719.

Kelleher R.J., III, Flanagan P.M., and Kornberg R.D. 1990. A novel mediator between activator proteins and the RNA polymerase II transcription apparatus. *Cell* **61:** 1209–1215.

Kim Y.J., Bjorklund S., Li Y., Sayre M.H., and Kornberg R.D. 1994. A multiprotein mediator of transcriptional activation and its interaction with the C-terminal repeat domain of RNA polymerase II. *Cell* **77:** 599–608.

Koleske A.J. and Young R.A. 1994. An RNA polymerase II holoenzyme responsive to activators. *Nature* **368:** 466–469.

Kruger W., Peterson C.L., Sil A., Coburn C., Arents G., Moudrianakis E.N., and Herskowitz I. 1995. Amino acid substitutions in the structured domains of histones H3 and H4 partially relieve the requirement of the yeast SWI/SNF complex for transcription. *Genes Dev.* **9:** 2770–2779.

Kwon H., Imbalzano A.N., Khavari P.A., Kingston R.E., and Green M.R. 1994. Nucleosome disruption and enhancement of activator binding by a human SW1/SNF complex. *Nature* **370:** 477–481.

Laurenson P. and Rine J. 1992. Silencers, silencing, and heritable transcriptional states. *Microbiol. Rev.* **56:** 543–560.

Ma J. and Ptashne M. 1988. Converting a eukaryotic transcriptional inhibitor into an activator. *Cell* **55:** 443–446.

Marcus G.A., Silverman N., Berger S.L., Horiuchi J., and Guarente L. 1994. Functional similarity and physical association between GCN5 and ADA2: Putative transcriptional adaptors. *EMBO J.* **13:** 4807–4815.

Matsui T., Segall J., Weil P.A., and Roeder R.G. 1980. Multiple factors required for accurate initiation of transcription by purified RNA polymerase II. *J. Biol. Chem.* **255:** 11992–11996.

Matsumoto K., Toh-e A., and Oshima Y. 1978. Genetic control of galactokinase synthesis in *Saccharomyces cerevisiae*: Evidence for constitutive expression of the positive regulatory gene *gal4*. *J. Bacteriol.* **134:** 446–457.

Meeks-Wagner D. and Hartwell L.H. 1986. Normal stoichiometry of histone dimer sets is necessary for high fidelity of mitotic chromosome transmission. *Cell* **44:** 43–52.

Moazed D. 2001. Common themes in mechanisms of gene silencing. *Mol. Cell* **8:** 489–498.

Myers L.C. and Kornberg R.D. 2000. Mediator of transcriptional regulation. *Annu. Rev. Biochem.* **69:** 729–749.

Nakajima N., Horikoshi M., and Roeder R.G. 1988. Factors involved in specific transcription by mammalian RNA polymerase II: Purification, genetic specificity, and TATA box-promoter interactions of TFIID. *Mol. Cell. Biol.* **8:** 4028–4040.

Neigeborn L. and Carlson M. 1984. Genes affecting the regulation of *SUC2* gene expression by glucose repression in *Saccharomyces cerevisiae*. *Genetics* **108:** 845–858.

Neigeborn L., Rubin K., and Carlson M. 1986. Suppressors of SNF2 mutations restore invertase derepression and cause temperature-sensitive lethality in yeast. *Genetics* **112:** 741–753.

Nonet M.L. and Young R.A. 1989. Intragenic and extragenic suppressors of mutations in the heptapeptide repeat domain of *Saccharomyces cerevisiae* RNA polymerase II. *Genetics* **123:** 715–724.

Perlman D. and Hopper J.E. 1979. Constitutive synthesis of the GAL4 protein, a galactose pathway regulator in *Saccharomyces cerevisiae*. *Cell* **16:** 89–95.

Peterson C.L., Dingwall A., and Scott M.P. 1994. Five SWI/SNF gene products are components of a large multisubunit complex required for transcriptional enhancement. *Proc. Natl. Acad. Sci.* **91:** 2905–2908.

Phelps C.B. and Brand A.H. 1998. Ectopic gene expression in *Drosophila* using GAL4 system. *Methods* **14:** 367–379.

Ptashne M. 1992. *A genetic switch: Phage λ and higher organisms*. Blackwell Science and Cell Press, Cambridge, Massachusetts.

Rine J. and Herskowitz I. 1987. Four genes responsible for a position effect on expression from HML and HMR in *Saccharomyces cerevisiae*. *Genetics* **116:** 9–22.

Rose M., Casadaban M.J., and Botstein D. 1981. Yeast genes fused to beta-galactosidase in *Escherichia coli* can be expressed normally in yeast. *Proc. Natl. Acad. Sci.* **78:** 2460–2464.

Rykowski M.C., Wallis J.W., Choe J., and Grunstein M. 1981. Histone H2B subtypes are dispensable during the yeast cell cycle. *Cell* **25:** 477–487.

Sadler J.R. and Novick A. 1965. The properties of repressor and the kinetics of its action. *J. Mol. Biol.* **12:** 305–327.

Saltzman A.G. and Weinmann R. 1989. Promoter specificity and modulation of RNA polymerase II transcription. *FASEB J.* **3:** 1723–1733.

Schmidt M.C., Kao C.C., Pei R., and Berk A.J. 1989. Yeast TATA-box transcription factor gene. *Proc. Natl. Acad. Sci.* **86:** 7785–7789.

Shore D. and Nasmyth K. 1987. Purification and cloning of a DNA binding protein from yeast that binds to both silencer and activator elements. *Cell* **51:** 721–732.

Smith C.L. and Peterson C.L. 2005. ATP-dependent chromatin remodeling. *Curr. Top. Dev. Biol.* **65:** 115–148.

Smith M., Leung D.W., Gillam S., Astell C.R., Montgomery D.L., and Hall B.D. 1979. Sequence of the gene for iso-1-cytochrome *c* in *Saccharomyces cerevisiae*. *Cell* **16:** 753–761.

Smith M.M. and Andresson O.S. 1983. DNA sequences of yeast H3 and H4 histone genes from two non-allelic gene sets encode identical H3 and H4 proteins. *J. Mol. Biol.* **169:** 663–690.

Smith M.M. and Murray K. 1983. Yeast H3 and H4 histone messenger RNAs are transcribed from two non-allelic gene sets. *J. Mol. Biol.* **169:** 641–661.

Stern M., Jensen R., and Herskowitz I. 1984. Five *SWI* genes are required for expression of the *HO* gene in yeast. *J. Mol. Biol.* **178:** 853–868.

Sternberg P.W., Stern M.J., Clark I., and Herskowitz I. 1987. Activation of the yeast *HO* gene by release from multiple negative controls. *Cell* **48:** 567–577.

St. John T.P. and Davis R.W. 1981. The organization and transcription of the galactose gene cluster of *Saccharomyces*. *J. Mol. Biol.* **152:** 285–315.

Strahl B.D. and Allis C.D. 2000. The language of covalent histone modifications. *Nature* **403:** 41–45.

Struhl K. 1984. Genetic properties and chromatin structure of the yeast gal regulatory element: An enhancer-like sequence. *Proc. Natl. Acad. Sci.* **81:** 7865–7869.

Struhl K. and Davis R.W. 1977. Production of a functional eukaryotic enzyme in *Escherichia coli:* Cloning and expression of the yeast structural gene for imidazole-glycerolphosphate dehydratase (his3). *Proc. Natl. Acad. Sci.* **74:** 5255–5259.

Sudarsanam P. and Winston F. 2000. The Swi/Snf family nucleosome-remodeling complexes and transcriptional control. *Trends Genet.* **16:** 345–351.

Taunton J., Hassig C.A., and Schreiber S.L. 1996. A mammalian histone deacetylase related to the yeast transcriptional regulator Rpd3p. *Science* **272:** 408–411.

Thompson C.M., Koleske A.J., Chao D.M., and Young R.A. 1993. A multisubunit complex associated with the RNA polymerase II CTD and TATA-binding protein in yeast. *Cell* **73:** 1361–1375.

Vidal M. and Gaber R.F. 1991. RPD3 encodes a second factor required to achieve maximum positive and negative transcriptional states in *Saccharomyces cerevisiae*. *Mol. Cell. Biol.* **11:** 6317–6327.

Vignali M., Hassan A.H., Neely K.E., and Workman J.L. 2000. ATP-dependent chromatin-remodeling complexes. *Mol. Cell. Biol.* **20:** 1899–1910.

Wallis J.W., Rykowski M., and Grunstein M. 1983. Yeast histone H2B containing large amino terminus deletions can function in vivo. *Cell* **35:** 711–719.

Wang W., Cote J., Xue Y., Zhou S., Khavari P.A., Biggar S.R., Muchardt C., Kalpana G.V., Goff S.P., Yaniv M., Workman J.L., and Crabtree G.R. 1996. Purification and biochemical heterogeneity of the mammalian SWI-SNF complex. *EMBO J.* **15:** 5370–5382.

Winston F. and Carlson M. 1992. Yeast SNF/SWI transcriptional activators and the SPT/SIN chromatin connection. *Trends Genet.* **8:** 387–391.

Winston F., Chaleff D.T., Valent B., and Fink G.R. 1984. Mutations affecting Ty-mediated expression of the HIS4 gene of *Saccharomyces cerevisiae*. *Genetics* **107:** 179–197.

Winston F., Dollard C., Malone E.A., Clare J., Kapakos J.G., Farabaugh P., and Minehart P.L. 1987. Three genes are required for trans-activation of Ty transcription in yeast. *Genetics* **115:** 649–656.

5

Translation

Alan G. Hinnebusch
Laboratory of Gene Regulation and Development
National Institute of Child Health and Human Development
Bethesda, Maryland 20892

Genetic analysis of the scanning mechanism of translation initiation, 88

Sherman F., Stewart J.W., and Schweingruber A.M. 1980. Mutants of yeast initiating translation of iso-1-cytochrome c within a region spanning 37 nucleotides. *Cell* **20:** 215–222.

Identification of *trans*-acting factors involved in AUG selection, 92

Donahue T.F., Cigan A.M., Pabich E.K., and Castilho Valavicius B. 1988. Mutations at a Zn(II) finger motif in the yeast eIF-2β gene alter ribosomal start-site selection during the scanning process. *Cell* **54:** 621–632.

Gene-specific translational control by phosphorylation of eIF2, 94

Mueller P.P. and Hinnebusch A.G. 1986. Multiple upstream AUG codons mediate translational control of GCN4. *Cell* **45:** 201–207.

Initiation factors working at both ends of the mRNA mediate ribosome binding and mRNA circularization, 98

Altmann M., Sonenberg N., and Trachsel H. 1989. Translation in *Saccharomyces cerevisiae*: Initiation factor 4E-dependent cell-free system. *Mol. Cell. Biol.* **9:** 4467–4472.

Multiple functions of poly(A) tail and Pab1 in translation initiation, 100

Tarun S.Z., Jr., Wells S.E., Deardorff J.A., and Sachs A.B. 1997. Translation initiation factor eIF4G mediates *in vitro* poly(A) tail-dependent translation. *Proc. Natl. Acad. Sci.* **94:** 9046–9051.

Note: The landmark papers listed above are those discussed in this chapter. Each landmark paper is preceded by the name of the section (with starting page number) where the paper is first discussed in detail.

TRANSLATIONAL CONTROL OF GENE EXPRESSION GENERALLY OCCURS at the initiation phase of protein synthesis. The translation initiation process brings together the small (40S) ribosomal subunit with the mRNA, positions it at the initiator codon of the open reading frame (ORF) base-paired with methionyl initiator tRNA (Met-tRNA$_i^{Met}$), and allows joining of the 60S ribosomal subunit. Formation of this 80S initiation complex in mammalian cells is dependent on 11 or more soluble initiation factors (eIFs), identified primarily through biochemical studies of translation in rabbit reticulocyte extracts (Hershey and Merrick 2000).

Note: Boldfaced references in the text denote landmark papers that are on the accompanying CD.

For translation of most eukaryotic mRNAs, it is thought that the process begins when a ternary complex (TC) of eIF2 bound to GTP and Met-tRNA$_i^{Met}$ associates with the 40S ribosomal subunit preloaded with eIF3, eIF1, and eIF1A. The resulting 43S complex binds to the 5′-terminal m^{7}GpppN structure (cap) of the mRNA (forming the 48S complex) and migrates into the 5′-untranslated region (UTR) of the mRNA, searching for an AUG start codon, in a reaction known as scanning (Fig. 1). Binding of the 43S complex to the mRNA is stimulated by the eIF4F complex, consisting of the cap-binding protein (eIF4E), a DEAD-box RNA helicase (eIF4A), and a scaffold polypeptide (eIF4G) that contains binding sites for eIF4E and eIF4A. The ATP-dependent helicase activity of eIF4A is thought to help unwind secondary structure in the 5′ UTR to facilitate movement of the ribosome along the mRNA during scanning (Hershey and Merrick 2000; Hinnebusch 2000). The poly(A)-binding protein (PABP), bound to the poly(A) tail, also interacts with eIF4G, so that both ends of the mRNA are tethered to eIF4G (Sachs and Varani 2000). It is believed that eIF3 bridges the interaction between this mRNP (ribonucleoprotein) complex and the ribosome by interacting simultaneously with eIF4G and the 40S subunit (Lamphear et al. 1995). Formation of a 48S complex with Met-tRNA$_i^{Met}$ base-paired to the start codon also requires eIF1 and eIF1A (Pestova et al. 1998). This assembly is then recognized by eIF5, which stimulates hydrolysis of GTP in the TC, leading to release of eIF2-GDP but leaving behind Met-tRNA$_i^{Met}$ in the P site (Hershey and Merrick 2000; Hinnebusch 2000). Finally, eIF5B catalyzes joining of the 60S subunit to form the 80S initiation complex, with hydrolysis of a second molecule of GTP (Fig. 1) (Pestova et al. 2000).

Genetic analysis of translation initiation in yeast had an important role in determining that AUG triplets are selected as start codons during the scanning process by their proximity to the cap and through complementarity to the anticodon of Met-tRNA$_i^{Met}$. It also provided the first evidence that eIFs 1, 5, and the three subunits of eIF2 contribute to the high fidelity with which only AUG triplets are selected as start sites. These studies are described in the first two sections below. In the third section, a translational control mechanism is described in which four "false" start codons in the mRNA leader are employed to inhibit translation of an mRNA encoding the transcriptional activator Gcn4. The four upstream AUGs waylay the ribosomes as they scan downstream from the cap in this mRNA. When the concentration of TC is reduced in the cell, which occurs in starvation conditions, ribosomes scanning the *GCN4* mRNA leader can skip over the upstream AUG codons and initiate at the authentic *GCN4* start site instead. The mechanism that was found to reduce TC levels in starved yeast cells, involving phosphorylation of eIF2, was first discovered in heme-starved rabbit reticulocytes. The studies on yeast *GCN4* revealed how a reduction in this key initiation intermediate could stimulate the translation of a specialized mRNA.

By combining phenotypic characterization of initiation factor mutants with biochemical analysis of translation defects in cell-free extracts, it is possible to evaluate the importance of a given factor in the various steps of the initiation pathway. Most eIFs are essential in yeast, so that null mutations in these factors are lethal. One way to study the effects of impairing an essential protein is to analyze a temperature-sensitive mutant that is inviable only at elevated temperatures. The first application of this strategy for studying protein synthesis in yeast involved the *prt1-1* mutation, now known to impair the second largest subunit of eIF3. Analysis of this mutant showed that Prt1 was essential for virtually all translation in vivo (Hartwell and McLaughlin 1969), and biochemical studies revealed a defect

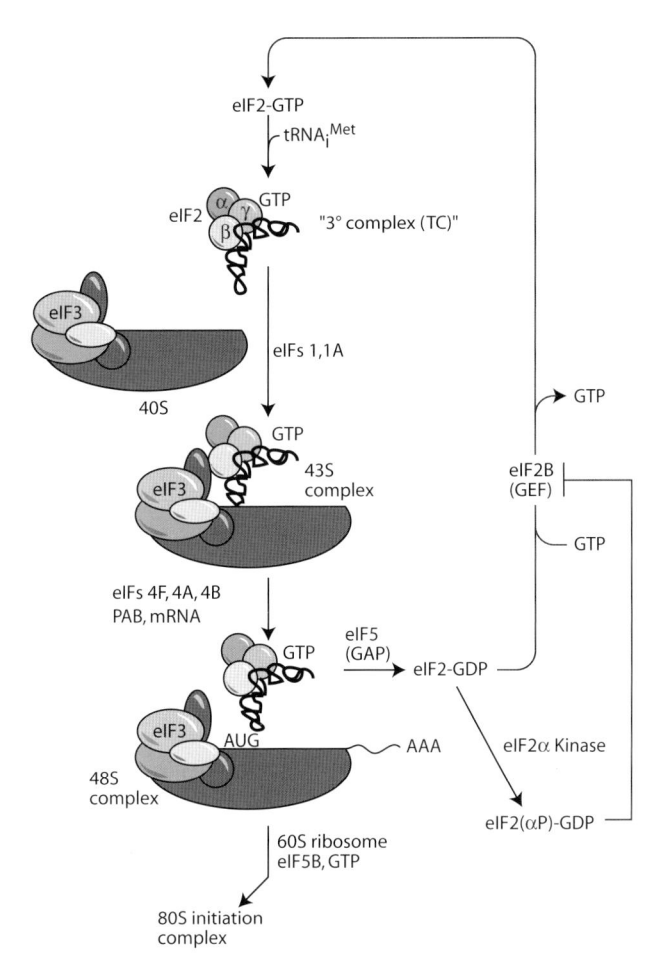

FIGURE 1. The eukaryotic translation initiation pathway. Conventional depiction of the pathway involving sequential binding of eIF3 and the ternary complex (TC), promoted by eIFs 1 and 1A to form the 43S preinitiation complex. The mRNA joins (facilitated by the eIF4 factors and PABP) to produce the 48S complex, which scans the mRNA for an AUG start codon. On AUG recognition (promoted by eIFs 1 and 1A), eIF5 binds and stimulates GTP hydrolysis, followed by release of factors and joining of the 60S subunit to produce the 80S initiation complex. Recycling of eIF2-GDP to eIF2-GTP is catalyzed by the GEF eIF2B. This reaction is inhibited by phosphorylation of eIF2-GDP on the α subunit, converting eIF2-GDP to a competitive inhibitor of eIF2B.

in stable association of the TC with 40S ribosomes in vitro (Feinberg et al. 1982), all of which was consistent with the role ascribed to mammalian eIF3 in recruiting TC to the ribosome (Hershey and Merrick 2000). Interestingly, application of a chemical cross-linking technique developed to measure 43S-48S complex assembly in living yeast cells showed recently that the rate-limiting defect in *prt1-1* cells lies downstream from 48S assembly, at scanning or AUG selection (Nielsen et al. 2004). In the future, this new tool should be combined with biochemical analysis of extracts or reconstituted systems to provide a more complete evaluation of initiation factor mutations.

Another way to characterize the null phenotype of an essential protein is to engineer the cognate gene to make its transcription dependent on a particular carbon source. The translational activity of cells or cell-free extracts can then be evaluated following depletion of the factor by culturing the mutant on the nonpermissive carbon source. In the fourth section below, this strategy was employed to show that the cap-binding protein eIF4E and the ATPase activity of the ATP-dependent RNA helicase eIF4A are both required in vitro for translation of most, if not all, yeast mRNAs.

Genetic analysis of the yeast PABP (Pab1) provided the first evidence that this protein participates in translation initiation in vivo (Sachs and Davis 1989), and biochemical studies showed that it functions in parallel with eIF4E to promote mRNA binding to 40S subunits (Tarun and Sachs 1995). A possible mechanism for its role in mRNA binding was prompted by the finding that Pab1, like eIF4E, binds to eIF4G, which in turn can interact with eIF3 in the 43S complex. The last section in this chapter describes genetic experiments that probed the relative importance of the eIF4E-eIF4G and Pab1-eIF4G interactions for translation initiation in vivo.

GENETIC ANALYSIS OF THE SCANNING MECHANISM OF TRANSLATION INITIATION

Formulation of the "First AUG Rule" from Classical Genetics of Cytochrome c Mutants

One of the fundamental questions concerning the mechanism of protein synthesis is how the ribosome can recognize the AUG start codon in the mRNA, distinguishing it from all other codons and from additional AUG triplets that occur near the 5´ end of the mRNA. Genetic analysis carried out in the laboratories of Fred Sherman and Thomas Donahue had a major role in determining how this discrimination occurs at the molecular level not only in yeast, but also in other eukaryotes (see, e.g., **Sherman et al. 1980;** Cigan et al. 1988b). It was known at the time that bacterial ribosomes recognize the start codon by its location approximately 7 nucleotides downstream from a short sequence closely related to 5´ GAGG 3´, known as the Shine-Dalgarno (SD) sequence (Shine and Dalgarno 1974). There was compelling evidence that this sequence base pairs with the 3´ end of the 16S rRNA in the small (30S) ribosomal subunit to position the ribosome with the initiation codon in the P site of the ribosome (for review, see Kozak 1984). However, it was unknown whether a similar mechanism operates in eukaryotes.

By the early 1970s, Sherman and colleagues had developed a large collection of mutants with lesions in the *CYC1* gene, encoding iso-1-cytochrome *c*, that provided a unique opportunity to identify important features of the initiation mechanism in yeast. More than 400 mutants lacking functional iso-1-cytochrome *c* were isolated, and the majority were located on a fine-structure genetic map of the gene. The position of the mutations on the map was determined by crossing each mutant to a panel of *cyc1* deletion mutants and analyzing the resulting diploids for the frequencies of X-ray-induced mitotic recombination that produced *CYC1*[+] recombinants. Of the 490 *cyc1* mutations analyzed, 17 were predicted to harbor mutations in the initiation codon, and these mutants contained no detectable *CYC1* product. Spontaneous revertants of certain initiation codon mutants were isolated in which iso-1-cytochrome *c* expression had been restored. The iso-1-cytochrome *c* was purified from each of the revertants and analyzed by mapping tryptic and chymotryptic peptides and by determining the amino acid sequence composition. From this information, Sherman et al. deduced that the revertants had acquired single-base-pair substitutions that reintroduced an AUG codon either at the normal start site (codon 1), at the triplet immediately preceding the normal start site (the –1 triplet), or at the 5th codon in the gene. Figure 2A depicts the three types of reversion events obtained for the *cyc1-13* start codon mutation. These findings indicated a degree of flexibility in the acceptable sequence context of the start codon that seemed to be inconsistent with a bacterial mode of initiation involving an SD-like sequence located at a defined distance from the AUG (Stewart et al. 1971; Sherman and Stewart 1982). The extraordinary factor about this work is that it occurred before the advent of yeast transformation and the techniques now used routinely to clone and sequence mutant alleles.

In the foregoing studies, reversion events at the other eight triplets within the first ten codons of *CYC1* were not obtained, even though the encoded amino acids were known to be dispensable for iso-1-cytochrome *c* function. Either an AUG at these other positions could not function as a start site or the frequency of the mutations required to create an AUG at these positions was prohibitively low. To distinguish between these possibilities, Sherman and colleagues used UV rays, X-rays, and γ-rays to increase the mutation frequency and employed a specially constructed allele for reversion analysis (*cyc1-345*) that contained the nonsense triplet UAA at the 3rd codon in addition to the start codon mutation (Fig. 2B). Introduction of the stop codon was designed to reduce the frequency of the reversion events at positions 1 and –1 obtained in their previous studies, as an AUG created at these two positions could not direct translation past the nonsense triplet at codon 3. The use of ionizing radiation as mutagen allowed for multiple base substitutions or deletions to arise in the same allele.

Analysis of the *cyc1-345* revertants showed that an AUG created at the –2, 4th, or 6th codons could direct normal levels of iso-1-cytochrome *c* expression, in addition to the –1 and 5th codon positions identified in the previous study. An AUG at the 10th codon also functioned as a start site, although possibly with reduced efficiency (Fig. 2B). Together, these data showed that translation could initiate with normal efficiency from AUG codons at any of six or seven locations within a span of 25 nucleotides near the 5′ end of the *CYC1*-coding region. This finding strongly implied that AUG is the only critical sequence required for initiation in yeast (Sherman et al. 1980).

Another important insight that emerged from revertant analysis of *cyc1* alleles was that an AUG triplet can function efficiently as a start codon only if it occurs as the 5′-proximal AUG in the mRNA. An allele containing the UAA nonsense triplet at codon 3 and AUG at the normal start site (*cyc1-9*) never reverted spontaneously by the introduction of an AUG downstream from the stop codon, even though it was clear from the earlier work that an AUG at codon 5 was a functional start site. The presumption was that an AUG downstream from the stop codon is preempted by the AUG at codon 1. This hypothesis was tested directly by creating the allele *cyc1-341* that contains AUG triplets at positions 1 and 5 plus the UAA terminator at position 3 (Fig. 2C). A revertant of this inactive allele was obtained (*CYC1-341-D*) in which the AUG at codon 1 was eliminated, making the AUG at codon 5 the first one in the mRNA (Fig. 2C). This result showed clearly that the AUG at codon 5 was perfectly functional as a start codon as long as it was not preceded by another AUG in the mRNA (Sherman and Stewart 1982).

Allele	Sequence
A CYC1	-2 -1 1 2 3 4 5 6 7 8 9 10 11 12 13 14 15 16 17 18 A UUA AUA AUG ACU GAA UUC AAG GCC GGU UCU GCU AAG AAA GGU GCU ACA CUU UUC AAG ACU (Met)Thr-Glu-Phe-Lys-Ala-Gly-Ser-Ala-Lys-Lys-Gly-Ala-Thr-Leu-Phe-Lys-Thr
cyc1-13 ↓	A UUA <u>AUA</u> <u>AUA</u> ACU GAA UUC <u>AAG</u> GCC GGU UCU GCU AAG AAA GGU GCU ACA CUU UUC AAG ACU no product
CYC1-13A	Met-*Ile*-Thr-Glu-Phe-Lys-Ala-Gly-Ser-Ala-Lys-Lys-Gly-Ala-Thr-Leu-Phe-Lys-Thr
CYC1-13P	(Met)Thr-Glu-Phe-Lys-Ala-Gly-Ser-Ala-Lys-Lys-Gly-Ala-Thr-Leu-Phe-Lys-Thr
CYC1-13S	(Met)Ala-Gly-Ser-Ala-Lys-Lys-Gly-Ala-Thr-Leu-Phe-Lys-Thr
B cyc1-345 ↓	<u>A UUA AUA AUA</u> ACU *UAA* <u>UUC</u> *CAA* <u>GCC</u> GGU UCU GCU <u>AAG</u> AAA GGU GCU ACA CUU UUC AAG ACU no product
CYC1-345-H	Met-*Asn*-*Asn*-*Asn*-Leu-*Ile*---Gln-Ala-Gly-Ser-Ala-Lys-Lys-Gly-Ala-Thr-Leu-Phe-Lys-Thr
CYC1-345-B	Met-*Ile*-Thr-*Tyr*-Phe-Gln-Ala-Gly-Ser-Ala-Lys-Lys-Gly-Ala-Thr-Leu-Phe-Lys-Thr
CYC1-345-A	(Met)Thr-*Gln*-Phe-Lys-Ala-Gly-Ser-Ala-Lys-Lys-Gly-Ala-Thr-Leu-Phe-Lys-Thr
CYC1-345-C	Met-Gln-Ala-Gly-Ser-Ala-Lys-Lys-Gly-Ala-Thr-Leu-Phe-Lys-Thr
CYC1-345-J	(Met)Gly-Ser-Ala-Lys-Lys-Gly-Ala-Thr-Leu-Phe-Lys-Thr
CYC1-345-F	*Met*-Lys-Gly-Ala-Thr-Leu-Phe-Lys-Thr
C cyc1-341 ↓	A UUA AUA <u>AUG</u> ACU *UAA* UUC *AUG* GCC GGU UCU GCU AAG AAA GGU GCU ACA CUU UUC AAG ACU (Met)Thr-TER
CYC1-341-D	A UUA AUA *AUA* ACU *UAA* UUC *AUG* GCC GGU UCU GCU AAG AAA GGU GCU ACA CUU UUC AAG ACU (Met)Ala-Gly-Ser-Ala-Lys-Lys-Gly-Ala-Thr-Leu-Phe-Lys-Thr

FIGURE 2. Summary of *cyc1* revertants that established the "first AUG rule" for translation initiation in yeast. The nucleotide sequence and corresponding amino acid sequences are shown for the beginning of the wild-type allele (*CYC1*). The Met residue encoded by the start codon (position 1) is in parentheses because it is removed by methionine aminopeptidase. The nucleotide sequences are shown for three mutant alleles that make no detectable iso-1-cytochrome *c* and were the parental alleles for isolating the functional revertants shown beneath each one. The italicized triplets in the parental alleles differ from the corresponding sequence of wild-type *CYC1*. It was surmised that the underlined triplets in the parental alleles sustained one or more base substitutions in order to produce the revertant proteins listed below. Residues italicized in the revertant proteins are novel residues not found at the corresponding positions in the wild-type protein. All three revertants of *cyc1-13* shown in *A* required only single base substitutions to introduce an AUG at codons −1, 1, or 5. The generation of *CYC1-345-H* from *cyc1-345* shown in *B* required a point mutation in the AUU triplet starting at the 7th base upstream of codon position 1, producing an AUG in the −1 reading frame. A base deletion in the UUC at codon position 4 was additionally required to shift translation back to the 0 reading frame. Generation of *CYC1-345-B* and *CYC1-345-A* required single-base mutations in codons −1 and 1, respectively, to produce AUG start codons at those sites, plus mutations in the UAA stop codon at position 3 to produce sense codons for Tyr or Gln, respectively. Generation of *CYC1-345-C* and *CYC1-345-J* required multiple substitutions in codons 4 and 6, respectively, to produce AUGs at those sites. Generation of *CYC1-345-F* required a single-base change in codon 10 to produce an AUG there. The mutational event that produced *CYC1-341-D* in *C* was identified by cloning the mutant allele and determining its DNA sequence, as shown.

Further Development of the Scanning Model

With the development of yeast transformation and the techniques for cloning mutant alleles, it was possible to obtain even stronger evidence for this "first AUG rule." The nonfunctional *cyc1-362* mutation had been mapped upstream of the normal initiation codon. Cloning and sequencing of this allele showed that it contains a single-base-pair substitution that generates a new AUG triplet 20 nucleotides upstream of the normal start site, initiating a short upstream ORF (uORF) of five codons that terminates two nucleotides before the AUG at codon 1. Presumably, the new AUG triplet preempted the ability of the wild-type start codon to function as an initiation site. The fact that the uORF in *cyc1-362* completely blocked initiation at the normal start codon also implied that ribosomes were incapable of resuming scanning and reinitiating downstream following termination of translation at the uORF (Stiles et al. 1981). This stood in sharp contrast to the ability of bacterial ribosomes to reinitiate translation efficiently on polycistronic mRNAs, particularly when the termination codon of the upstream cistron was very close to the start codon of the downsteam cistron. As shown below, the prohibition on reinitiation is not absolute. The process can be enhanced by increasing the intercistronic distance between the uORF and the downstream start site and modulated by sequences surrounding the termination codon of the uORF.

An elegant genetic analysis of initiation codon mutations at *HIS4* showed that for this gene as well, the 5´-proximal location was far more important than the surrounding sequence context in determining whether an AUG triplet could function as a start codon (Donahue and Cigan 1988). The lessons learned from genetic analysis of *CYC1* and *HIS4* initiation mutants complemented Marilyn Kozak's previous observations that the great majority of eukaryotic mRNAs are monocistronic and lack AUG codons upstream of their initiation sites. Combining these facts with the results of her experiments showing that eukaryotic ribosomes cannot bind to circular mRNAs (Kozak 1979), Kozak proposed the scanning model for translation initiation. According to this hypothesis, the small (40S) ribosomal subunit binds to the mRNA at the 5´ end and threads along the mRNA until an AUG codon is encountered, whereupon an 80S initiation complex is assembled (Kozak 1978). The hypothesis was modified after the discovery that sequences immediately surrounding the start codon can have a strong effect on initiation frequency in mammalian cells, so that the first AUG can be bypassed if it occurs in an unsuitable sequence context. This exception to the first AUG rule was called "leaky scanning" (Kozak 1986). Subsequent work on *CYC1* (Baim and Sherman 1988) and *HIS4* (Cigan et al. 1988b) showed that nucleotides present at the –3 and +4 base positions relative to the AUG (where A is designated the +1 base) can affect the initiation frequency in yeast cells, but the effect is considerably smaller in magnitude than that observed in mammals. Moreover, evidence was obtained that a poor sequence context can lead to leaky scanning in yeast (Werner et al. 1987).

The phenomenon of leaky scanning has several interesting consequences for gene expression. First, translation of an mRNA can be down-regulated by the presence of a naturally occurring upstream AUG codon in the mRNA leader. If ribosomes can leaky-scan past the upstream AUG, then the downstream cistron will be translated at a frequency that is directly in proportion to the probability of leaky scanning upstream. Leaky scanning can also allow for production of multiple proteins from the same mRNA. Interesting examples of this phenomenon involve proteins that function in both mitochondria and the cytoplasm and require a leader peptide for import into mitochondria. In some cases, it appears that longer and shorter forms of the same protein, containing or lacking the leader peptide, respectively, can be translated from one mRNA. Ribosomes initiating at the first AUG produce the longer protein, whereas the shorter protein is translated by ribosomes that leaky-scan past the first AUG and initiate at the second start site (Wolfe et al. 1994). Reinitiation following translation of a short uORF is another way to circumvent an upstream AUG, and the efficiency of reinitiation can be regulated to control gene expression (e.g., *GCN4*).

A final point to consider in this section is that an increasing number of mRNAs in mammalian cells, or their viruses, seem to violate the first-AUG rule entirely by employing an alternative to the scanning mechanism known as internal initiation. In these mRNAs, the 40S ribosome bypasses the cap and binds to an internal ribosome entry site (IRES) located in the 5′ UTR. In bypassing the cap, an IRES can dispense with the requirement for some or all mRNA-binding eIFs (eIF4E, eIF4A, eIF4B, eIF4G, and Pab1) (Hellen and Sarnow 2001). In an extreme case, an IRES found in cricket paralysis virus (CrPV) functions independently of all eIFs, and even of Met-tRNA$_i^{Met}$. It seems that the CrPV IRES occupies the ribosomal P site, and the first codon of the gene (specifying alanine) is decoded in the A site by alanyl-tRNA (Wilson et al. 2000).

The presence of IRES elements with reduced factor requirements can permit viral protein synthesis to proceed even when some or all of the host-cell eIFs have been inactivated and host protein synthesis is impaired in the infected cells. Likewise, IRESs may exist in cellular mRNAs to permit their translation under starvation or stress conditions where eIF function and the canonical scanning mechanism has been down-regulated. At present, in yeast, only a few cases of naturally occurring IRESs in the mRNAs for *YAP1, TIF4631* (encoding eIF4G1), and *URE2* have been described (Zhou et al. 2001; Komar et al. 2003). Interestingly, the CrPV IRES has been shown to function in yeast as long as eIF2 function is down-regulated by phosphorylation (Thompson et al. 2001).

IDENTIFICATION OF *TRANS*-ACTING FACTORS INVOLVED IN AUG SELECTION

If AUG is the only critical sequence required for initiation, how is it recognized by the scanning 40S ribosome? In a series of elegant genetic experiments, Donahue and colleagues showed that the anticodon of tRNA$_i^{Met}$ has a key role in this process, along with contributions from eIFs 1, 2, and 5. In one approach, they altered the anticodon of tRNA$_i^{Met}$ from 3′ UAC 5′ to 3′ UCC 5′ by mutating one of the *IMT* genes encoding this tRNA in yeast. Overproduction of the mutant initiator from a high-copy-number plasmid restored expression from a *his4* allele in which the AUG start codon was changed to AGG, a triplet complementary to the anticodon of the mutant tRNA$_i^{Met}$(UCC). Other *his4* alleles with different mutant start codons were not suppressed by tRNA$_i^{Met}$(UCC), showing that codon-anticodon base pairing was required for suppression. Moreover, introduction of an extra AGG codon upstream of the AGG start site abolished suppression, indicating that the upstream AGG was recognized preferentially by the mutant tRNA$_i^{Met}$ (UCC)—a hallmark of the scanning process (Fig. 2) (Cigan et al. 1988a). This work established that perfect base pairing between the anticodon of the initiator and the start codon in mRNA, regardless of their sequences, is a fundamental requirement for efficient initiation in yeast.

As shown in Figure 1, Met-tRNA$_i^{Met}$ is transferred to the 40S ribosome in a TC with eIF2 and GTP. After the resulting 43S complex binds to the mRNA and locates the AUG start codon, base pairing between Met-tRNA$_i^{Met}$ and AUG triggers hydrolysis of the GTP in the TC, in a manner stimulated by eIF5. Genetic studies in Donahue's lab implicated all three subunits of eIF2, eIF5, and a third factor, eIF1, in stringent recognition of AUG as the start codon. To reach this goal, they selected mutations in *trans*-acting factors that would increase expression of a defective *his4* allele containing AUU in place of AUG at the start codon. Suppressor mutations obtained in three different genes, *SUI1*, *SUI2*, and *SUI3* (for suppressor of initiation codon mutation), overcame the histidine auxotrophy of the starting *his4(AUU)* mutant (Castilho-Valavicius et al. 1990). Amino-terminal amino acid sequencing of purified His4-LacZ fusion proteins expressed from a *his4(AUU)-lacZ* fusion was conducted to determine what codon was being used as the start site in the Sui⁻ mutants. The results showed that the *sui* mutations allowed a UUG present at the third codon in *HIS4* to be recognized as the start codon by Met-tRNA$_i^{Met}$, despite the predicted U-U mismatch at the first position of the codon–anticodon duplex (Fig. 3). The *SUI2* and *SUI3* genes were cloned and sequenced and were found to encode the α and β subunits of eIF2 (**Donahue et al. 1988**; Cigan et al. 1989). *SUI1* was later found to encode the yeast homolog of eIF1 (Yoon and Donahue 1992).

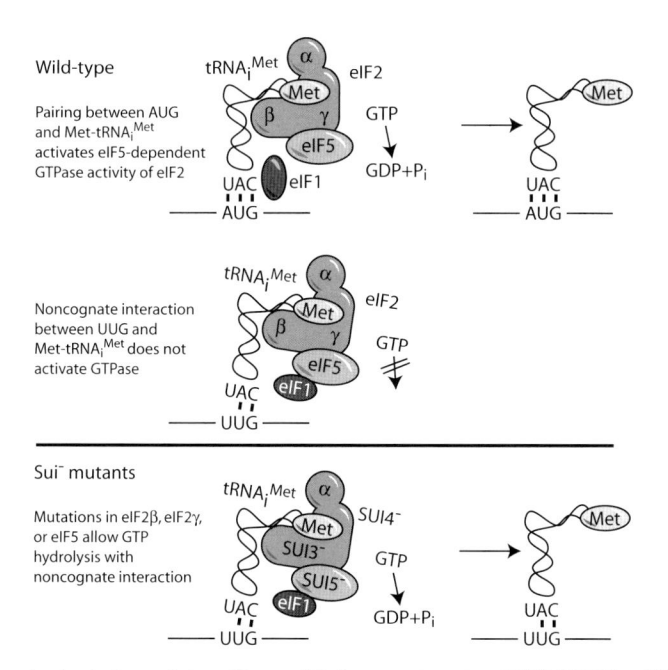

FIGURE 3. Schematic depiction of the effects of Sui⁻ mutations in eIF2β (*SUI3*), eIF2γ (*SUI4*), or eIF5 (*SUI5*) in allowing UUG to be selected as start codon. In wild-type cells, perfect pairing between the anticodon loop of Met-tRNA$_i^{Met}$ and AUG start codon is required to trigger hydrolysis of GTP by eIF2 in the ternary complex, dependent on eIF5 and modulated by eIF1. This leads to dissociation of eIF2-GDP, eIF5, and eIF1 from the 48S complex and leaves only Met-tRNA$_i^{Met}$ base-paired to the start codon. In the Sui⁻ mutants, it is thought that imperfect pairing of the anticodon loop with a UUG codon triggers GTP hydrolysis at a much higher rate than occurs in wild-type cells. See text for further details.

SUI1, *SUI2*, and *SUI3* were found to be essential, as deleting each gene from the chromosome had a lethal phenotype. Thus, it appears that the Sui⁻ mutations alter the functions of eIF1 or eIF2 in a way that reduces the accuracy of initiation without greatly impairing its efficiency. In a subsequent screen, dominant Sui⁻ mutations in *SUI3* and in two other genes, *SUI4* and *SUI5*, were isolated in a diploid strain and found to be lethal in haploid cells (Castilho-Valavicius et al. 1992). The *SUI4* mutations were identified as alleles of *GCD11* (Huang et al. 1997), encoding the γ subunit of eIF2 (Hannig et al. 1993), and *SUI5* was identical to *TIF5* (Huang et al. 1997), encoding eIF5 (Chakravarti and Maitra 1993).

Biochemical analysis of purified mutant proteins showed that two Sui⁻ mutations in *SUI3* (*S264Y* and *L254P*) led to increased GTP hydrolysis in the purified TC independently of eIF5. The *S264Y* mutation also led to increased dissociation of Met-tRNA$_i^{Met}$ from the TC independently of GTP hydrolysis. The *SUI4/GCD11-N135K* mutation in eIF2γ alters its GTP-binding (G) domain that is highly conserved with bacterial translation elongation factor EF-Tu. This lesion destabilized the TC by increasing both the rate of spontaneous GTP hydrolysis and the off-rate of Met-tRNA$_i^{Met}$ from eIF2. It was suggested that all of these defects might increase the probability that the TC dissociates inappropriately and leaves Met-tRNA$_i^{Met}$ base-paired with a UUG start codon, accounting for the Sui⁻ phenotypes of the mutations. The same mechanism could account for the Sui⁻ phenotype of the *SUI5/TIF5-G31R* allele, as it increases the rate of eIF5-stimulated GTP hydrolysis by the TC in a model assay for eIF5 GAP function (Huang et al. 1997). These findings represent an impressive convergence of genetic and biochemical data suggesting that GTP hydrolysis by the TC is triggered by perfect base pairing between Met-tRNA$_i^{Met}$ and the AUG start codon (Fig. 3). Subsequent biochemical studies conducted by Pestova, Borukhov, and Hellen using purified mammalian eIFs, Met-tRNA$_i^{Met}$, mRNA, and 40S ribosomes, showed that eIF1 is essential to form a stable 48S initiation complex with TC positioned precisely at the start codon (Pestova et al. 1998). Pestova and colleagues further showed that mammalian eIF1 enables discrimination against non-AUG triplets and AUGs in a poor sequence context by the scanning 43S complex in this reconstituted system (Pestova et al. 1998; Pestova and Kolupaeva 2002). These last results confirm the genetic findings from Donahue et al. implicating eIF1 in stringent AUG selection. Recent findings from biochemical (Unbehaun et al. 2004), biophysical (Maag et al. 2005), and genetic (Valasek et al. 2004) experiments indicate that both yeast and mammalian eIF1 impede eIF5-dependent GTP hydrolysis at non-AUG triplets, thus serving a "gatekeeper" function in AUG selection beyond its role in promoting ribosomal scanning.

GENE-SPECIFIC TRANSLATIONAL CONTROL BY PHOSPHORYLATION OF eIF2

As discussed above, binding of TC to the 40S ribosome and AUG recognition during scanning are critical steps involved in the translation of all mRNAs. Here, we turn to genetic studies of the *GCN4* gene in yeast that have revealed that these fundamental reactions can be modulated to control the translation of specific mRNAs that harbor specialized regulatory sequences in their mRNA leaders. Gerald Fink and colleagues established that Gcn4 is a transcriptional activator of multiple amino acid biosynthetic genes that are induced by starvation for any amino acid, a regulatory response known as general amino acid control. Because *gcn4* mutants are defective for this response (a general control nonderepressible [Gcn⁻] phenotype), they cannot grow in the presence of inhibitors of amino acid biosynthetic enzymes, e.g., 3-aminotriazole (3AT, a competitive inhibitor of His3), which causes histidine limitation. Mutations in *GCN2* and *GCN3* also confer 3AT sensitivity, and it was shown that these gene products are required for induction of Gcn4 in amino-acid-starved cells. Negative regulatory factors were identified by the isolation of *gcd* (general control derepressed) mutants in which *GCN4* and its target genes are expressed at high levels constitutively. Genetic epistasis studies suggested that Gcn2 and Gcn3 function by antagonizing the negatively regulatory functions of the *GCD* factors, which in turn down-regulate Gcn4 (Fig. 4A) (Hinnebusch and Fink 1983; Bushman et al. 1993). For example, *gcn2 gcd1* double mutants are 3AT-resistant, whereas *gcn4 gcd1* double mutants are 3AT-sensitive. Indeed, a number of *gcd* mutations were isolated as suppressors of the 3AT-sensitive phenotype of a *gcn2 gcn3* double mutant (Harashima and Hinnebusch 1986).

The *GCN* and *GCD* products were shown to regulate *GCN4* expression at the translational level by modulating the effects of four short uORFs in the *GCN4* mRNA leader. As described above, uORFs inhibit translation initiation at downstream AUG codons because scanning ribosomes reach the uORF start codons first, and reinitiation downstream following translation of the uORF is generally inefficient. Thus, it was not surprising to find that simultaneous removal of all four upstream AUG codons by point mutations led to constitutive derepression of β-galactosidase activity expressed from a *GCN4-lacZ* fusion construct, without increasing its fusion mRNA level (**Mueller and Hinnebusch 1986**). When introduced into the *GCN4* gene, these AUG mutations also derepressed transcription of *HIS4*, a gene under Gcn4 control. By analyzing triple mutations that left only a single uORF intact, it was deduced that uORF3 and uORF4 (the two closest to the authentic start site) were responsible for the strong inhibition of *GCN4* translation in nonstarved cells, whereas uORF1 was a relatively weak translational barrier. Further mutational analysis led to the unexpected conclusion that uORF1 was required for the induction of *GCN4* translation in starved cells and that it functioned by allowing ribosomes to overcome the strong translational barrier at uORF3–uORF4. Because *gcn2* and *gcd1* mutations affected *GCN4* expression only when uORF1 was present together with uORF3 or uORF4, it was concluded that the GCD factors block the positive effect of uORF1 in overcoming the inhibitory effects of uORF3–uORF4 on *GCN4* translation (Fig. 4A) (Mueller and Hinnebusch 1986). It is striking that this regulatory model, constructed primarily on the basis of genetic epistasis studies, is still an accurate description of the interplay among the *cis*- and *trans*-acting regulatory elements in *GCN4* translational control.

Subsequent mutational analysis of the uORFs showed that the large differences between uORF1 and uORF4 as translational barriers, when examined as solitary uORFs, were dependent on sequences surrounding their termination codons (Miller and Hinnebusch 1989; Grant and Hinnebusch 1994). This led to the idea that ribosomes would remain attached to the mRNA, resume scanning, and reinitiate translation downstream at a high frequency following translation of uORF1, whereas termination at uORF4 would generally lead to dissociation of ribosomes from the mRNA (Fig. 4B).

Kozak (1987) had shown that in mammalian cells, a uORF introduced into the leader of an mRNA became progressively less inhibitory as it was moved further upstream from the protein-coding ORF, and she proposed that ribosomes must rebind a factor following termination at the uORF before they can reinitiate downstream. When the uORF is further upstream, it provides more time to rebind the requisite factor, increasing the probability of reinitiation. Applying this concept to *GCN4* led to the prediction that the approxi-

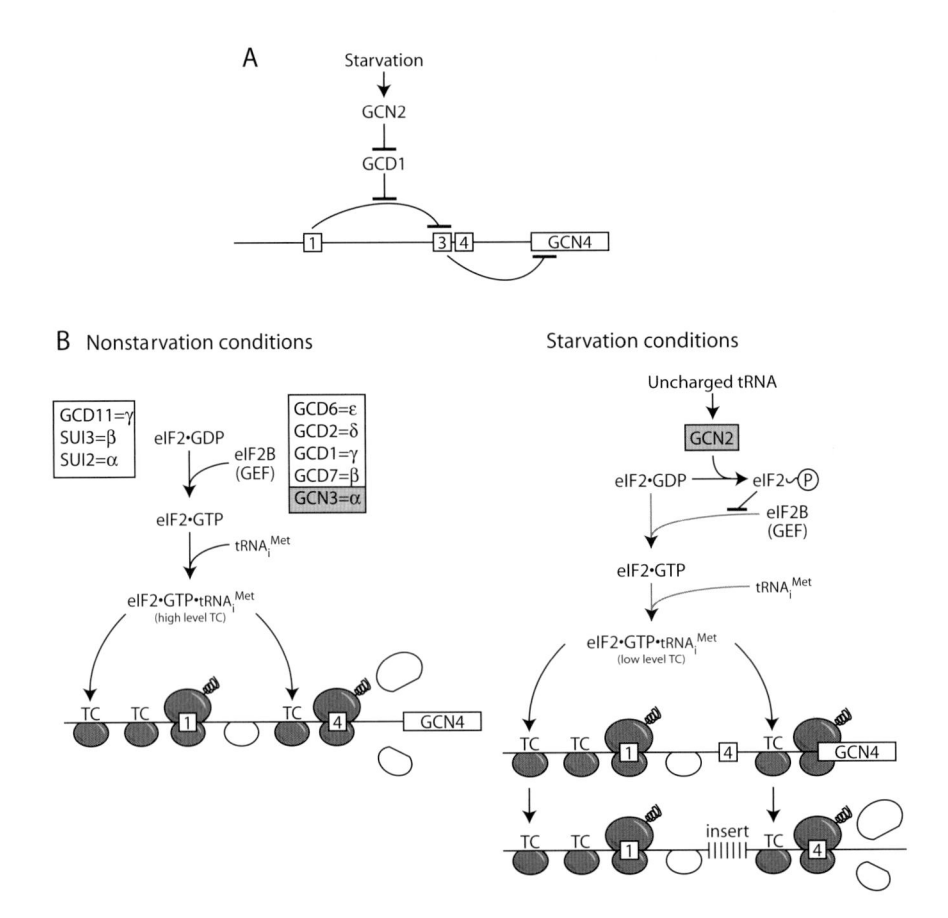

FIGURE 4. Schematic depiction of the mechanism of *GCN4* translational control. (*A*) Regulatory scheme deduced from genetic epistasis analysis. *GCN4* mRNA is depicted with uORF1, uORF3, and uORF4 and the *GCN4*-coding sequences as boxes (uORF2 is removed to simplify the analysis). Epistasis analysis of mutations in the uORFs and the *trans*-acting factors *GCD1* and *GCN2* indicated that uORF3–uORF4 are potent negative elements, whereas uORF1 functions positively by overcoming the negative effects of uORF3–uORF4. *GCD1* is a negative regulator that blocks the stimulatory effect of uORF1, and *GCN2* is a positive regulatory factor that antagonizes *GCD1*. (*B*) Molecular model for *GCN4* translational control. For simplicity, uORF2 and uORF3 were omitted because their functions are redundant with uORF4. The 40S ribosomal subunits are shaded when associated with the ternary complex (TC) and thus are competent to reinitiate at the next start codon they encounter. 80S ribosomes are shown translating uORF1, uORF4, or *GCN4* with the synthesized peptides depicted as coils. Free 40S and 60S subunits are shown dissociating from the mRNA following translation of uORF4. The three subunits of eIF2 and the five subunits of eIF2B are listed in the boxes on the left panel. Following translation of uORF1, the 40S ribosome remains attached to the mRNA and resumes scanning. Under nonstarvation conditions, the 40S quickly rebinds the TC and reinitiates at uORF4 because the TC concentration is high. Under amino acid starvation conditions, many 40S ribosomes fail to rebind the TC until scanning past uORF4, because the TC concentration is low, and reinitiate at *GCN4* instead. TC levels are reduced in starved cells due to phosphorylation of eIF2 by the kinase Gcn2, converting eIF2 from substrate to inhibitor of its guanine nucleotide exchange factor eIF2B. The bottom construct in the right panel contains an insertion that increases the spacing between uORF1 and uORF4. This forces all ribosomes to reinitiate at uORF4, even under starvation conditions, thereby preventing the induction of *GCN4* translation in starved cells. See text for further details.

mately 200 nucleotides separating uORF1 and uORF4 are sufficient to allow nearly all ribosomes to rebind the requisite factor after terminating at uORF1 and reinitiate at uORF4 under nonstarvation conditions. After translating uORF4, these ribosomes would dissociate from the mRNA and fail to reach the *GCN4* start codon. In amino-acid-starved cells, in contrast, rebinding of the requisite factors would be delayed, so that a fraction of ribosomes that translated uORF1 would not be competent to reinitiate until after scanning past uORF4, but before reaching the *GCN4* start codon. This pool of ribosomes would bypass uORF2–uORF4 and reinitiate at *GCN4* instead (Fig. 4B). Strong evidence for this hypothesis was provided by the demonstration that progressively increasing the distance between uORF1 and uORF4 led to a progressive decrease in *GCN4* translation. This result was predicted by the model because a larger fraction of ribosomes should be competent to reinitiate before reaching uORF4, and be excluded from reinitiation at *GCN4,* in starved cells when the scanning time between uORF1 and uORF4 was increased (Fig. 4B). Other experiments showed that translation of uORF4 decreased in starved cells when *GCN4* translation is induced and that ribosomes which translate *GCN4* have not previously translated uORF4. These observations fit with the idea that ribosomes must bypass uORF4 to reach *GCN4* in starved cells (Abastado et al. 1991).

A delay in the rebinding of a key initiation factor to ribosomes scanning downstream from uORF1 in starved cells could arise from a reduction in the concentration of that factor in the cell. Given that Met-tRNA$_i^{Met}$ directs the scanning ribosome to the start codon in yeast (Cigan et al. 1988a), a reduction in TC levels was an attractive mechanism for inducing *GCN4* translation. Supporting this possibility, it was found that a *gcd1* mutation (which exhibits derepressed translation of *GCN4* mRNA) reduces binding of Met-tRNA$_i^{Met}$ to 40S ribosomes in vivo (Tzamarias and Thireos 1988). Moreover, *sui2* and *sui3* mutations in eIF2α and eIF2β were shown to constitutively derepress *GCN4* translation (Williams et al. 1989). It was known that TC levels in rabbit reticulocyte lysates are diminished by phosphorylation of Ser-51 in the α subunit of eIF2. The phosphorylated form of the protein (eIF2α[P]) acts as a competitive inhibitor of the guanine nucleotide exchange factor, eIF2B, responsible for recycling eIF2-GDP to eIF2-GTP for regenerating the TC after each round of initiation (Fig. 1) (Hinnebusch 2000). By this time, it was established that Gcn2 is a protein kinase (Roussou et al. 1988; Wek et al. 1989) and that its kinase domain is closely related in sequence to that of a mammalian kinase that phosphorylates eIF2α on Ser-51, known as DAI (now called PKR) (Ramirez et al. 1991; Chong et al. 1992). The stage was set to determine whether *GCN4* translation is induced in amino-acid-starved cells by phosphorylation of eIF2α on Ser-51 by Gcn2 (Fig. 4B).

Convincing evidence for this model was provided by showing that eIF2α phosphorylation increases in amino-acid-starved cells and is abolished by deleting *GCN2* or by a *sui2* mutation that replaces Ser-51 in eIF2α with alanine (*SUI2-S51A*). These measurements involved isoelectric focusing polyacrylamide gel electrophoresis to resolve the phosphorylated and nonphosphorylated forms of eIF2α, which were visualized by western blot analysis. Importantly, the *SUI2-S51A* mutation impaired *GCN4* translation in starved cells to the same extent as did deletion of *GCN2* (Dever et al. 1992). Subsequently, a large body of genetic and biochemical evidence was obtained indicating that phosphorylation of eIF2α inhibits eIF2B function and reduces TC levels in yeast cells, just as it does in rabbit reticulocyte lysates. The work in yeast further revealed that eIF2B contains a regulatory subcomplex of three subunits (Gcd2, Gcd7, and Gcn3) that mediates the inhibitory effect of eIF2(αP) on the ability of the remaining two subunits (Gcd6 and Gcd1) to catalyze GDP-GTP exchange on eIF2 (Hinnebusch 2000). This finally explained why a deletion of Gcn3 (the only nonessential subunit in eIF2B) impairs *GCN4* translation (Fig. 4A,B). It remains to be determined what essential function of eIF2B is carried out by Gcd2 and Gcd7.

A mammalian homolog of Gcn2 was identified and found to phosphorylate Ser-51 in eIF2α in cultured cells in response to amino acid starvation. As expected, this led to a general reduction in the rate of translation initiation; but interestingly, it also stimulated translation of *ATF4* mRNA (encoding a transcription factor) in a manner dependent on two uORFs in the *ATF4* leader (Harding et al. 2000). Remarkably, genetic analysis indicates that induction of *ATF4* translation occurs by essentially the same reinitiation mechanism established for *GCN4* and described above, with the 3-codon uORF1 acting positively to suppress the inhibitory effect of uORF2 on *ATF4* translation in a manner dependent on the proper spacing between the two uORFs (Lu et al. 2004; Vattem and Wek 2004). It is striking that eIF2α phosphorylation induces the translation of uORF-containing mRNAs in both yeast and mammalian cells.

Before leaving the topic of translational control by uORFs, it should be noted that an uORF can regulate translation by a mechanism distinct from that just described for *GCN4*. A single uORF in the mRNA of *CPA1*, encoding an arginine biosynthetic enzyme in yeast, represses *CPA1* translation in the presence of excess arginine. Because the amino acid sequence of the uORF is critical for regulation, the uORF-encoded peptide has a role in attenuating *CPA1* translation (Werner et al. 1987; Delbecq et al. 1994; Wang et al. 1999). This regulatory mechanism was reconstituted in a cell-free translation system and, using a physical technique called ribosomal toe-printing, it was possible to map the locations of ribosomes translating the *CPA1* uORF. Interestingly, the ribosomes were found to stall after completing translation of the uORF peptide, with the uORF stop codon in the ribosomal A site (Wang et al. 1999). Presumably, the stalled ribosomes prevent 48S complexes that have leaky-scanned past the uORF AUG codon from reaching the *CPA1* start site downstream. Recently, an elegant modification of the toe-printing technique was used to confirm that ribosomes cannot reinitiate following translation of the *CPA1* uORF and, hence, must reach the *CPA1* start site by leaky-scanning. Under the same conditions, reinitiation at the *GCN4* start codon was readily detected for ribosomes that had previously translated uORF1 (but not uORF4) in *GCN4* mRNA. Thus, the *CPA1* uORF and uORF1 of *GCN4* regulate translation by completely different mechanisms (Gaba et al. 2001).

The regulatory mechanism elucidated for *CPA1* has also been described in mammalian cells for the second uORF of the cytomegalovirus gene encoding gpUL4 (Cao and Geballe 1996) and for the uORF at the gene encoding S-adenosylmethionine decarboxylase (Mize et al. 1998).

INITIATION FACTORS WORKING AT BOTH ENDS OF THE mRNA MEDIATE RIBOSOME BINDING AND mRNA CIRCULARIZATION

Cap-binding Factor and RNA Helicases Are Required for Ribosome Binding to mRNA

Once the TC binds to the 40S ribosome, the resulting 43S complex can interact with the 5´ end of the mRNA to initiate the scanning process (Fig. 1). Biochemical analysis of the mammalian system had shown that this interaction is strongly dependent on a factor known as eIF4F, which associates with the m⁷G(5′)ppp(5′)X cap structure at the 5′ end of the mRNA. In mammals, eIF4F is a complex of three proteins: (1) eIF4E, the cap-binding protein; (2) eIF4A, an ATP-dependent RNA helicase; and (3) eIF4G, a scaffold protein that contains binding sites for the other two components in eIF4F and for the multisubunit eIF3 complex (Fig. 5). As eIF3 can bind directly to 40S subunits, it is generally assumed that the eIF3-eIF4G interaction serves to bridge association between the 40S ribosome and the mRNA/eIF4F complex (Hershey and Merrick 2000). The importance of the subunits of eIF4F in vivo was shown in yeast by the fact that all three are required for cell viability (Altmann et al. 1987; Linder and Slonimski 1989; Goyer et al. 1993). In the case of eIF4E and eIF4A, Hans Trachsel, Nahum Sonenberg, and colleagues established an essential role in translation for these factors by examining the translational activity of extracts prepared from specially constructed yeast strains in which the wild-type proteins could be eliminated from cells, leaving behind either no factor or a heat-labile version encoded by a temperature-sensitive allele (**Altmann et al. 1989**; Blum et al. 1989). In these studies, the wild-type proteins were depleted by placing their structural genes under the *GAL* promoter, which is inactivated when cells are shifted from galactose to glucose as the carbon source in the medium.

The yeast strain in which eIF4E expression was made dependent on galactose facilitated the isolation of temperature-sensitive (Ts⁻) mutations in the gene encoding this factor (*CDC33*). After mutagenizing the gene and introducing it into the strain harboring the galactose-regulated *CDC33* allele, random transformants were screened for those that failed to grow at high temperature on glucose medium, but showed no growth defect on galactose medium. This criterion ensured that the temperature-sensitive phenotype of the mutant was conferred by a defect in *CDC33* (being complemented by expression of the wild-type protein) rather than by an unrelated mutation arising during transformation. One Ts⁻*cdc33* allele isolated in this fashion (*cdc33-4-2*) was analyzed for its effects on translation in extracts prepared by growing the cells on glucose to deplete wild-type eIF4E, as described above. Following a limited micrococcal nuclease treatment of the extract to digest

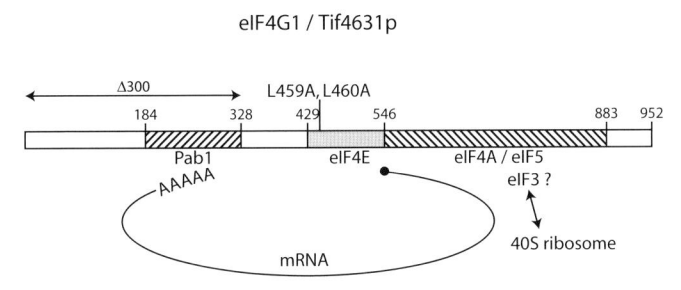

FIGURE 5. The postulated role of the eIF4G subunit of eIF4F in bridging interaction between the mRNA and the 40S ribosome. The 952-amino-acid sequence of eIF4G1 is depicted as a rectangle with the binding domains for Pab1, eIF4E, eIF4A, and eIF5 shown with shading or hatching. The locations of the binding domains were compiled from several publications (Mader et al. 1995; Tarun and Sachs 1996; Dominguez et al. 1999; Asano et al. 2001). Mammalian eIF4G interacts with eIF3 through the domain indicated in the schematic, but this has not been demonstrated for the yeast factors. The presence of Pab1- and eIF4E-binding domains in eIF4G leads to the prediction that simultaneous interaction of these factors with eIF4G and with the polyadenylated and capped ends of the mRNA would lead to circularization of the mRNA, as depicted here. This mRNP complex could be recruited to the 40S ribosome via the eIF4G-eIF3 interaction, as eIF3 interacts directly with the ribosome.

endogenous mRNAs, and a brief heat treatment to inactivate the mutant protein, it was found that translation of virtually all exogenous yeast mRNAs added back to the extract was impaired. Importantly, translation was rescued by the addition of purified eIF4E to the extract (Altmann et al. 1989). This ruled out the possibility that depletion of wild-type eIF4E during growth of the cells on glucose impaired protein synthesis indirectly by altering the expression or activity of some other critical factor. These results provided compelling evidence that eIF4E is required for translation of the vast majority of mRNAs.

The same strategy was used to study the effects on translation of an unconditionally lethal point mutation in eIF4A that impaired the ATPase activity of this protein. The mutation was introduced by site-directed mutagenesis of Ala-66, located in one of the conserved sequence motifs found in the "DEAD" family of RNA helicases, of which eIF4A is the founding member (Linder et al. 1989). (These proteins received their name from the Asp-Glu-Ala-Asp [D-E-A-D] motif that they all contain.) Ala-66 is present in the **A**XXGXGKT motif believed to be part of the ATP-binding motif. Mutant eIF4A containing a valine at position 66 was expressed in *Escherichia coli* and purified along with the wild-type protein. Biochemical analysis showed that the Val-66 mutant protein had a fivefold higher K_m for ATP than did wild-type eIF4A, signifying reduced binding affinity, and was nearly inactive as an RNA helicase. In addition, the mutant protein did not support translation of total yeast mRNAs in an eIF4A-dependent extract. In vivo, the Val-66 mutation was lethal. This study proved that ATP hydrolysis by eIF4A is required for translation initiation in vivo. Interestingly, a synthetic mRNA containing the start codon only eight nucleotides 3′ of the cap also was not translated in the mutant extract containing eIF4A-A66V. A 43S complex that binds to this mRNA at the cap should not have to scan in order to reach the AUG codon; hence, these results suggest that the RNA helicase activity of eIF4A is required for one or more functions besides the unwinding of secondary structure ahead of the ribosome to facilitate scanning (Blum et al. 1992). It may be required for the initial binding of a ribosome near the cap, or perhaps for dissociating nonproductive interactions of the mRNA with Met-tRNA$_i^{Met}$ or rRNA during the process of AUG selection.

The RNA helicase activity of eIF4A is stimulated by eIF4B (Hershey and Merrick 2000), and Altmann and colleagues showed that both mammalian eIF4B and its yeast homolog (encoded by *TIF3*) have RNA annealing activity that may facilitate these functions of eIF4A (Altmann et al. 1995). Another essential DEAD-box RNA helicase in yeast, encoded by *DED1*, also is required for translation initiation and may have overlapping functions with eIF4A/eIF4B (Chuang et al. 1997; de la Cruz et al. 1997).

Multiple Functions of Poly(A) Tail and Pab1 in Translation Initiation

The 3′ ends of eukaryotic mRNAs also are modified posttranscriptionally by addition of the poly(A) tail. Studies using in vitro translation systems had suggested that the poly(A) tail can stimulate translation and pointed to the involvement of the poly(A)-binding protein (Pab1) in mediating the enhancement (Sachs 2000). Evidence from Allan Jacobson's laboratory indicated that the poly(A) tail stimulates joining of the 60S subunit to the 48S initiation complex, i.e., the last phase of initiation (Munroe and Jacobson 1990). However, the importance of poly(A) and Pab1 for translation in vivo was unknown. Alan Sachs showed that yeast cells carrying the Ts⁻ lethal allele *pab1-F363L* contain a much reduced polysome content when incubated at the restrictive temperature, with a commensurate increase in the amount of 80S ribosomes. Most of the 80S species that accumulated dissociated in high salt (0.8 M KCl), identifying them as inactive "80S couples" rather than translating 80S monosomes (which are salt resistant). This is the classical phenotype of a mutation that reduces translation initiation without impeding elongation or termination: Ribosomes already engaged in elongation can finish translation, terminate, and dissociate from the mRNA, but cannot initiate a new round of translation at the restrictive temperature. Interestingly, it was found that the lethal phenotype of *PAB1* null mutations can be suppressed by mutations that reduce the concentration of 60S subunits (Sachs and Davis 1989, 1990). Although the mechanism of this suppression was unclear, it bolstered the conclusion that Pab1 has an important function in translation initiation in vivo.

Using yeast cell extracts in which both cap and poly(A) tail contribute synergistically to the translation of exogenous mRNAs, it was possible to show that the stimulatory effect of the poly(A) tail on translation was independent of eIF4E but dependent on Pab1. These experiments involved translation of uncapped, polyadenylated (C–A+) or capped, non-polyadenylated (C+A–) reporter mRNAs in extracts in which eIF4E was inactivated (by addition of the cap analog [m^{7}GpppG] that sequesters eIF4E) or in which Pab1 was immunodepleted. Translation of the C+A– mRNA was inhibited much more severely than that of the C–A+ mRNA by cap analog, whereas immunodepletion of Pab1 had the opposite effect, inhibiting translation of C–A+ mRNA more so than that of C+A– mRNA. Analysis of the initiation intermediates present in these extracts revealed that Pab1 stimulates the binding of mRNA to the 40S subunit, the same function ascribed to eIF4E (Tarun and Sachs 1995). In view of this finding, it may be possible to explain why the lethal phenotype of a *pab1Δ* mutation is suppressed by a reduction in 60S subunit abundance: The attendant increase in the concentration of free 40S subunits may drive 48S complex formation to completion by mass action.

Subsequent biochemical analysis revealed that Pab1, in addition to eIF4E, is physically associated with each of the two isoforms of eIF4G, the scaffold component of eIF4F, in yeast cell extracts. Furthermore, recombinant forms of Pab1 and the eIF4G isoforms (encoded by *TIF4631* and *TIF4632*) could interact directly in vitro. The binding sites for Pab1 and eIF4E were mapped to adjacent regions in the amino-terminal half of eIF4G2, between residues 201–315 and 400–515, respectively (Mader et al. 1995; Tarun and Sachs 1996). eIF4G1 contains a Pab1-binding site in a similar location that is necessary and sufficient for the interaction in vitro (Tarun and Sachs 1996). An intriguing prediction of these results is that mutual association of eIF4E and Pab1 with eIF4G should mediate circularization of a capped, polyadenylated mRNA (Fig. 5). This prediction was confirmed with purified components using atomic force microscopy (Wells et al. 1998).

How critical is the interaction of Pab1 with eIF4G for translation initiation? To answer this question, Sachs and colleagues prepared extracts from yeast strains containing only one of the two eIF4G isoforms, in which the amino-terminal 300 amino acids spanning the Pab1-binding site had been deleted (*ΔN300* mutation). These extracts were nearly incapable of translating a C–A+ reporter mRNA that depends on the poly(A) tail and Pab1 for initiation, but were perfectly capable of translating a C+A– mRNA in which the cap mediates ribosome binding to mRNA. Consistently, translation of a C+A+ mRNA was reduced by a factor of 20 to 50 in the mutant extracts, to roughly the level given by the C+A– mRNA. Thus, the amino terminus of eIF4G containing the Pab1-binding site was critical for the stimulatory effect of poly(A) on translation in vitro, consistent with the functional importance of Pab1 binding to eIF4G in translation initiation (**Tarun et al. 1997**).

The situation in vivo is rather different. A strain deleted for *TIF4632* and containing the *TIF4631-ΔN300* allele showed only a slight growth defect, implying that the Pab1-eIF4G interaction is not crucial for protein synthesis and cell growth. On the other hand, the *ΔN300* mutation was lethal when introduced into a mutant allele of *TIF4631* harboring two point mutations in the eIF4E-binding site of eIF4G1 (*L459A, L460A*) (Fig. 5). The latter mutations alone were not lethal, but conferred slow growth and Ts⁻ phenotypes. The synthetic-lethal phenotype produced by combining the mutations in *TIF4631* was taken as evidence that Pab1-eIF4G interaction is crucial in vivo when the eIF4E-eIF4G interaction is impaired. In this view, the Pab1-eIF4G and eIF4E-eIF4G interactions provide partly redundant means of recruiting the 40S ribosome to the mRNA, with the latter interaction having a more prominent role than the Pab1-eIF4G interaction (Fig. 5). It was speculated that Pab1 may contact other regions of eIF4G, or other eIFs, in vivo and that such interactions may not occur in vitro, making Pab1 association with the amino terminus of eIF4G more crucial in vitro than in vivo (Tarun et al. 1997). This last study further illustrates the importance of being able to analyze the effects of mutations in initiation factors on translation in living cells in addition to cell-free extracts. It may be easier to observe a defect conferred by a given mutation in vitro where partially redundant functions or interactions can be compromised. Whether the function impaired by the mutation is critical in vivo can be decided only by analyzing translation initiation in mutant cells. Yeast is currently the eukaryotic organism best suited for this integrated approach.

Reed Wickner and colleagues have examined the function of the poly(A) tail by using the approach of electroporating mRNAs into living yeast cells and measuring expression of the protein products over time. The synergistic effects of the cap and poly(A) tail in stimulating expression were observed with electroporated mRNAs in yeast, and mutations in subunits of eIF2 or eIF3 impair translation of such mRNAs, regardless of the presence or absence of the cap or poly(A) tail. Interestingly, a deletion of *FUN12*, encoding eIF5B, eliminated the stimulatory effect of the poly(A) tail but not that of cap (Searfoss et al. 2001). The eIF5B is a nonessential factor implicated in the last step of initiation where the 60S subunit joins with the 48S initiation complex to produce an 80S complex (Pestova et al. 2000). The *fun12Δ* mutation reduced translation of C+A+ and C–A+ mRNAs by tenfold, but reduced that of C+A– mRNAs by only 30%. One way to interpret the fact that *fun12Δ* masks the deleterious effect on translation of removing the poly(A) tail is to propose that poly(A)/Pab1 stimulates initiation by promoting the function of eIF5B in subunit joining, as originally suggested by Jacobson (Munroe and Jacobson 1990).

ACKNOWLEDGMENTS

The author is grateful to members of his laboratory for many helpful suggestions.

STUDY QUESTIONS

1. Whereas reticulocyte and wheat germ in vitro translation systems have contributed to the early knowledge of translation initiation, yeast had an important role in determin-

ing the role of many individual factors. Why could the yeast system complement in an ideal manner the in vitro translation systems and what were the two main approaches used in this analysis?

2. Whereas in bacteria the ribosome is guided to the initiator codon by the Shine-Dalgarno (SD) sequence, the majority of cellular mRNAs are translated from the first AUG codon. Cite two major contributions that helped to establish this "first AUG rule" notion. Detail the contribution of the yeast system.

3. Which type of results indicated that eukaryotic translation initiation generally does not use an SD-type positioning mechanism?

4. Although early genetic experiments in yeast clearly contributed to the "first AUG rule," the introduction of yeast transformation and molecular biology further helped to establish today's knowledge of initiator codon selection and the inefficiency of reinitiation. How?

5. What translation initiation factors participate in accurate initiator codon selection and how has this been established by the isolation of novel mutants? How is it possible that in these mutants, the bulk of mRNA can still be translated accurately?

6. Explain the experimental approach that showed the importance of base pairing of the initiator tRNA in initiator codon selection.

7. Several mutants affecting initiator codon selection are in genes encoding subunits of a heterotrimeric GTPase or its activating protein. What general biological mechanism is affected in these mutants and how would you explain this?

8. Explain how the phenotypes of single and double mutants harboring *gcn2*, *gcn3*, and *gcd1* mutations showed that Gcn2 and Gcn3 act as positive regulators in general amino acid control by antagonizing the negative regulatory function of Gcd1.

9. Summarize the genetic results showing that (1) ribosomes must translate uORF1 and resume scanning in order to translate the *GCN4*-coding sequence and (2) ribosomes bypass uORFs 3–4 because the distance from uORF1 to uORF4 is too small to recover the factors needed for reinitiation under starvation conditions.

10. Describe (1) the biochemical evidence that *GCN2* phosphorylates the α subunit eIF2; (2) the genetic evidence (from mutations in *GCN2* and *SUI2*) that this phosphorylation event is critical for inducing *GCN4* translation; (3) the explanation of why reducing eIF2 activity through phosphorylation would cause ribosomes which have translated uORF1 to reinitiate at the *GCN4* start codon instead of at uORF4.

11. What is the fundamental difference in the regulation of *GCN4* versus *CPA1*? What were the crucial experiments that show this difference?

12. The poly(A) tail of mRNA has been shown to have an important role in translation initiation. Give the in vitro arguments for its involvement in the initiation process and compare the results to the in vivo data.

REFERENCES

Abastado J.P., Miller P.F., Jackson B.M., and Hinnebusch A.G. 1991. Suppression of ribosomal reinitiation at upstream open reading frames in amino acid-starved cells forms the basis for GCN4 translational control. *Mol. Cell. Biol.* **11:** 486–496.

Altmann M., Handschin C., and Trachsel H. 1987. mRNA cap-binding protein: Cloning of the gene encoding protein synthesis initiation factor eIF-4E from *Saccharomyces cerevisiae*. *Mol. Cell. Biol.* **7:** 998–1003.

Altmann M., Sonenberg N., and Trachsel H. 1989. Translation in *Saccharomyces cerevisiae:* Initiation factor 4E-dependent cell-free system. *Mol. Cell. Biol.* **9:** 4467–4472.

Altmann M., Wittmer B., Methot N., Sonenberg N., and Trachsel H. 1995. The *Saccharomyces cerevisiae* translation initiation factor Tif3 and its mammalian homologue, eIF-4B, have RNA annealing activity. *EMBO J.* **14:** 3820–3827.

Asano K., Shalev A., Phan L., Nielsen K., Clayton J., Valasek L., Donahue T.F., and Hinnebusch A.G. 2001. Multiple roles for the carboxyl terminal domain of eIF5 in translation initiation complex assembly and GTPase activation. *EMBO J.* **20:** 2326–2337.

Baim S.B. and Sherman F. 1988. mRNA structures influencing translation in the yeast *Saccharomyces cerevisiae*. *Mol. Cell. Biol.* **8:** 1591–1601.

Blum S., Mueller M., Schmid S.R., Linder P., and Trachsel H. 1989. Translation in *Saccharomyces cerevisiae:* Initiation factor 4A-dependent cell-free system. *Proc. Natl. Acad. Sci.* **86:** 6043–6046.

Blum, S., Schmid S.R., Pause A., Buser P., Linder P., Sonenberg N., and Trachsel H. 1992. ATP hydrolysis by initiation factor 4A is required for translation initiation in *Saccharomyces cerevisiae*. *Proc. Natl. Acad. Sci.* **89:** 7664–7668.

Bushman J.L., Asuru A.I., Matts R.L., and Hinnebusch A.G. 1993. Evidence that GCD6 and GCD7, translational regulators of *GCN4* are subunits of the guanine nucleotide exchange factor for eIF-2 in *Saccharomyces cerevisiae*. *Mol. Cell. Biol.* **13:** 1920–1932.

Cao J. and Geballe A.P. 1996. Inhibition of nascent-peptide release at translation termination. *Mol. Cell. Biol.* **16:** 7109–7114.

Castilho-Valavicius B., Thompson G.M., and Donahue T.F. 1992. Mutation analysis of the $Cys-X_2-Cys-X_{19}-Cys-X_2-Cys$ motif in the α subunit of eukaryotic translation factor 2. *Gene Expr.* **2:** 297–309.

Castilho-Valavicius B., Yoon H., and Donahue T.F. 1990. Genetic characterization of the *Saccharomyces cerevisiae* translational initiation suppressors *sui1*, *sui2* and *SUI3* and their effects on *HIS4* expression. *Genetics* **124:** 483–495.

Chakravarti D. and Maitra U. 1993. Eukaryotic translation initiation factor 5 from *Saccharomyces cerevisiae* cloning, characterization, and expression of the gene encoding the 45,346-Da protein. *J. Biol. Chem.* **268:** 10524–10533.

Chong K.L., Feng L., Schappert K., Meurs E., Donahue T.F., Friesen J.D., Hovanessian A.G., and Williams B.R.G. 1992. Human p68 kinase exhibits growth suppression in yeast and homology to the translational regulator GCN2. *EMBO J.* **11:** 1553–1562.

Chuang R.Y., Weaver P.L., Liu Z., and Chang T.H. 1997. Requirement of the DEAD-Box protein Ded1p for messenger RNA translation. *Science* **275:** 1468–1471.

Cigan A.M., Feng L., and Donahue T.F. 1988a. $tRNA_1^{Met}$ functions in directing the scanning ribosome to the start site of translation. *Science* **242:** 93–97.

Cigan A.M., Pabich E.K., and Donahue T.F. 1988b. Mutational analysis of the *HIS4* translational initiator region in *Saccharomyces cerevisiae*. *Mol. Cell. Biol.* **8:** 2964–2975.

Cigan A.M., Pabich E.K., Feng L., and Donahue T.F. 1989. Yeast translation initiation suppressor *sui2* encodes the α subunit of eukaryotic initiation factor 2 and shares identity with the human α subunit. *Proc. Natl. Acad. Sci.* **86:** 2784–2788.

de la Cruz J., Iost I., Kressler D., and Linder P. 1997. The p20 and Ded1 proteins have antagonistic roles in eIF4E-dependent translation in *Saccharomyces cerevisiae*. *Proc. Natl. Acad. Sci.* **94:** 5201–5206.

Delbecq P., Werner M., Feller A., Filipkowski R.K., Messenguy F., and Pierard A. 1994. A segment of mRNA encoding the leader peptide of the *CPA1* gene confers repression by arginine on a heterologous yeast gene transcript. *Mol. Cell. Biol.* **14:** 2378–2390.

Dever T.E., Feng L., Wek R.C., Cigan A.M., Donahue T.D., and Hinnebusch A.G. 1992. Phosphorylation of initiation factor 2α by protein kinase GCN2 mediates gene-specific translational control of GCN4 in yeast. *Cell* **68:** 585–596.

Dominguez D., Altmann M., Benz J., Baumann U., and Trachsel H. 1999. Interaction of translation initiation factor eIF4G with eIF4A in the yeast *Saccharomyces cerevisiae*. *J. Biol. Chem.* **274:** 26720–26726.

Donahue T.F. and Cigan A.M. 1988. Genetic selection for mutations that reduce or abolish ribosomal recognition of the *HIS4* translational initiator region. *Mol. Cell. Biol.* **8:** 2955–2963.

Donahue T.F., Cigan A.M., Pabich E.K., and Castilho Valavicius B. 1988. Mutations at a Zn(II) finger motif in the yeast eIF-2β gene alter ribosomal start-site selection during the scanning process. *Cell* **54:** 621–632.

Feinberg B., McLaughlin C.S., and Moldave K. 1982. Analysis of temperature-sensitive mutant *ts*187 of *Saccharomyces cerevisiae* altered in a component required for the initiation of protein synthesis. *J. Biol. Chem.* **257:** 10846–10851.

Gaba A., Wang Z., Krishnamoorthy T., Hinnebusch A.G., and Sachs M.S. 2001. Physical evidence for distinct mechanisms of translational control by upstream open reading frames. *EMBO J.* **20:** 6453–6463.

Goyer C., Altmann M., Lee H.S., Blanc A., Deshmukh M., Woolford J.L., Trachsel H., and Sonenberg N. 1993. *TIF4631* and *TIF4632:* Two yeast genes encoding the high-molecular-weight subunits of the cap-binding protein complex (eukaryotic initiation factor 4F) contain an RNA recognition motif-like sequence and carry out an essential function. *Mol. Cell. Biol.* **13:** 4860–4874.

Grant C.M. and Hinnebusch A.G. 1994. Effect of sequence context at stop codons on efficiency of reinitiation in *GCN4* translational control. *Mol. Cell. Biol.* **14:** 606–618.

Hannig E.M., Cigan A.M., Freeman B.A., and Kinzy T.G. 1993. *GCD11,* a negative regulator of *GCN4* expression, encodes the γ subunit of eIF-2 in *Saccharomyces cerevisiae*. *Mol. Cell. Biol.* **13:** 506–520.

Harashima S. and Hinnebusch A.G. 1986. Multiple *GCD* genes required for repression of *GCN4,* a transcriptional activator of amino acid biosynthetic genes in *Saccharomyces cerevisiae*. *Mol. Cell. Biol.* **6:** 3990–3998.

Harding H.P., Novoa I., Zhang Y., Zeng H., Wek R., Schapira M., and Ron D. 2000. Regulated translation initiation controls stress-induced gene expression in mammalian cells. *Mol. Cell* **6:** 1099–1108.

Hartwell L.H. and McLaughlin C.S. 1969. A mutant of yeast apparently defective in the initiation of protein synthesis. *Proc. Natl. Acad. Sci.* **62:** 468–474.

Hellen C.U. and Sarnow P. 2001. Internal ribosome entry sites in eukaryotic mRNA molecules. *Genes Dev.* **15:** 1593–1612.

Hershey J.W.B. and Merrick W.C. 2000. Pathway and mechanism of initiation of protein synthesis. In *Translational control of gene expression* (ed. N. Sonenberg et al.), pp. 33–88. Cold Spring Harbor Laboratory Press, Cold Spring Harbor, New York.

Hinnebusch A.G. 2000. Mechanism and regulation of initiator methionyl-tRNA binding to ribosomes. In *Translational control of gene expression* (ed. N. Sonenberg et al.), pp. 185–243. Cold Spring Harbor Laboratory Press, Cold Spring Harbor, New York.

Hinnebusch A.G. and Fink G.R. 1983. Positive regulation in the general amino acid control of *Saccharomyces cerevisiae*. *Proc. Natl. Acad. Sci.* **80:** 5374–5378.

Huang H., Yoon H., Hannig E.M., and Donahue T.F. 1997. GTP hydrolysis controls stringent selection of the AUG start codon during translation initiation in *Saccharomyces cerevisiae*. *Genes Dev.* **11:** 2396–2413.

Komar A.A. Lesnik T., Cullin C., Merrick W.C., Trachsel H., and Altmann M. 2003. Internal initiation drives the synthesis of Ure2 protein lacking the prion domain and affects [URE3] propagation in yeast cells. *EMBO J.* **22:** 1199–1209.

Kozak M. 1978. How do eucaryotic ribosomes select initiation regions in messenger RNA? *Cell* **15:** 1109–1123.

———. 1979. Inability of circular mRNA to attach to eukaryotic ribosomes. *Nature* **280:** 82-85.

———. 1984. Selection of initiation sites by eukaryotic ribosomes: Effect of inserting AUG triplets upstream from the coding sequence for preproinsulin. *Nucleic Acids Res.* **12:** 3873–3893.

———. 1986. Point mutations define a sequence flanking the AUG initiator codon that modulates translation by eukaryotic ribosomes. *Cell* **44:** 283–292.

———. 1987. Effects of intercistronic length on the efficiency of reinitiation by eukaryotic ribosomes. *Mol. Cell. Biol.* **7:** 3438–3445.

Lamphear B.J., Kirchweger R., Skern T., and Rhoads R.E. 1995. Mapping of functional domains in eukaryotic protein synthesis initiation factor 4G (eIF4G) with picornaviral proteases. *J. Biol. Chem.* **270:** 21975–21983.

Linder P. and Slonimski P.P. 1989. An essential yeast protein, coded by duplicated genes *TIF1* and *TIF2*, and homologues to the mammalian translation initiation factor eIF-4E can suppress a mitochondrial missense mutation. *Proc. Natl. Acad. Sci.* **86:** 2286–2290.

Linder P., Lasko P.F., Ashburner M., Leroy P., Nielsen P., Nishi K., Schnier J., and Slonimski P.P. 1989. Birth of the D-E-A-D box. *Nature* **337:** 121–122.

Lu P.D., Harding H.P., and Ron D. 2004. Translation reinitiation at alternative open reading frames regulates gene expression in an integrated stress response. *J. Cell. Biol.* **167:** 27–33.

Maag D., Fekete C.A., Gryczynski Z., and Lorsch J.R. 2005. A conformational change in eukaryotic translation preinitiation complex and release of eIF1 signal recognition of the start codon. *Mol. Cell* **17:** 265–275.

Mader S., Lee H., Pause A., and Sonenberg N. 1995. The translation initiation factor eIF-4E binds to a common motif shared by the translation factor eIF-4γ and the translational repressors 4E-binding proteins. *Mol. Cell. Biol.* **15:** 4990–4997.

Miller P.F. and Hinnebusch A.G. 1989. Sequences that surround the stop codons of upstream open reading frames in *GCN4* mRNA determine their distinct functions in translational control. *Genes Dev.* **3:** 1217–1225.

Mize G.J., Ruan H., Low J.J., and Morris D.R. 1998. The inhibitory upstream open reading frame from mammalian S-adenosylmethionine decarboxylase mRNA has a strict sequence specificity in critical positions. *J. Biol. Chem.* **273:** 32500–32505.

Mueller P.P. and Hinnebusch A.G. 1986. Multiple upstream AUG codons mediate translational control of *GCN4*. *Cell* **45:** 201–207.

Munroe D. and Jacobson A. 1990. mRNA poly(A) tail, a 3′ enhancer of translational initiation. *Mol. Cell. Biol.* **10:** 3441–3455.

Nielson K.H., Szamecz B., Valasek L., Jivotovskaya A., Shin B.S., and Hinnebusch A.G. 2004. Functions of eIF3 downstream of 48S assembly impact AUG recognition and GCN4 translational control. *EMBO J.* **23:** 1166–1177.

Pestova T.V. and Kolupaeva V.G. 2002. The roles of inidividual eukaryotic translation initiation factors in ribosomal scanning and initiation codon selection. *Genes Dev.* **16:** 2906–2922.

Pestova T.V., Borukhov S.I., and Hellen C.U.T. 1998. Eukaryotic ribosomes require initiation factors 1 and 1A to locate initiation codons. *Nature* **394:** 854–859.

Pestova T.V., Lomakin I.B., Lee J.H., Choi S.K., Dever T.E., and Hellen C.U.T. 2000. The joining of ribosomal subunits in eukaryotes requires eIF5B. *Nature* **403:** 332–335.

Ramirez M., Wek R.C., and Hinnebusch A.G. 1991. Ribosome-association of GCN2 protein kinase, a translational activator of the *GCN4* gene of *Saccharomyces cerevisae*. *Mol. Cell. Biol.* **11:** 3027–3036.

Roussou I., Thireos G., and Hauge B.M. 1988. Transcriptional-translational regulatory circuit in *Saccharomyces cerevisiae* which involves the *GCN4* transcriptional activator and the GCN2 protein kinase. *Mol. Cell. Biol.* **8:** 2132–2139.

Sachs A. 2000. Physical and functional interactions between the mRNA cap structure and the poly(A) tail. In *Translational control of gene expression* (ed. N. Sonenberg et al.), pp. 447–465. Cold Spring Harbor Laboratory Press, Cold Spring Harbor, New York.

Sachs A.B. and Davis R.W. 1989. The poly(A) binding protein is required for poly(A) shortening and 60S ribosomal subunit-dependent translation initiation. *Cell* **58:** 857–867.

———. 1990. Translation initiation and ribosomal biogenesis: Involvement of a putative rRNA helicase and RPL46. *Science* **247:** 1077–1079.

Sachs A.B. and Varani G. 2000. Eukaryotic translation initiation: There are (at least) two sides to every story. *Nat. Struct. Biol.* **7:** 356–361.

Searfoss A., Dever T.E., and Wickner R. 2001. Linking the 3′ poly(A) tail to the subunit joining step of translation initiation: Relations of Pab1p, eukaryotic translation initiation factor 5b (Fun12p), and Ski2p-Slh1p. *Mol. Cell. Biol.* **21:** 4900–4908.

Sherman F. and Stewart J.W. 1982. Mutations altering initiation of translation of yeast iso-1-cytochrome c; contrasts between the eukaryotic and prokaryotic initiation process. In *The molecular biology of the yeast Saccharomyces: Metabolism and gene expression* (ed. J.N. Strathern et al.), pp. 301–334. Cold Spring Harbor Laboratory, Cold Spring Harbor, New York.

Sherman F., Stewart J.W., and Schweingruber A.M. 1980. Mutants of yeast initiating translation of iso-1-cytochrome c within a region spanning 37 nucleotides. *Cell* **20:** 215–222.

Shine J. and Dalgarno L. 1974. The 3′-terminal sequence of *Escherichia coli* 16S ribosomal RNA: Complementarity to nonsense triplets and ribosome binding sites. *Proc. Natl. Acad. Sci.* **71:** 1342–1346.

Stewart J.W., Sherman F., Shipman N.A., and Jackson M. 1971. Identification and mutational relocation of the AUG codon initiating translation of iso-1-cytochrome *c* in yeast. *J. Biol. Chem.* **246:** 7429–7445.

Stiles J.I., Szostak J.W., Young A.T., Wu R., Consaul S., and Sherman F. 1981. DNA sequence of a mutation in the leader region of the yeast iso-1-cytochrome c mRNA. *Cell* **25:** 277–284.

Tarun S.Z. and Sachs A.B. 1995. A common function for mRNA 5′ and 3′ ends in translation initiation in yeast. *Genes Dev.* **9:** 2997–3007.

———. 1996. Association of the yeast poly(A) tail binding protein with translation initiation factor eIF-4G. *EMBO J.* **15:** 7168–7177.

Tarun S.Z., Jr., Wells S.E., Deardorff J.A., and Sachs A.B. 1997. Translation initiation factor eIF4G mediates *in vitro* poly(A) tail-dependent translation. *Proc. Natl. Acad. Sci.* **94:** 9046–9051.

Thompson S.R., Gulyas K.D., and Sarnow P. 2001. Internal initiation in *Saccharomyces cerevisiae* mediated by an initiator tRNA/eIF2-independent internal ribosome entry site element. *Proc. Natl. Acad. Sci.* **98:** 12972–12977.

Tzamarias D. and Thireos G. 1988. Evidence that the *GCN2* protein kinase regulates reinitiation by yeast ribosomes. *EMBO J.* **7:** 3547–3551.

Unbehaun A., Borukhov S.I., Hellen C.U., and Pestova T.V. 2004. Release of initiation factors from 48S complexes during ribosomal subunit joining and the link between establishment of codon-anticodon base-pairing and hydrolysis of eIF2-bound GTP. *Genes Dev.* **18:** 3078–3093.

Valasek L., Nielsen K.H., Zhang F., Fekete C.A., and Hinnebusch A.G. 2004. Interactions of eukaryotic translation initiation factor 3 (eIF3) subunit NIP1/c with eIF1 and eIF5 promote preinitiation complex assembly and regulate start codon selection. *Mol. Cell. Biol.* **24:** 9437–9455.

Vattem K.M. and Wek R.C. 2004. Reinitiation involving upstream ORFs regulates ATF4 mRNA translation in mammalian cells. *Proc. Natl. Acad. Sci.* **101:** 11269–11274.

Wang, Z., Gaba A., and Sachs M.S. 1999. A highly conserved mechanism of regulated ribosome stalling mediated by fungal arginine attenuator peptides that appears independent of the charging status of arginyl-tRNAs. *J. Biol. Chem.* **274:** 37565–37574.

Wek R.C., Jackson B.M., and Hinnebusch A.G. 1989. Juxtaposition of domains homologous to protein kinases and histidyl-tRNA synthetases in GCN2 protein suggests a mechanism for coupling *GCN4* expression to amino acid availability. *Proc. Natl. Acad. Sci.* **86:** 4579–4583.

Wells S.E., Hillner P.E., Vale R.D., and Sachs A.B. 1998. Circularization of mRNA by eukaryotic translation initiation factors. *Mol. Cell* **2:** 135–140.

Werner M., Feller A., Messenguy F., and Pierard A. 1987. The leader peptide of yeast gene *CPA1* is essential for the translational repression of its expression. *Cell* **49:** 805–813.

Williams N.P., Hinnebusch A.G., and Donahue T.F. 1989. Mutations in the structural genes for eukaryotic initiation factors 2α and 2β of *Saccharomyces cerevisiae* disrupt translational control of *GCN4* mRNA. *Proc. Natl. Acad. Sci.* **86:** 7515–7519.

Wilson J.E., Pestova T.V., Hellen C.U., and Sarnow P. 2000. Initiation of protein synthesis from the A site of the ribosome. *Cell* **102:** 511–520.

Wolfe C.L., Lou Y.C., Hopper A.K., and Martin N.C. 1994. Interplay of heterogeneous transcriptional start sites and translational selection of AUGs dictate the production of mitochondrial and cytosolic/nuclear tRNA nucleotidyltransferase from the same gene in yeast. *J. Biol. Chem.* **269:** 13361–13366.

Yoon H.J. and Donahue T.F. 1992. The *sui1* suppressor locus in *Saccharomyces cerevisiae* encodes a translation factor that functions during tRNA$_i^{Met}$ recognition of the start codon. *Mol. Cell. Biol.* **12:** 248–260.

Zhou W., Edelman G.M., and Mauro V.P. 2001. Transcript leader regions of two *Saccharomyces cerevisiae* mRNAs contain internal ribosome entry sites that function in living cells. *Proc. Natl. Acad. Sci.* **98:** 1531–1536.

6

Cell Division

Kim Nasmyth
Research Institute of Molecular Pathology
Vienna, A-1030, Austria

Once and only once: Regulating the initiation of DNA replication, 110
Diffley J.F.X., Cocker J.H., Dowell S.J., and Rowley A. 1994. Two steps in the assembly of complexes at yeast replication origins in vivo. *Cell* **78:** 303–316.

The anaphase-promoting complex (APC) and sister-chromatid separation, 112
Irniger S., Piatti S., Michaelis C., and Nasmyth K. 1995. Genes involved in sister chromatid separation are needed for B-type cyclin proteolysis in budding yeast. *Cell* **81:** 269–277.

The mitotic trigger: Cyclin-dependent kinases, 116
Nurse P. and Thuriaux P. 1980. Regulatory genes controlling mitosis in the fission yeast *Schizosaccharomyces pombe*. *Genetics* **96:** 627–637.

Surveillance mechanisms that halt the Cdk oscillation, 119
Weinert T.A. and Hartwell L.H. 1988. The *RAD9* gene controls the cell cycle response to DNA damage in *Saccharomyces cerevisiae*. *Science* **241:** 317–322.

Note: The landmark papers listed above are those discussed in this chapter. Each landmark paper is preceded by the name of the section (with starting page number) where the paper is first discussed in detail.

UNDERSTANDING THE CHROMOSOME CYCLE

The process by which cells grow and divide to produce two daughter cells resembling their parents is known as the cell cycle. Instructions for cell duplication are largely encoded by the genome, which is present in two copies in diploids and one copy in haploids. The duplication of chromosomal DNA and its segregation to opposite poles of the cell prior to division must therefore be performed with a particularly high fidelity. This process is known as the chromosome cycle and is the most intensively studied aspect of the cell cycle.

There are several reasons why yeast has proved to be a powerful model system to study the chromosome cycle. The ability to isolate mutants facilitated the identification of many, if not most, cell cycle proteins with conserved functions, whereas conditional mutations (temperature-sensitive proteins or galactose-dependent gene expression) enabled the rigorous analysis of the consequences of interfering with the function of such proteins. The ability to obtain cultures in which cells progress through the cell cycle in a semisynchronous manner was crucial to being able to compare carefully wild-type and mutant cells. Finally, the ability to grow large quantities of cells from different stages of the cell cycle made possible biochemical studies that address molecular mechanisms. With the advent of epitope tagging and green fluorescent protein (GFP), cytological analyses have greatly improved and are an important aspect of many modern studies.

This chapter attempts to pinpoint key papers in four areas where experiments using yeast have led to major discoveries about the chromosome cycle. My criteria for selection were that the work was original and led to a genuine understanding about a mechanism subsequently found in all eukaryotic cells. The four topics concern how the genome is duplicated and segregated with high fidelity and how these two processes are coupled. Some of these topics are also treated by other articles in this volume but nevertheless are mentioned here because of the insight they have shed on the cell cycle. The first two topics deal with the clearly defined processes of DNA replication and chromatid segregation, whereas the last two focus on cyclin-dependent kinases (Cdks) and surveillance mechanisms (checkpoints), which serve to coordinate these two processes.

ONCE AND ONLY ONCE: REGULATING THE INITIATION OF DNA REPLICATION

It is thought that the primary event in DNA replication initiation in all living organisms is the loading of a helicase, whose role is to melt origins (see Chapter 3). Although the proteins involved in this process share little or no sequence identity in eukaryotes and eubacteria, the principles appear to be similar. Nevertheless, fundamental differences between bacteria and eukaryotes in the coupling between replication and segregation of DNA necessitate major differences in the mode by which initiation is regulated.

Most if not all bacterial genomes are replicated from a single origin that must fire as frequently as the cell doubles its mass. In rich medium, cells duplicate their mass faster than they can complete replication of the genome and therefore re-fire origins long before the previous round of replication has been completed, which rarely occurs in eukaryotic cells (Sherratt et al. 2001). Soon after their replication, nascent origins move toward opposite poles of the cell, and in rod-shaped bacteria, they occupy the front and rear positions where they are progressively joined by all other newly replicated DNA sequences. The segregation of sister DNA molecules therefore commences long before the two replication forks converge at a unique terminus. Although the mechanism by which bacteria segregate the two nascent genomes from each other is not understood, the process starts soon after the initiation of DNA replication and may even be linked to it. If the diverging/converging forks remain associated with each other, then the stationary "replisome" would push nascent chromatids away from their site of synthesis (Lemon and Grossman 2001), although how sisters are pushed in opposite directions remains unclear.

In most eukaryotic cells, on the other hand, chromosome duplication and segregation take place during separate periods, segregation is rarely initiated until replication is complete, and origins do not re-fire until chromatids produced by a previous round of DNA replication have been segregated away from each other during mitosis (Nasmyth 2001a). Rereplication without an intervening mitosis is the exception rather than the rule in eukaryotic cells and is known as endoduplication. Unlike bacteria, eukaryotic cells that undergo endoduplication are seldom if ever capable of returning to a proliferative state.

Endoduplication is rare in animals but more frequent in plants, which often require large cells (Edgar and Orr-Weaver 2001). The general rule that origins fire once and only once between rounds of chromosome segregation is all the more remarkable when one considers that different origins on the same chromosome fire at different times during the replicative "S" period (see also Chapter 3).

The molecular mechanism behind this "once and only once" phenomenon is still not fully understood. Nevertheless, investigation of the state of origins during the chromosome cycle in yeast has shed important insight into the principles involved (Fig. 1). Work on *Xenopus* extracts raised the possibility that initiation might be a two-step process, the first step being the assembly of replication competent origins (sometimes referred to licensing) and the second step being their actual firing (Blow and Laskey 1988). Further understanding of this phenomenon required focusing on events at individual replication origins, which had been identified in yeast but not in *Xenopus* or other multicellular eukaryotes. Autonomous replication sequence (ARS) elements, identified by their ability to promote the autonomous replication of plasmids, had been shown to function as chromosomal replication origins (Fangman and Brewer 1991; see Chapter 3). ARS elements contain a binding site for the multisubunit replication initiator protein, ORC (origin-recognition complex) (Bell and Stillman 1992). Upon binding DNA in vitro, ORC protects a region within the ARS element from digestion by DNase I, producing a characteristic "footprint" that can also be detected in chromatin of isolated yeast nuclei. The ORC-specific footprint was found at all stages of the chromosome cycle, but it was accompanied by protection of adjacent sequences in G_1 cells but not G_2 cells, i.e., only in cells about to undergo replication (**Diffley et al. 1994**). The G_1-specific footprint was subsequently shown to be due to the recruitment by ORC of the Cdc6 (Cocker et al. 1996) and Cdt1 (Maiorano et al. 2000; Nishitani et al.

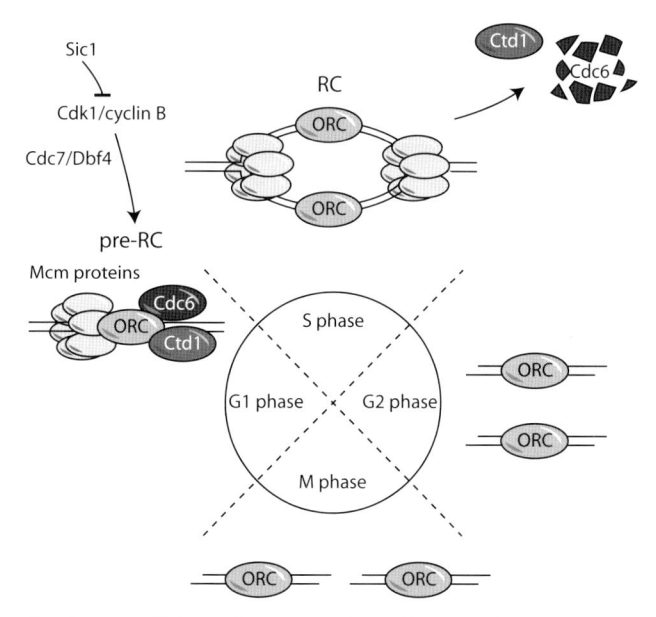

FIGURE 1. Once and only once. Schematic presentation of the origin recognition complex (ORC) on a replication origin during the cell cycle and its activation as discussed in the text.

2000), and finally, the hexameric Mcm2-7 complex (Aparicio et al. 1997; Tanaka et al. 1997) currently thought to be the helicase that actually melts origins (*MCM* genes were originally identified as minichromosome maintenance genes) (Maine et al. 1984). The assembly of these "prereplication" complexes (pre-RCs) is what makes origins competent to initiate replication, but their actual firing depends on activation of the cyclin B/Cdk1 (Schwob et al. 1994) and the Cdc7/Dbf4 kinases (Jackson et al. 1993) during late G_1 (Fig. 1).

The Cdks that trigger initiation from competent origins remain active from late G_1 until the onset of anaphase; therefore, their regulation cannot be responsible for the block to reinitiation. Origins fail to fire during G_2, not because of any lack of S-phase promoting Cdks, but because of a failure to reassemble pre-RCs after their destruction during the course of initiation. Two pieces of evidence suggested that Cdks not only trigger replication from competent origins, but also have a crucial role in regulating the formation of pre-RCs. First, formation of pre-RCs coincided with the inactivation of Cdks during anaphase (Diffley et al. 1994); second, mutants defective in mitotic Cdks (Hayles et al. 1994) or mutants defective in maintaining the activity of S-phase Cdks were found to undergo endoduplication (Dahmann et al. 1995). This raised the possibility that Cdks not only trigger "competent" origins that have previously assembled pre-RCs to fire, but also block de novo formation of pre-RCs. A clinching experiment was to inactivate Cdks in cells blocked in mitosis by inducing high levels of Sic1, a Cdk inhibitory protein. Under these conditions, new pre-RCs were formed, and when Cdks were reactivated by turning Sic1 off again, a new round of DNA replication ensued (Dahmann et al. 1995). The detailed mechanisms by which Cdks both trigger initiation and block formation of pre-RCs have yet to be elucidated. It involves, but does not critically depend on, the phosphorylation of ORC and of Cdc6, with the latter causing Cdc6's rapid proteolysis (Drury et al. 2000). Exclusion of the MCM complex from nuclei is also involved (Nguyen et al. 2001).

By using the same Cdk to trigger initiation and block formation of pre-RCs, yeast cells ensure that origin re-firing depends on a Cdk cycle: a period of low Cdk activity that is permissive for pre-RC formation, followed by a period of high Cdk activity that both triggers competent origins to fire and blocks any further assembly of pre-RCs. The transition between low and high Cdk1 kinase states is largely effected by proteolytic mechanisms (see below).

By coupling the inactivation of Cdks to the trigger for chromatid segregation (see below under Anaphase Promoting Complex), cells ensure that reassembly of pre-RCs cannot take place until mitosis has been completed or at least initiated. Reinitiation of DNA replication is thereby linked to chromosome segregation. The fundamental principle is similar to that of all machines that do work using a cyclical mechanism. The work stroke must be preceded by a phase in which the machine reverts to a ground state, i.e., one competent to perform a work stroke. Eukaryotic cells have evolved a remarkably economical system for switching their replication machine between competent and work states. Although some of the details may differ, similar principles are found to govern the replication cycle of animal cells (Coverley et al. 2002).

THE ANAPHASE-PROMOTING COMPLEX (APC) AND SISTER-CHROMATID SEPARATION

Sister DNA molecules (sister chromatids) produced by DNA replication must be dragged toward opposite poles of the cell by microtubules before cell division if both daughters are to inherit a complete copy of the genome. How chromatids attach to microtubules with opposing orientations, which is known as biorientation, is a key question. Moreover, connections holding sisters together long after their production during DNA replication have a crucial role.

The two chromatids produced by the DNA replication machinery are not only intercatenated (Sundin and Varshavsky 1981), but also connected by a multiprotein complex called cohesin (Hirano 2000; Nasmyth 2001a), which is essential for holding sister chromatids together until their separation at the onset of anaphase. This cohesion complex is essential for the biorientation process (Tanaka et al. 2000; Sonoda et al. 2001). Chromosomes missegregate if sister chromatids attach to the same spindle pole. Eukaryotic cells therefore possess a mechanism for avoiding this situation, which is known as syntelic attachment. A protein kinase called Ipl1 (in yeast) or Aurora B (in mammals) is thought to destabilize microtubule-kinetochore connections that do not give rise to tension within centromeric chromatin (Tanaka et al. 2002). Cohesin not only makes this tension possible, and hence favors the bioriented state, but also resists spindle forces until all chromatid pairs have bioriented. The climax of the mitotic process is therefore a balance of forces between microtubules attempting to pull sister chromatids apart and cohesin resisting the force.

Chromatid separation is finally triggered in a synchronous manner, only when all chromosomes have bioriented, by activation of a cysteine protease called separase, which destroys sister-chromatid cohesion by cleaving the cohesin Scc1/Mcd1 subunit (Uhlmann et al. 1999, 2000). Separase is kept inactive for most of the cell cycle by its association with an inhibitory chaperone called securin, which is destroyed through the action of the same ubiquitin protein ligase (the APC/C, sometimes also called the cyclosome; hence, its abbreviation as the APC/C; see below) that also mediates destruction of Cdk1's regulatory cyclin subunits. The discovery of the APC/C preceded that of the biorientation kinase Ipl1, cohesin, and separase and therefore opened the doors not only to understanding how cells switch from high to low Cdk1 states, but also to the elucidation of the role of separase in triggering the onset of anaphase.

The APC/C was discovered by studying the mechanism by which Cdk1 is inactivated at the end of mitosis. Work on sea urchins and *Xenopus* extracts had demonstrated that Cdk1 inactivation is accompanied by proteolysis of its regulatory cyclin subunit (Evans et al. 1983), which coincided with ubiquitination of cyclin. An amino-terminal sequence, called the destruction box, was found to be essential for both the ubiquitination and proteolysis of cyclin (Glotzer et al. 1991). Expression of cyclins lacking their destruction box prevented Cdk1 inactivation and caused cells to arrest in a mitotic state (Murray et al. 1989). However, work in yeast showed that cyclin destruction was not needed for chromatid separation at the onset of anaphase (Surana et al. 1993), as was thought to be the case in *Xenopus* extracts (Murray et al. 1989). Fractionation of egg extracts (both *Xenopus* and clam) showed that the transfer of ubiquitin from its conjugating enzymes to cyclin was mediated by a large particle with a sedimentation velocity of around 12S (King et al. 1995; Sudakin et al. 1995). The constituents of this ubiquitin protein ligase were discovered through genetic studies in yeast, which revealed that the cyclin ubiquitin machinery has a key role in regulating the onset of anaphase as well as the inactivation of Cdk1.

The identification of proteins involved in cyclin proteolysis was made possible by the finding that the cyclin destruction machinery in yeast remains active in daughter cells long after the initiation of anaphase. It persists throughout G_1 and is only shut off shortly before the onset of S phase (Amon et al. 1994). Why was this important? When hunting for mutants defective in cyclin proteolysis, it was reasonable to assume that the proteolytic machinery would be essential for proliferation and that mutant screens would therefore have to search for "conditional" mutants whose growth was temperature sensitive. However, merely looking for mutants that failed to destroy cyclins upon shift to the restrictive temperature would not have led to the real culprits because the destruction machinery is physiologically inactive throughout S, G_2, and M phases. Any mutant that arrested in these stages of the cell cycle, for example, mutants defective in DNA replication, would arrest in a state where cyclins are stable. Thus, very few mutants that arrested with high cyclin levels would have any genuine defect in cyclin proteolysis.

This problem was circumvented by first arresting cells in early G_1, when the cyclin machinery was known to be active, and then measuring the ability of cells to produce high levels of cyclin upon its induction from the galactose-dependent *GAL1-10* promoter (**Irniger et al. 1995**). The latter was measured by observing the ability of G_1-arrested colonies to accumulate a cyclin–β-galactosidase fusion protein whose abundance could be detected using the chromogenic substrate X-gal (5-bromo-4-chloro-3-indolyl-β-D-galactoside). Colonies that turned blue in the presence of X-gal among a sea of unaffected white colonies were considered candidates.

By this means, several gene products specifically required for cyclin destruction, but not for the destruction of other unstable proteins, were identified. Two such proteins were the products of the *CDC16* and *CDC23* genes, which had long been known to be required for anaphase onset (Culotti and Hartwell 1971), and were now shown to be necessary for sister-chromatid separation (Irniger et al. 1995). Fortunately, antibodies against the vertebrate homolog of Cdc16 protein and one of its partners, Cdc27, had already been produced (Tugendreich et al. 1995) and were used to show that both proteins were subunits of the *Xenopus* cyclin ubiquitin protein ligase complex (King et al. 1995). This ligase contains up to a dozen subunits and is known as the anaphase-promoting complex (APC) because of its role in triggering anaphase. It is also sometimes called the cyclosome (C), hence its abbreviation as the APC/C.

The finding that the cyclin ubiquitination machinery was required for anaphase onset in yeast (Irniger et al. 1995; Zachariae and Nasmyth 1996) was surprising because it was known that expression of nondegradable cyclins blocked exit from mitosis but not sister-chromatid separation (Surana et al. 1993). This finding provided the first incontrovertible evidence for the notion that the cyclin destruction machinery might target for destruction not only cyclins, but also inhibitors of the chromatid separation apparatus. This hypothesis was first proposed because of the metaphase arrest induced in *Xenopus* extracts by high levels of a cyclin destruction box peptide (Holloway et al. 1993). It remained unclear whether this phenomenon was due to interference with the APC/C or with the ubiquitin proteolytic system in general (Yamano et al. 1998).

The hunt for this inhibitor did not last long as a candidate (the Pds1 protein) quickly emerged from a screen for budding yeast mutants defective in arresting chromatid separation when treated with spindle poisons (Yamamoto et al. 1996). Pds1 is degraded by the APC/C shortly before the onset of anaphase, and nondegradable mutants block sister-chromatid separation (Cohen-Fix et al. 1996). Even more compelling, deletion of the *PDS1*

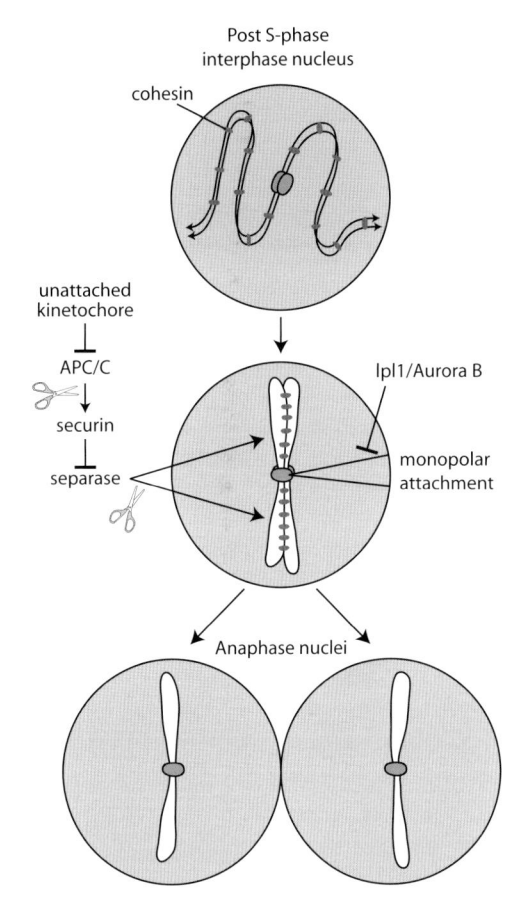

FIGURE 2. Holding together and separation of sister chromatids. After replication, the sister chromatids are held together by cohesion (Scc1/Mcsd1). The APC/C is inhibited by unattached kinetochores. For separation to occur, APC/C inactivates securin (Pds1/Cut2), which in turn activates the separase (Esp1/Cut1) and thereby allows cleavage of the cohesin. See text for details.

gene permits cells lacking APC/C activity to undergo anaphase (Yamamoto et al. 1996; Ciosk et al. 1998). However, what eventually transpired to be Pds1's homolog in fission yeast was identified around the same time, surprisingly in a search for anaphase-promoting proteins, not for inhibitors (Funabiki et al. 1996b). Because of the almost nonexistent sequence homology between Cut2 and Pds1 and because of the very different phenotypes caused by their deletion, it was not recognized that the two proteins were homologs until they were discovered to bind to homologous proteins: Cut1 and Esp1, respectively (Fig. 2).

Purification of Pds1 showed it to be tightly associated with the product of the *ESP1* gene, which had previously been shown to be necessary for chromosome segregation (McGrew et al. 1992), and was also found to be essential for sister separation (Ciosk et al. 1998). This led to the notion that the APC/C promoted sister separation only indirectly, by activating Esp1 through the destruction of Pds1. Previous work had already shown that Esp1's fission yeast homolog, Cut1, associated with Cut2 (Funabiki et al. 1996a). Although their primary sequences are not highly conserved, Pds1- and Cut2-like proteins are found in all eukaryotic cells (Zou et al. 1999) and are now known as securins. Cut1- and Esp1-like proteins are found in all eukaryotes and are now known as separases (see below).

How the APC/C-separase pathway actually triggered sister-chromatid separation had to await the discovery of cohesins, which were identified through the isolation of mutants in which sister chromatids separated even in the absence of securin destruction, e.g., in APC/C mutants (Michaelis et al. 1997) or in cells treated with spindle poisons (Guacci et al. 1997), which block APC/C activation. Four of the genes identified in this manner encode subunits of a cohesin complex (Losada et al. 1998; Toth et al. 1999), which is essential for sister-chromatid cohesion. Cohesin was found to bind tightly to chromosomes from S phase until the onset of anaphase (Michaelis et al. 1997), whereupon proteolytic cleavage of its Scc1 subunit by the Esp1 separase triggers its dissociation from chromosomes and the migration of sisters to opposite poles of the cell (Uhlmann et al. 2000). Cleavage of Scc1 by Esp1/Cut1-like proteins is also essential for sister separation in fission yeast (Tomonaga et al. 2000) and in human cells (Hauf et al. 2001).

It has transpired that by binding separase, securins not only inhibit its protease activity, but also prepare it for activation once securin has been removed by the APC/C. This role in activation explains why the genes for securin are essential in some organisms such as fission yeast (Funabiki et al. 1996a) and *Drosophila* (Stratmann and Lehner 1996) but are nonessential, although important, for optimal separase activity in yeast and mammals (Wang et al. 2001). It is ironic that securin's nonessentiality in budding yeast was crucial in establishing its credentials as the prime target of the APC/C.

Use of the APC/C to trigger destruction of securin, cyclins, and an inhibitor of pre-RC formation called geminin (McGarry and Kirschner 1998; Wohlschlegel et al. 2000) provides at least one mechanism by which eukaryotic cells ensure that preparations for a new round of chromosome duplication cannot occur before chromatids produced in the previous S phase have been segregated. The APC/C therefore has a key role in ordering S and M phases. Other mechanisms must also exist because, remarkably, deletion of the *PDS1* and the *CLB5* cyclin genes permits yeast cells that overproduce the Cdk inhibitor Sic1 to proliferate in the complete absence of the APC/C (Shirayama et al. 1999; Thornton and Toczyski 2003).

THE MITOTIC TRIGGER: CYCLIN-DEPENDENT KINASES

The sister-chromatid cohesion apparatus appears to be unique to eukaryotic cells and makes it possible for them to segregate chromosomes long after DNA replication has been completed. Without cohesion, the S and M phases of the eukaryotic cell cycle could not be separated by an interval known as G_2. G_2 is sometimes very long. Fission yeast cells spend most of their cell cycle in G_2 waiting to enter mitosis, and the equivalent interval can last up to 50 years in the case of human oocytes. How cells awake from this slumber and initiate mitosis has long been a key question in the cell cycle field. Because many other eukaryotic cells, including *Saccharomyces cerevisiae*, spend most of their cycle in G_1, a similar question can be posed about entry into S phase. Although cell fusion experiments were key to defining S- and M-phase triggers, it was genetic studies in yeast that actually identified the triggers (Nasmyth 2001b). Remarkably, both S and M phases were found to be triggered by the same molecule, a protein kinase called Cdk1.

The potential of yeast for identifying cell cycle regulatory genes was first recognized by Hartwell and his colleagues who had embarked on a search for budding yeast mutants defective in macromolecular synthesis, i.e., cells unable to synthesize protein, RNA, or DNA. These authors soon realized that many temperature-sensitive lethal mutants arrested at defined stages of the cell cycle upon being shifted to the restrictive temperature (Hartwell et al. 1970). Such "*cdc*" mutants were particularly easy to recognize in *S. cerevisiae* because bud size serves as a measure to determine the position in the cell cycle. By this means, more than 40 genes required for cell division were identified, and the genetic analysis of the cell cycle was launched (Hartwell et al. 1974). The original *cdc* collection included not only the gene for Cdk1 (Reid and Hartwell 1977), but also those encoding some of the APC/C subunits (Culotti and Hartwell 1971), as well as subunits of SCF (Skp1/Cul1/F-box protein complex) (Hereford and Hartwell 1974), another key ubiquitin protein ligase involved in cell cycle control, whose ability to target for destruction the Cdk1 inhibitory protein Sic1 is a crucial step in the initiation of S phase (Schwob et al. 1994).

It was not, however, apparent which if any of the *S. cerevisiae CDC* genes might encode proteins whose activation might trigger S or M phase as opposed to simply being required for these events. None of the mutants arrested in G_2, and as a consequence, there was not a single candidate for the M-phase trigger. Several mutants did arrest in G_1, and one of these, the *CDC28* gene, which encodes Cdk1 in *S. cerevisiae*, emerged as a potential candidate (Reid and Hartwell 1977). Nevertheless, a mere failure to enter S phase or initiate other early cell cycle events could have been caused by many defects besides inactivation of the S-phase trigger. There was no compelling reason at the time to think that *CDC28* encoded a protein that triggered G_1 cells to initiate DNA replication.

This problem was solved by the identification of mutants that underwent key cell cycle transitions not less but more readily than wild type. Furthermore, it was the M-phase trigger that emerged first. Inspired by Hartwell, Nurse and his colleagues isolated a large collection of *cdc* genes in the fission yeast *Schizosaccharomyces pombe*, where a failure to divide gives rise to very long elongated cells (Nurse et al. 1976). Quite by chance, a mutant with precisely the opposite characteristics was also isolated. A temperature-sensitive allele of the *wee1* gene (initially called *cdc50-1*) caused *S. pombe* cells not to cease division upon shift to the restrictive temperature, but instead to divide at half the normal cell size (Nurse 1975). Inactivation of Wee1 caused most G_2 cells to enter mitosis almost immediately. Wee1, which was subsequently found to encode a protein kinase (Russell and Nurse 1987), is an inhibitor of the G_2-to-M-phase transition. Among a large collection of small-sized mutants (Thuriaux et al. 1978), most were recessive to wild type and were alleles of *wee1*, but one was partially dominant and resided at a second locus, initially called *wee2*. This raised the possibility that the *wee2-1* mutation might be a hyperactive allele of a gene required for mitotic entry. Crosses between *wee2-1* and representative alleles of all genes required for mitotic entry revealed that the *wee2* and *cdc2* loci were very tightly linked. Fine-structure mapping suggested that *wee2-1* was situated between two different temperature-sensitive *cdc2* alleles and must therefore be a mutation in the *cdc2* (the homolog of the budding yeast *CDC28* gene itself) (**Nurse and Thuriaux 1980**).

This then was the first unambiguous evidence for the existence of a gene whose role was to trigger the G_2-to-M-phase transition. Subsequent work revealed that *cdc2* was also required for the G_1-to-S-phase transition (Nurse and Bissett 1981), that it shared sequence identity and function with the *CDC28* gene (Beach et al. 1982), that homologous genes exist in all eukaryotic cells (Lee and Nurse 1987), and that these genes encode a specific class of protein kinases (cyclin-dependent kinases or CDKs) whose activity depends on regulatory subunits called cyclins. The accumulation of cyclins during G_2 creates a potentially active cyclin/Cdk1 complex whose activity is inhibited through phosphorylation of a tyro-

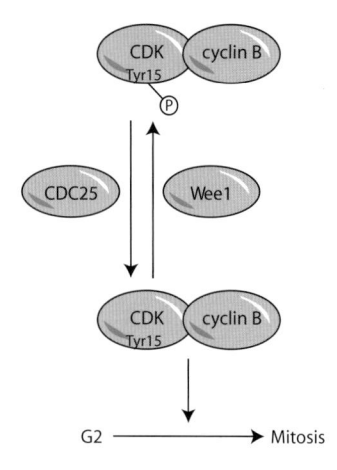

FIGURE 3. Triggering entry into mitosis. Schematic presentation of the interplay of the Cdc25 phosphatase and the Wee1 kinase in activation of the Cdk/cyclin complex triggering the G_2 to mitosis transition. For details, see text.

sine residue close to Cdk1's active site (Tyr-15) by the Wee1 protein kinase (Gould and Nurse 1989). The precise mechanism by which Cdk1 is suddenly activated when fission yeast cells reach a certain cell size remains to be elucidated, but involves activation of the Cdc25 phosphatase, which removes the inhibitory phosphorylation on Tyr-15 (Fig. 3).

One of the surprises to emerge from this work is that the same protein kinase triggers S and M phases. How does Cdk1 trigger S phase and not M phase when activated in late G_1 and M and not S phase when activated in G_2? I think that it is fair to say we do not yet fully understand this paradox. Cdk1's association with different types of cyclins to create S- and M-phase-specific variants of the kinase suggested that S- and M-phase initiation might require kinases with different substrate specificities. In budding yeast, S phase is triggered by Cdk1 associated with Clb5 and Clb6 cyclins (Kuhne and Linder 1993; Schwob and Nasmyth 1993), whereas M phase is triggered by Cdk1 associated with Clbs1, 2, 3, and 4 (Fitch et al. 1992). However, it soon emerged that M as well as S Clb/Cdk1 kinases might be able to trigger S phase in *S. cerevisiae* (Amon et al. 1994). Furthermore, Cdk1 associated with the B-type cyclin encoded by *cdc13* can trigger both S and M phases in *S. pombe* (Fisher and Nurse 1996). Whether S or M phase is triggered by Cdk1 cannot therefore be determined by the type of cyclin associated with Cdk1.

The outlines of a solution to this paradox emerged from the discovery that Cdk1 will only trigger DNA replication from origins that have already assembled pre-RCs (Cocker et al. 1996; Piatti et al. 1996). Activation of Cdk1 triggers S phase in G_1 cells and not in G_2 cells because the former and not the latter have assembled pre-RCs at origins. By analogy, might Cdk1 not trigger M phase in G_1 cells because they have not yet assembled structures required for building a mitotic spindle? One of the preconditions for the formation of bipolar spindles, a key mitotic event in yeast, is the duplication of spindle pole bodies, which takes place during late G_1 in *S. cerevisiae* (Byers and Goetsch 1974). As a result, wild-type *S. cerevisiae* cells do indeed build spindles rather soon after the initiation of S phase. Furthermore, in cells lacking the S-phase cyclins Clb5 and Clb6, where S phase is initiated by Cdk1 associated with Clb1–Clb4, formation of bipolar spindles quite possibly does coincide with the initiation of S phase. The situation is more complicated in organisms like *S. pombe* where there is usually a long gap between S and M phases. *S. pombe* cells emerge from S phase with Cdk1 heavily phosphorylated on Tyr-15, which clearly prevents Cdk1 from triggering entry into mitosis. Activation of Cdk1 during late G_1 in *S. pombe* fails to trigger M phase because spindle pole bodies have not yet duplicated. Cdk1 becomes rapidly phosphorylated on Tyr-15 as soon as cells enter S phase and this prevents mitotic entry (Zarzov et al. 2002) until cells grow to a critical cell size.

Although Cdks were identified as molecules whose activity triggers S or M phase, it is clear that they have an equally important role in inhibiting cell cycle events as they do in promoting them. This duality of Cdk function, promoting one or more processes while simultaneously blocking others, appears to be fundamental to their ability to coordinate cell cycle events. A good analogy would be the role of traffic lights at a road intersection. In this regard, traffic moving along one road represents a biological process like the initiation of DNA replication, whereas traffic moving along an intersecting road represents another process, e.g., the formation of pre-RCs, whose simultaneous occurrence with the first process is incompatible with it. Just as traffic from one road must be prevented from crossing an intersection while traffic from the other is doing so, so must formation of pre-RCs never coincide with the initiation of DNA replication. The consequences would be reduplication of at least certain segments of the genome before segregation of sister chromatids from the previous round of duplication, which is incompatible with the orderly segregation of chromosomes during mitosis. Cytokinesis must likewise never coincide with chromosome alignment during metaphase, and it is therefore inhibited by the Cdks which trigger entry into mitosis. Cdks are therefore analogous to traffic lights, which, when green for traffic moving along one road, are simultaneously red for traffic attempting to move along an intersecting one. An active Cdk "shines" as it were green in one direction but red in the other. This metaphor has of course its limitations because unlike traffic lights, which only regulate two potentially conflicting flows of traffic, Cdks both promote and block multiple processes in parallel. If Cdks are analogous to the traffic lights themselves, their regulators, in particular the APC/C and SCF ubiquitin protein ligases, are analogous to the switching mechanisms that turn these lights on and off.

SURVEILLANCE MECHANISMS THAT HALT THE Cdk OSCILLATION

Given that an oscillation in Cdk activity is a key aspect of the eukaryotic cell cycle, then the mechanism by which cells create this oscillation becomes of central interest. At least two kinds of mechanisms capable of generating Cdk oscillations have been described. The first stems from work on extracts from *Xenopus* eggs, suggesting that activation of cyclin B/Cdk1 promotes activity of APC/C activity mediated by Cdc20 (King et al. 1996). The mechanistic basis for this phenomenon is not well understood, but it is thought to involve the APC/C's phosphorylation by Cdk1. A negative feedback loop in which Cdk1 promotes its own destruction in this manner could alone give rise to a limit cycle (oscillation) if there were a sufficient time lag between Cdk1's activation and the subsequent activation of APC/C. This mechanism may also exist in yeast, because some APC/C subunits, for example, Cdc16, Cdc23, and Cdc27, appear to be phosphorylated in a Cdk1-dependent fashion and mutants in which these subunits can no longer be phosphorylated by Cdk1 are delayed in the onset of anaphase (Rudner and Murray 2000).

Even if this negative feedback mechanism does exist, and the case is by no means proven, it cannot be solely responsible for the yeast cell's Cdk1 oscillation. There is abundant evidence of an oscillator of a very different nature. Several factors ensure that both Cdk1 states, the low one characteristic of G_1 cells and the high one characteristic of G_2/M cells, are self-reinforcing. This largely arises because Cdk1 promotes ubiquitination and hence proteolysis of the Cdk1 inhibitor Sic1 (Verma et al. 1997), whereas it inhibits the activity of Cdh1, which takes over from Cdc20, the activator for the APC, when cells enter G_1 (Zachariae et al. 1998). Cdk1 performs both tasks by directly phosphorylating Sic1 and Cdh1. As a consequence, Cdh1 is active and Sic1 accumulates when Cdk1 is low, whereas Cdh1 is inactive and Sic1 is destroyed when Cdk1 is high. Because the low and high Cdk1 states are both stable states, the yeast cell needs special mechanisms to trigger the transitions between these states. The low-to-high Cdk1 state is catalyzed by the production in a growth-dependent manner of a form of Cdk1 that is refractory to inhibition by Sic1 and resistant to proteolysis mediated by Cdh1 (Nasmyth 1996). The high-to-low Cdk1 transition, on the other hand, is catalyzed by activation shortly after anaphase onset of the Cdc14 phosphatase, which is thought to dephosphorylate both Sic1 and Cdh1 (Visintin et al. 1998).

Such oscillators appear to be capable of driving chromosome replication and segregation cycles in a highly autonomous fashion. The Cdk1 oscillations that drive the cleavage divisions of early *Xenopus* embryos persist even when microtubule damage is severe enough to abolish chromosome segregation, when chromosomal DNA badly damaged by irradiation, or even when inhibitors of DNA replication prevent the production of chromatids to be segregated. Nevertheless, in most somatic cells and in most microorganisms, numerous highly specialized surveillance mechanisms exist that detect DNA damage (Zhou and Elledge 2000) or failure of kinetochores to attach to microtubules (Amon 1999) and block activation of either Cdk1 or APC/C. The first of these mechanisms to be discovered was one that blocks activation of the APC/C in *S. cerevisiae* when chromosomes are broken by irradiation.

It had long been known that the induction of double-strand breaks by γ-irradiation prevents the onset of M phase in G_2 cells. For many years, before the factors that trigger M phase had been discovered, it was assumed that mitosis failed to take place because the chromosomes were too broken to be effectively segregated. An indication that this might not be so was the finding that cells from patients with ataxia-telangiectasia (AT) entered mitosis even when chromosomes were damaged (Painter and Young 1980). This raised the possibility that the cell cycle arrest induced by irradiation might be an active process that depended on specific gene products. A good precedent was the SOS control system that triggers not only the production of DNA repair enzymes, but also a block to septation in bacteria (Walker 1984). In budding yeast, double-strand breaks do not block entry into M phase but rather the onset of anaphase. Thus, cells with damaged chromosomes arrest in a metaphase-like state.

Weinert and Hartwell (1988) reasoned that yeast cells must also possess SOS-like surveillance mechanisms. They further postulated that by blocking the segregation of damaged chromatids, cell cycle arrest would facilitate the repair of double-strand breaks by holding cells in a cell cycle stage in which they possessed a sister chromatid to be used as a template for repair. This reasoning predicted that the sensitivity to γ-irradiation of some "radiation repair" or *rad* mutants might be due to their inability to arrest the cell cycle. These authors therefore tested whether any of these mutants might, like the mammalian AT mutant cells, be defective in irradiation-induced cell cycle arrest. Remarkably, they found that *rad9* mutants were completely defective in arrest and suggested (although this has never been satisfactorily proven) that this and not any defect in repair per se was responsible for their inviability after irradiation. They named such surveillance mechanisms "checkpoints."

Subsequent work has unearthed a vast network of proteins necessary for arresting the cell cycle in response to different types of DNA damage. It is now recognized that cell cycle arrest is just one of many responses mediated by these proteins (Zhou and Elledge 2000). In addition to controlling cell cycle arrest, these proteins control activation of DNA-repair pathways, the composition of telomeric chromatin, the movement of DNA-repair proteins to sites of DNA damage, and, in some cells, the induction of apoptosis.

In most eukaryotic cells, the cell cycle arm of these DNA-damage surveillance mechanisms (known as checkpoints) blocks entry of G_2 cells into mitosis, largely by blocking dephosphorylation of Cdk1, possibly by inhibiting the Cdc25 phosphatase (Zhou and Elledge 2000). In budding yeast, in contrast, entry into M phase is rarely if ever blocked by DNA damage surveillance mechanisms, which instead block ubiquitination mediated by APC/C-Cdc20 and hence the onset of anaphase (Cohen-Fix and Koshland 1997). All eukaryotic cells including *S. cerevisiae* possess another set of surveillance mechanisms that monitors the attachment of kinetochores to microtubules and interferes with Cdc20's interaction with the APC/C (Hoyt et al. 1991; Li and Murray 1991; Amon 1999). As a consequence of this control mechanism, known variously as the mitotic or spindle checkpoint, cells delay activation of separase and inactivation of Cdk1 until every chromosome has bioriented on the mitotic spindle. The value of such a system is obvious because the destruction of sister-chromatid cohesion by separase is a global process that affects all chromosomes, whether or not they have bioriented. It is therefore vital that separase not be activated until every single chromosome has achieved this state. How a single unattached kinetochore manages to inhibit APC/C-Cdc20 throughout the cell remains largely a mystery. This system is possibly the best example of a "checkpoint" in the sense of a surveillance mechanism whose primary role is to regulate the chromosome cycle. There is little or no evidence, for example, that spindle checkpoint proteins ameliorate spindle damage or directly strengthen kinetochore-spindle attachments.

STUDY QUESTIONS

1. What are the main differences between eukaryotic and prokaryotic replication and what could be the reason for these differences?
2. How did Diffley and colleagues infer that ORC and Abf1 binding alone are not sufficient for DNA replication origin activation?
3. Why are Cdks proposed to have a dual role in replication initiation?
4. Explain the roles of the ubiquitination system in sister-chromatid segregation and the experiments that allowed the genetic characterization of this system.
5. How do you explain that the budding yeast Pds1 has been found to be a inhibitor of sister-chromatin separation, whereas the fission yeast homolog has been found in a search for factors that promote anaphase?

6. On what basis were the first *cdc* mutants identified as such and how would you go about ordering these mutants within the cell cycle and among each other?

7. How did Nurse and collaborators, in the absence of a characteristic budding pattern in *S. pombe* that would indicate a cell's position within the cell cycle, isolate *cdc* mutants?

8. What is the phenotype of a *rad9* surveillance mutant? Can you imagine other explanations for the observed phenotype?

9. Can you propose distinct stages or functional categories that might be common to most or all "checkpoint" systems? How might you assign specific checkpoint factors to these different roles?

REFERENCES

Amon A. 1999. The spindle checkpoint. *Curr. Opin. Genet. Dev.* **9:** 69–75.

Amon A., Irniger S., and Nasmyth K. 1994. Closing the cell cycle circle in yeast: G2 cyclin proteolysis initiated at mitosis persists until the activation of G1 cyclins in the next cycle. *Cell* **77:** 1037–1050.

Aparicio O.M., Weinstein D.M., and Bell S.P. 1997. Components and dynamics of DNA replication complexes in *S. cerevisiae*: Redistribution of MCM proteins and Cdc45p during S phase. *Cell* **91:** 59–69.

Beach D., Durkacz B., and Nurse P. 1982. Functionally homologous cell cycle control genes in budding and fission yeast. *Nature* **300:** 706–709.

Bell S.P. and Stillman B. 1992. ATP-dependent recognition of eukaryotic origins of DNA replication by a multiprotein complex. *Nature* **357:** 128–134.

Blow J.J. and Laskey R.A. 1988. A role for the nuclear envelope in controlling DNA replication within the cell cycle. *Nature* **332:** 546–548.

Byers B. and Goetsch L. 1974. Duplication of spindle plaques and integration of the yeast cell cycle. *Cold Spring Harbor Symp. Quant. Biol.* **38:** 123–131.

Ciosk R., Zachariae W., Michaelis C., Shevchenko A., Mann M., and Nasmyth K. 1998. An ESP1/PDS1 complex regulates loss of sister chromatid cohesion at the metaphase to anaphase transition in yeast. *Cell* **93:** 1067–1076.

Cocker J.H., Piatti S., Santocanale C., Nasmyth K., and Diffley J.F. 1996. An essential role for the Cdc6 protein in forming the pre-replicative complexes of budding yeast. *Nature* **379:** 180–182.

Cohen-Fix O. and Koshland D. 1997. The anaphase inhibitor of *Saccharomyces cerevisiae* Pds1p is a target of the DNA damage checkpoint pathway. *Proc. Natl. Acad. Sci.* **94:** 14361–14366.

Cohen-Fix O., Peters J.M., Kirschner M.W., and Koshland D. 1996. Anaphase initiation in *Saccharomyces cerevisiae* is controlled by the APC-dependent degradation of the anaphase inhibitor Pds1p. *Genes Dev.* **10:** 3081–3093.

Coverley D., Laman H., and Laskey R.A. 2002. Distinct roles for cyclins E and A during DNA replication complex assembly and activation. *Nat. Cell Biol.* **4:** 523–528.

Culotti J. and Hartwell L.H. 1971. Genetic control of the cell division cycle in yeast. 3. Seven genes controlling nuclear division. *Exp. Cell Res.* **67:** 389–401.

Dahmann C., Diffley J.F., and Nasmyth K.A. 1995. S-phase-promoting cyclin-dependent kinases prevent re-replication by inhibiting the transition of replication origins to a pre-replicative state. *Curr. Biol.* **5:** 1257–1269.

Diffley J.F.X., Cocker J.H., Dowell S.J., and Rowley A. 1994. Two steps in the assembly of complexes at yeast replication origins in vivo. *Cell* **78:** 303–316.

Drury L.S., Perkins G., and Diffley J.F. 2000. The cyclin-dependent kinase Cdc28p regulates distinct modes of Cdc6p proteolysis during the budding yeast cell cycle. *Curr. Biol.* **10:** 231–240.

Edgar B.A. and Orr-Weaver T.L. 2001. Endoreplication cell cycles: More for less. *Cell* **105:** 297–306.

Evans T., Rosenthal E.T., Youngblom J., Distel D., and Hunt T. 1983. Cyclin: A protein specified by maternal mRNA in sea urchin eggs that is destroyed at each cleavage division. *Cell* **33:** 389–396.

Fangman W.L. and Brewer B.J. 1991. Activation of replication origins within yeast chromosomes. *Annu. Rev. Cell Biol.* **7:** 375–402.

Fisher D.L. and Nurse P. 1996. A single fission yeast mitotic cyclin B p34cdc2 kinase promotes both S-phase and mitosis in the absence of G1 cyclins. *EMBO J.* **15:** 850 860.

Fitch I., Dahmann C., Surana U., Amon A., Nasmyth K., Goetsch L., Byers B., and Futcher B. 1992. Characterization of four B-type cyclin genes of the budding yeast *Saccharomyces cerevisiae*. *Mol. Biol. Cell* **3:** 805–818.

Funabiki H., Kumada K., and Yanagida M. 1996a. Fission yeast Cut1 and Cut2 are essential for sister chromatid separation, concentrate along the metaphase spindle and form large complexes. *EMBO J.* **15:** 6617–6628.

Funabiki H., Yamano H., Kumada K., Nagao K., Hunt T., and Yanagida M. 1996b. Cut2 proteolysis required for sister-chromatid separation in fission yeast. *Nature* **381:** 438–441.

Glotzer M., Murray A.W., and Kirschner M.W. 1991. Cyclin is degraded by the ubiquitin pathway. *Nature* **349:** 132–138.

Gould K.L. and Nurse P. 1989. Tyrosine phosphorylation of the fission yeast cdc2+ protein kinase regulates entry into mitosis. *Nature* **342:** 39–45.

Guacci V., Koshland D., and Strunnikov A. 1997. A direct link between sister chromatid cohesion and chromosome condensation revealed through the analysis of MCD1 in *S. cerevisiae*. *Cell* **91:** 47–57.

Hartwell L.H., Culotti J., and Reid B. 1970. Genetic control of the cell-division cycle in yeast. I. Detection of mutants. *Proc. Natl. Acad. Sci.* **66:** 352–359.

Hartwell L.H., Culotti J., Pringle J.R., and Reid B.J. 1974. Genetic control of the cell division cycle in yeast. *Science* **183:** 46–51.

Hauf S., Waizenegger I.C., and Peters J.M. 2001. Cohesin cleavage by separase required for anaphase and cytokinesis in human cells. *Science* **293:** 1320–1323.

Hayles J., Fisher D., Woollard A., and Nurse P. 1994. Temporal order of S phase and mitosis in fission yeast is determined by the state of the p34cdc2-mitotic B cyclin complex. *Cell* **78:** 813–822.

Hereford L.M. and Hartwell J.H. 1974. Sequential gene function in the initiation of *Saccharomyces cerevisiae* DNA synthesis. *J. Mol. Biol.* **85:** 445–461.

Hirano T. 2000. Chromosome cohesion, condensation, and separation. *Annu. Rev. Biochem.* **69:** 115–144.

Holloway S.L., Glotzer M., King R.W., and Murray A.W. 1993. Anaphase is initiated by proteolysis rather than by the inactivation of maturation-promoting factor. *Cell* **73:** 1393–1402.

Hoyt M.A., Totis L., and Roberts B.T. 1991. *S. cerevisiae* genes required for cell cycle arrest in response to loss of microtubule function. *Cell* **66:** 507–517.

Irniger S., Piatti S., Michaelis C., and Nasmyth K. 1995. Genes involved in sister chromatid separation are needed for B-type cyclin proteolysis in budding yeast. *Cell* **81:** 269–277.

Jackson A.L., Pahl P.M., Harrison K., Rosamond J., and Sclafani R.A. 1993. Cell cycle regulation of the yeast Cdc7 protein kinase by association with the Dbf4 protein. *Mol. Cell. Biol.* **13:** 2899–2908.

King R.W., Deshaies R.J., Peters J.M., and Kirschner M.W. 1996. How proteolysis drives the cell cycle. *Science* **274:** 1652–1659.

King R.W., Peters J.M., Tugendreich S., Rolfe M., Hieter P., and Kirschner M.W. 1995. A 20S complex containing CDC27 and CDC16 catalyzes the mitosis-specific conjugation of ubiquitin to cyclin B. *Cell* **81:** 279–288.

Kuhne C. and Linder P. 1993. A new pair of B-type cyclins from *Saccharomyces cerevisiae* that function early in the cell cycle. *EMBO J.* **12:** 3437–3447.

Lee M.G. and Nurse P. 1987. Complementation used to clone a human homologue of the fission yeast cell cycle control gene cdc2. *Nature* **327:** 31–35.

Lemon K.P. and Grossman A.D. 2001. The extrusion-capture model for chromosome partitioning in bacteria. *Genes Dev.* **15:** 2031–2041.

Li R. and Murray A.W. 1991. Feedback control of mitosis in budding yeast. *Cell* **66:** 519–531.

Losada A., Hirano M., and Hirano T. 1998. Identification of *Xenopus* SMC protein complexes required for sister chromatid cohesion. *Genes Dev.* **12:** 1986–1997.

Maine G.T., Sinha P., and Tye B.K. 1984. Mutants of *S. cerevisiae* defective in the maintenance of minichromosomes. *Genetics* **106:** 365–385.

Maiorano D., Moreau J., and Mechali M. 2000. XCDT1 is required for the assembly of pre-replicative com-

plexes in *Xenopus laevis*. *Nature* **404:** 622–625.
McGarry T.J. and Kirschner M.W. 1998. Geminin, an inhibitor of DNA replication, is degraded during mitosis. *Cell* **93:** 1043–1053.
McGrew J.T., Goetsch L., Byers B., and Baum P. 1992. Requirement for ESP1 in the nuclear division of *Saccharomyces cerevisiae*. *Mol. Biol. Cell* **3:** 1443–1454.
Michaelis C., Ciosk R., and Nasmyth K. 1997. Cohesins: Chromosomal proteins that prevent premature separation of sister chromatids. *Cell* **91:** 35–45.
Murray A.W., Solomon M.J., and Kirschner M.W. 1989. The role of cyclin synthesis and degradation in the control of maturation promoting factor activity. *Nature* **339:** 280–286.
Nasmyth K. 1996. At the heart of the budding yeast cell cycle. *Trends Genet.* **12:** 405–412.
———. 2001a. Disseminating the genome: Joining, resolving, and separating sister chromatids during mitosis and meiosis. *Annu. Rev. Genet.* **35:** 673–745.
———. 2001b. A prize for proliferation. *Cell* **107:** 689–701.
Nguyen V.Q., Co C., and Li J.J. 2001. Cyclin-dependent kinases prevent DNA re-replication through multiple mechanisms. *Nature* **411:** 1068–1073.
Nishitani H., Lygerou Z., Nishimoto T., and Nurse P. 2000. The Cdt1 protein is required to license DNA for replication in fission yeast. *Nature* **404:** 625–628.
Nurse P. 1975. Genetic control of cell size at cell division in yeast. *Nature* **256:** 547–551.
Nurse P. and Bissett Y. 1981. Gene required in G1 for commitment to cell cycle and in G2 for control of mitosis in fission yeast. *Nature* **292:** 558–560.
Nurse P. and Thuriaux P. 1980. Regulatory genes controlling mitosis in the fission yeast *Schizosaccharomyces pombe*. *Genetics* **96:** 627–637.
Nurse P., Thuriaux P., and Nasmyth K. 1976. Genetic control of the cell division cycle in the fission yeast *Schizosaccharomyces pombe*. *Mol. Gen. Genet.* **146:** 167–178.
Painter R.B. and Young B.R. 1980. Radiosensitivity in ataxia-telangiectasia: A new explanation. *Proc. Natl. Acad. Sci.* **77:** 7315–7317.
Piatti S., Bohm T., Cocker J.H., Diffley J.F., and Nasmyth K. 1996. Activation of S-phase-promoting CDKs in late G1 defines a "point of no return" after which Cdc6 synthesis cannot promote DNA replication in yeast. *Genes Dev.* **10:** 1516–1531.
Reid B.J. and Hartwell L.H. 1977. Regulation of mating in the cell cycle of *Saccharomyces cerevisiae*. *J. Cell Biol.* **75:** 355–365.
Rudner A.D. and Murray A.W. 2000. Phosphorylation by Cdc28 activates the Cdc20-dependent activity of the anaphase-promoting complex. *J. Cell Biol.* **149:** 1377–1390.
Russell P. and Nurse P. 1987. Negative regulation of mitosis by wee1+, a gene encoding a protein kinase homolog. *Cell* **49:** 559–567.
Schwob E. and Nasmyth K. 1993. CLB5 and CLB6, a new pair of B cyclins involved in DNA replication in *Saccharomyces cerevisiae*. *Genes Dev.* **7:** 1160–1175.
Schwob E., Bohm T., Mendenhall M.D., and Nasmyth K. 1994. The B-type cyclin kinase inhibitor p40SIC1 controls the G1 to S transition in *S. cerevisiae*. *Cell* **79:** 233–244.
Sherratt D.J., Lau I.F., and Barre F.X. 2001. Chromosome segregation. *Curr. Opin. Microbiol.* **4:** 653–659.
Shirayama M., Toth A., Galova M., and Nasmyth K. 1999. APC(Cdc20) promotes exit from mitosis by destroying the anaphase inhibitor Pds1 and cyclin Clb5. *Nature* **402:** 203–207.
Sonoda E., Matsusaka T., Morrison C., Vagnarelli P., Hoshi O., Ushiki T., Nojima K., Fukagawa T., Waizenegger I.C., Peters J.M., Earnshaw W.C., and Takeda S. 2001. Scc1/Rad21/Mcd1 is required for sister chromatid cohesion and kinetochore function in vertebrate cells. *Dev. Cell* **1:** 759–770.
Stratmann R. and Lehner C.F. 1996. Separation of sister chromatids in mitosis requires the *Drosophila* pimples product, a protein degraded after the metaphase/anaphase transition. *Cell* **84:** 25–35.
Sudakin V., Ganoth D., Dahan A., Heller H., Hershko J., Luca F.C., Ruderman J.V., and Hershko A. 1995. The cyclosome, a large complex containing cyclin-selective ubiquitin ligase activity, targets cyclins for destruction at the end of mitosis. *Mol. Biol. Cell* **6:** 185–197.
Sundin O. and Varshavsky A. 1981. Arrest of segregation leads to accumulation of highly intertwined catenated dimers: Dissection of the final stages of SV40 DNA replication. *Cell* **25:** 659–669.
Surana U., Amon A., Dowzer C., McGrew J., Byers B., and Nasmyth K. 1993. Destruction of the CDC28/CLB

mitotic kinase is not required for the metaphase to anaphase transition in budding yeast. *EMBO J.* **12:** 1969–1978.
Tanaka T., Knapp D., and Nasmyth K. 1997. Loading of an Mcm protein onto DNA replication origins is regulated by Cdc6p and CDKs. *Cell* **90:** 649–660.
Tanaka T., Fuchs J., Loidl J., and Nasmyth K. 2000. Cohesin ensures bipolar attachment of microtubules to sister centromeres and resists their precocious separation. *Nat. Cell Biol.* **2:** 492–499.
Tanaka T.U., Rachidi N., Janke C., Pereira G., Galova M., Schiebel E., Stark M.J., and Nasmyth K. 2002. Evidence that the Ipl1-Sli15 (Aurora kinase-INCENP) complex promotes chromosome bi-orientation by altering kinetochore-spindle pole connections. *Cell* **108:** 317–329.
Thornton B.R. and Toczyski D.P. 2003. Securin and B-cyclin/CDK are the only essential targets of the APC. *Nat. Cell Biol.* **5:** 1090–1094.
Thuriaux P., Nurse P., and Carter B. 1978. Mutants altered in the control co-ordinating cell division with cell growth in the fission yeast *Schizosaccharomyces pombe*. *Mol. Gen. Genet.* **161:** 215–220.
Tomonaga T., Nagao K., Kawasaki Y., Furuya K., Murakami A., Morishita J., Yuasa T., Sutani T., Kearsey S.E., Uhlmann F., Nasmyth K., and Yanagida M. 2000. Characterization of fission yeast cohesin: Essential anaphase proteolysis of Rad21 phosphorylated in the S phase. *Genes Dev.* **14:** 2757–2770.
Toth A., Ciosk R., Uhlmann F., Galova M., Schleiffer A., and Nasmyth K. 1999. Yeast cohesin complex requires a conserved protein, Eco1p(Ctf7), to establish cohesion between sister chromatids during DNA replication. *Genes Dev.* **13:** 320–333.
Tugendreich S., Tomkiel J., Earnshaw W., and Hieter P. 1995. CDC27Hs colocalizes with CDC16Hs to the centrosome and mitotic spindle and is essential for the metaphase to anaphase transition. *Cell* **81:** 261–268.
Uhlmann F., Lottspeich F., and Nasmyth K. 1999. Sister-chromatid separation at anaphase onset is promoted by cleavage of the cohesin subunit Scc1. *Nature* **400:** 37–42.
Uhlmann F., Wernic D., Poupart M.A., Koonin E.V., and Nasmyth K. 2000. Cleavage of cohesin by the CD clan protease separin triggers anaphase in yeast. *Cell* **103:** 375–386.
Verma R., Annan R.S., Huddleston M.J., Carr S.A., Reynard G., and Deshaies R.J. 1997. Phosphorylation of Sic1p by G1 Cdk required for its degradation and entry into S phase. *Science* **278:** 455–460.
Visintin R., Craig K., Hwang E.S., Prinz S., Tyers M., and Amon A. 1998. The phosphatase Cdc14 triggers mitotic exit by reversal of Cdk-dependent phosphorylation. *Mol. Cell* **2:** 709–718.
Walker G.C. 1984. Mutagenesis and inducible responses to deoxyribonucleic acid damage in *Escherichia coli*. *Microbiol. Rev.* **48:** 60–93.
Wang Z., Yu R., and Melmed S. 2001. Mice lacking pituitary tumor transforming gene show testicular and splenic hypoplasia, thymic hyperplasia, thrombocytopenia, aberrant cell cycle progression, and premature centromere division. *Mol. Endocrinol.* **15:** 1870–1879.
Weinert T.A. and Hartwell L.H. 1988. The *RAD9* gene controls the cell cycle response to DNA damage in *Saccharomyces cerevisiae*. *Science* **241:** 317–322.
Wohlschlegel J.A., Dwyer B.T., Dhar S.K., Cvetic C., Walter J.C., and Dutta A. 2000. Inhibition of eukaryotic DNA replication by geminin binding to Cdt1. *Science* **290:** 2309–2312.
Yamamoto A., Guacci V., and Koshland D. 1996. Pds1p, an inhibitor of anaphase in budding yeast, plays a critical role in the APC and checkpoint pathway(s). *J. Cell Biol.* **133:** 99–110.
Yamano H., Tsurumi C., Gannon J., and Hunt T. 1998. The role of the destruction box and its neighbouring lysine residues in cyclin B for anaphase ubiquitin-dependent proteolysis in fission yeast: Defining the D-box receptor. *EMBO J.* **17:** 5670–5678.
Zachariae W. and Nasmyth K. 1996. TPR proteins required for anaphase progression mediate ubiquitination of mitotic B-type cyclins in yeast. *Mol. Biol. Cell* **7:** 791–801.
Zachariae W., Schwab M., Nasmyth K., and Seufert W. 1998. Control of cyclin ubiquitination by CDK-regulated binding of Hct1 to the anaphase promoting complex. *Science* **282:** 1721–1724.
Zarzov P., Decottignies A., Baldacci G., and Nurse P. 2002. G(1)/S CDK is inhibited to restrain mitotic onset when DNA replication is blocked in fission yeast. *EMBO J.* **21:** 3370–3376.
Zhou B.B. and Elledge S.J. 2000. The DNA damage response: Putting checkpoints in perspective. *Nature* **408:** 433–439.
Zou H., McGarry T.J., Bernal T., and Kirschner M.W. 1999. Identification of a vertebrate sister-chromatid separation inhibitor involved in transformation and tumorigenesis. *Science* **285:** 418–422.

7

Cell Growth

James R. Broach
Department of Molecular Biology
Princeton University
Princeton, New Jersey 08544

The cell cycle depends on cell growth, 128
Johnston G.C., Pringle J.R., and Hartwell L.H. 1977. Coordination of growth with cell division in the yeast *Saccharomyces cerevisiae*. *Exp. Cell Res.* **105:** 79–98.

Ras controls cell growth in response to nutrients, 130
Toda T., Uno I., Ishikawa T., Powers S., Kataoka T., Broek D., Cameron S., Broach J., Matsumoto K., and Wigler M. 1985. In yeast, *RAS* proteins are controlling elements of adenylate cyclase. *Cell* **40:** 27–36.

Ras is not alone in controlling cell growth in response to nutrients, 132
Cameron S., Levin L., Zoller M., and Wigler M. 1988. cAMP-independent control of sporulation, glycogen metabolism, and heat shock resistance in *S. cerevisiae*. *Cell* **53:** 555–566.

Tor has a role in the growth response to nutrients, 134
Barbet N.C., Schneider U., Helliwell S.B., Stansfield I., Tuite M.F., and Hall M.N. 1996. TOR controls translation initiation and early G1 progression in yeast. *Mol. Biol. Cell* **7:** 25–42.

Note: The landmark papers listed above are those discussed in this chapter. Each landmark paper is preceded by the name of the section (with starting page number) where the paper is first discussed in detail.

How does a yeast cell control its growth? Growth of yeast, or of any cell for that matter, comprises two independent but loosely coupled processes: the discontinuous cell cycle events directing duplication and segregation of the cell's genetic material (the cell cycle) and the continuous increase in cell mass doubling the nongenetic components of the cell (the growth cycle). The former process has received enormous attention, which has resulted in an understanding sufficient to allow precise, inclusive, and predictive models of cell cycle progression (Chen et al. 2000), but our understanding of the growth cycle is woefully lacking. In 1977, **Johnston et al. (1977)** bemoaned the fact that "the definition of the events that comprise the growth cycle is little better today than at the time of Swann's review" (Swann first formulated the distinction between the growth cycle and been determined, although as noted below, Ras and Tor signaling has a central role in this process. Thus, these remain fundamental questions to be resolved in the future.

Ras CONTROLS CELL GROWTH IN RESPONSE TO NUTRIENTS

The second set of papers all deal with the means by which the cell coordinates its synthetic apparatus, and thereby the cell cycle machinery, in response to nutrient availability. The first two papers in this set bookend the era of Ras, whereas the third ushers in the era of Tor.

The paper by **Toda et al. (1985)** represents the convergence of two independent lines of investigation. The first line encompassed studies on the role of two genes in yeast that have substantial homology with each other and with the mammalian proto-oncogene *ras*. Initial analysis of these genes indicated that they redundantly performed an essential function, and, since cells lacking both these *RAS* genes arrested as unbudded cells, that Ras's function was cell-cycle-related and required for the transition from G_1 to S. The identification of these genes in yeast raised the attractive prospect of using a genetically tractable microorganism to decipher what role this oncogene has in mammalian tumorigenesis. The prospect was invigorated by the observation that mammalian *ras*-encoded protein could substitute for the yeast Ras proteins in executing their essential functions in the cell. In addition, cells containing a mutant allele of *RAS2* ($RAS2^{G19V}$), analogous to the oncogenic mutation of human *ras*, had growth properties strikingly similar to those of cancer cells. Namely, such yeast cells fail to respond appropriately to growth signals elicited by nutrient availability. Although wild-type haploid cells arrest in G_1 when starved for nutrients, $RAS2^{G19V}$ cells starved for nutrients fail to arrest at G_1 and accumulate at random stages of the cell cycle. In addition, wild-type diploid cells sporulate on nutrient starvation, but $RAS2^{G19V}$ fail to do so. Thus, like cancer cells, $RAS2^{G19V}$ cells fail to heed signals for growth arrest or differentiation.

The second independent line of investigation that preceded the paper by Toda et al. (1985) included work from Matsumoto, Uno, and Ishikawa, who had been studying the role of cAMP in yeast. By a clever series of selections, this group had isolated yeast strains whose growth was dependent on exogenous cAMP and thereby identified the structural gene for adenylyl cyclase, which they designated *CYR1*. Their analysis indicated that the gene was essential for yeast viability and, furthermore, that this essential function was required for the transition from G_1 to S. By additional genetic tricks, they identified the gene, *BCY1* (*b*ypass of *cy*clic AMP requirement), encoding the regulatory subunit of the cAMP-dependent protein kinase (PKA). Since inactivation of *BCY1* suppressed the cell cycle defect of *cyr1* mutants, these authors concluded that cAMP was required for cell cycle progression through its regulation of PKA. This group further demonstrated that loss of *CYR1* function induced meiosis in diploid cells even in the presence of nutrients (a condition that precludes meiosis in wild-type cells) and that inactivation of *BCY1* prevented meiosis even under starvation conditions.

The realization of the remarkable correspondence of phenotypes between *ras1 ras2* and *cyr1* mutants, on the one hand, and between $RAS2^{G19V}$ and *bcy1* mutants, on the other hand, was the "Eureka" moment that spawned the study by Toda et al. (1985). The work reported in this paper thoroughly and definitively demonstrated that the primary essential function of Ras proteins in yeast was to activate adenylyl cyclase. Thus, in one fell swoop, the role of the proto-oncogene homolog in yeast was revealed.

Although this report had a major impact on analysis of cell growth in yeast, it came as a substantial disappointment to those who hoped that yeast would hold the key to understanding the role of Ras in cancer. Although data at the time were inconclusive, they generally indicated that cAMP had little connection to tumorigenic transformation. So, the role of Ras in cancer was unlikely to involve regulation of cAMP production. Indeed, subsequent studies, prompted by genetic analysis in two other model organisms—*Drosophila* and *Caenorhabditis elegans*—demonstrated that *ras* affects growth in human cells primarily through activation of the mitogen-activated protein (MAP) kinase pathway via direct interaction with Raf protein kinase (Marshall 1996). Further analysis has expanded Ras's interacting partners to include phosphatidylinositol (PI) 3-kinase, the guanine nucleotide exchange factor for the small G-protein Ral, and possibly others (Shields et al. 2000). However, no connection to adenylyl cyclase has ever been seen. Thus, in this case, analysis in yeast failed to reveal the function of a highly conserved homolog.

This is not to say that the studies with yeast had no impact on the analysis of mammalian Ras pathway. Although the pathways in which Ras functions have not been conserved from yeast to humans, most of the structural features, regulation of nucleotide binding and biosynthetic routes of the protein, have been conserved. As a consequence, identification of Cdc25 as the exchange factor and Ira1 and Ira2 as the GTPase-activating proteins for yeast Ras allowed identification of the corresponding factors in metazoans. In this way, Sos in *Drosophila* (and mSos in mammals) was recognized as the exchange factor, an observation that helped formulate the tyrosine kinase connection to Ras. Similarly, the sequence relationship of *NF1*, the gene responsible for neurofibromatosis, to Ira1 led to a fundamental understanding of the etiology of this hyperproliferative disease. Furthermore, genetic studies on Ras in yeast yielded innumerable insights into the nature and mechanism of posttranslational processing required for production of functional Ras protein. These insights ultimately yielded targets for novel anticancer therapies that are still being evaluated in the clinic (Johnston 2001; Ohkanda et al. 2002). Thus, despite the initial disappointment attendant on this report, the analysis of Ras in yeast has fulfilled much of the expectation that yeast studies could provide a shortcut in understanding the biology of a human protein.

In the broader context of the control of cell growth, the results of this study highlighted the connection between the growth cycle and the cell cycle, i.e., previous work had indicated that nutrient restriction resulted in cell cycle arrest prior to commitment to the cell cycle. Substantial discussion ensued as to whether the arrest state of the cell under nutrient limitation represented a step in the normal cell cycle progression or a state outside the normal progression accessed during G_1 but distinct from it, referred to as either G_0 or stationary phase. As reinforced in this study, Ras and cAMP mutants are epistatic to the effects of nutrient availability on cell cycle progress (or entry into G_0), thus formally placing these components on the pathway that links nutrient availability to this critical step in the cell cycle. This observation in theory provides a mechanistic underpinning for the G_0 transition, placing PKA-directed phosphorylation at a central position in this transition. However, as we noted previously, we still do not know whether a single master regulator of cell cycle entry (the yeast equivalent of retinoblastoma [Rb]) is the target of PKA activity or, conversely, whether the transition to G_0 occurs simply as the accumulated changes in a number of different metabolic, physiological, and transcriptional processes, each sensitive to regulation by cAMP, heat shock, nutrient availability, etc. Furthermore, as noted in Chapter 8, despite the beautiful work sider the pathway in which product of gene X stimulates product of gene Y to produce a measurable output. We then use two different mutations of X, one that is inactive and one that is hyperactive, and two similar mutations of Y. If we combine the hyperactive mutation of Y with the inactive mutation of X, then we would expect to obtain the output. The other combination—hyperactive X and inactive Y—would give no output. On the basis of these two experiments, we would conclude that X precedes Y in the pathway. However, if a pathway is branched, the genetics becomes more complicated. Consider a signaling pathway as follows:

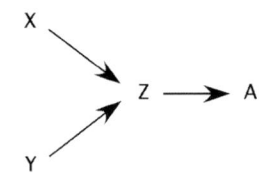

$$X \searrow$$
$$Z \rightarrow A$$
$$Y \nearrow$$

What would be the output (production of A)
- in a strain combining an inactive mutation of Z and a hyperactive mutation of Y?
- of a strain containing a hyperactive mutation of Z and an inactive mutation of X?
- of a strain containing a hyperactive mutation of X and an inactive mutation of Y?

If you know the members of a signal pathway but do not know whether they lie on a branched or linear pathway, can you devise a genetic test (using hyperactive and inactive mutations) that allows you to determine the topology of the pathway (branched or linear)?

REFERENCES

Barbet N.C., Schneider U., Helliwell S.B., Stansfield I., Tuite M.F., and Hall M.N. 1996. TOR controls translation initiation and early G1 progression in yeast. *Mol. Biol. Cell* **7:** 25–42.

Broach J.R. and Deschenes R.J. 1990. The function of ras genes in *Saccharomyces cerevisiae*. *Adv. Cancer Res.* **54:** 79–139.

Brunn G.J., Hudson C.C., Sekulic A., Williams J.M., Hosoi H., Houghton P.J., Lawrence J.C., Jr., and Abraham R.T. 1997. Phosphorylation of the translational repressor PHAS-I by the mammalian target of rapamycin. *Science* **277:** 99–101.

Burnett P.E., Barrow R.K., Cohen N.A., Snyder S.H., and Sabatini D.M. 1998. RAFT1 phosphorylation of the translational regulators p70 S6 kinase and 4E-BP1. *Proc. Natl. Acad. Sci.* **95:** 1432–1437.

Cameron S., Levin L., Zoller M., and Wigler M. 1988. cAMP-independent control of sporulation, glycogen metabolism, and heat shock resistance in *S. cerevisiae*. *Cell* **53:** 555–566.

Chen K.C., Csikasz-Nagy A., Gyorffy B., Val J., Novak B., and Tyson J.J. 2000. Kinetic analysis of a molecular model of the budding yeast cell cycle. *Mol. Biol. Cell* **11:** 369–391.

Dennis P.B., Jaeschke A., Saitoh M., Fowler B., Kozma S.C., and Thomas G. 2001. Mammalian TOR: A homeostatic ATP sensor. *Science* **294:** 1102–1105.

Gingras A.C., Raught B., and Sonenberg N. 2001. Regulation of translation initiation by FRAP/mTOR. *Genes Dev.* **15:** 807–826.

Herman P.K. and Rine J. 1997. Yeast spore germination: A requirement for Ras protein activity during re-entry into the cell cycle. *EMBO J.* **16:** 6171–6181.

Jacinto E. and Hall M.N. 2003. Tor signalling in bugs, brain, and brawn. *Nat. Rev. Mol. Cell Biol.* **4:** 117–126.

Jiang Y. and Broach J.R. 1999. Tor proteins and protein phosphatase 2A reciprocally regulate Tap42 in controlling cell growth in yeast. *EMBO J.* **18:** 2782–2792.

Jiang Y., Davis C., and Broach J.R. 1998. Efficient transition to growth on fermentable carbon sources in *Saccharomyces cerevisiae* requires signaling through the Ras pathway. *EMBO J.* **17:** 6942–6951.

Johnston G.C., Pringle J.R., and Hartwell L.H. 1977. Coordination of growth with cell division in the yeast *Saccharomyces cerevisiae*. *Exp. Cell Res.* **105:** 79–98.

Johnston M. 1999. Feasting, fasting and fermenting. Glucose sensing in yeast and other cells. *Trends Genet.* **15:** 29–33.

Johnston S.R. 2001. Farnesyl transferase inhibitors: A novel targeted therapy for cancer. *Lancet Oncol.* **2:** 18–26.

Jorgensen P., Nishikawa J.L., Breitkrentz B.J., and Tyers M. 2002. Systematic identification of pathways that couple cell growth and division in yeast. *Science* **297:** 295–400.

Loewith R., Jacinto E., Wullschleger S., Lorberg A., Crespo J.L., Bonenfant D., Oppliger W., Jenoe P., and Hall M.N. 2002. Two TOR complexes, only one of which is rapamycin sensitive, have distinct roles in cell growth control. *Mol. Cell* **10:** 457–468.

Marshall C.J. 1996. Ras effectors. *Curr. Opin. Cell Biol.* **8:** 197–204.

Mitchison J.M. 1971. *The biology of the cell cycle*. Cambridge University Press, Cambridge, United Kingdom.

Ohkanda J., Knowles D.B., Blaskovich M.A., Sebti S.M., and Hamilton A.D. 2002. Inhibitors of protein farnesyltransferase as novel anticancer agents. *Curr. Top. Med. Chem.* **2:** 303–323.

Pringle J.R. and Hartwell L.H. 1981. The *Saccharomyces cerevisiae* cell cycle. In *The molecular biology of the yeast* Saccharomyces: *Life cycle and inheritance* (ed. J.N. Strathern et al.), pp. 97–142. Cold Spring Harbor Laboratory, Cold Spring Harbor, New York.

Schmidt A., Bickle M., Beck T., and Hall M.N. 1997. The yeast phosphatidylinositol kinase homolog TOR2 activates RHO1 and RHO2 via the exchange factor ROM2. *Cell* **88:** 531–542.

Shields J.M., Pruitt K., McFall A., Shaub A., and Der C.J. 2000. Understanding Ras: "It ain't over 'til it's over." *Trends Cell Biol.* **10:** 147–154.

Swann M.M. 1957. The control of cell division —A review. 1. General mechanisms. *Cancer Res.* **17:** 727–757.

Thompson-Jaeger S., Francois J., Gaughran J.P., and Tatchell K. 1991. Deletion of SNF1 affects the nutrient response of yeast and resembles mutations which activate the adenylate cyclase pathway. *Genetics* **129:** 697–706.

Toda T., Uno I., Ishikawa T., Powers S., Kataoka T., Broek D., Cameron S., Broach J., Matsumoto K., and Wigler M. 1985. In yeast, *RAS* proteins are controlling elements of adenylate cyclase. *Cell* **40:** 27–36.

8

Differentiation: Mating and Filamentation

George F. Sprague, Jr.
Institute of Molecular Biology
University of Oregon
Eugene, Oregon 97403-1229

Cell type: Genetics as the entrée, 142

Strathern J., Hicks J., and Herskowitz I. 1981. Control of cell type in yeast by the mating type locus: The α1-α2 hypothesis. *J. Mol. Biol.* **147:** 357–372.

Cell type: Molecular redux, 145

Bender A. and Sprague G.F., Jr. 1987. MATα1 protein, a yeast transcription activator, binds synergistically with a second protein to a set of cell-type-specific genes. *Cell* **50:** 681–691.

Cell type: Lessons from pleiotropy, 148

Keleher C.A., Redd M.J., Schultz J., Carlson M., and Johnson A.D. 1992. Ssn6-Tup1 is a general repressor of transcription in yeast. *Cell* **68:** 709–719.

Mobile mating-type information: The cassette hypothesis, 148

Hicks J.B. and Herskowitz I. 1977. Interconversion of yeast mating types. II. Restoration of mating ability to sterile mutants in homothallic and heterothallic strains. *Genetics* **85:** 373–393.

Filamentation: Hunting for food, 152

Gimeno C.J., Ljungdahl P.O., Styles C.A., and Fink G.R. 1992. Unipolar cell divisions in the yeast *S. cerevisiae* lead to filamentous growth: Regulation by starvation and *RAS*. *Cell* **68:** 1077–1090.

Note: The landmark papers listed above are those discussed in this chapter. Each landmark paper is preceded by the name of the section (with starting page number) where the paper is first discussed in detail.

YEAST CELLS HAVE THE CAPACITY TO ADOPT DIFFERENT DEVELOPMENTAL FORMS. Vegetative cells growing in the presence of abundant nutrients are ovoid and produce new buds in a predictable pattern. However, if these vegetative cells receive the appropriate extracellular cue, they are induced to differentiate and express new cellular phenotypes. One such cue governs the sexual reproductive cycle. Vegetative haploid **a** or α cells are induced to become mating-proficient gametes when exposed to the pheromone produced by the opposite mating type. This pheromone-induced differentiation event has three facets: the transcription of genes whose products participate in mating increases, budding

---

Note: Boldfaced references in the text denote landmark papers that are on the accompanying CD.

and progression through the cell cycle ceases, and cell growth orients toward the perceived mate, resulting in the formation of a pear-shaped cell called a "shmoo." The diploid zygote grows vegetatively as long as abundant nutrients are present, again with a characteristic ovoid morphology and a predictable budding pattern. However, when starved for nitrogen and provided with a nonfermentable carbon source, diploid cells are induced to undergo a differentiation program that culminates in meiosis and spore development.

Another cue, nutrient limitation, governs the budding pattern and overall pattern of growth in both haploid and diploid cells. The precise cue—i.e., the relevant limiting nutrient—may be different for haploid and diploid cells, but the differentiation program is very similar. The budding pattern changes, the new cells are elongated, and they adhere to one another. The result is a filamentous string of cells growing away from the original nutrient-challenged cell as if foraging for nutrients.

In this chapter, I focus on two of the differentiation programs outlined above, namely, the mating program and the filamentation program, and discuss seminal papers that define the problem and set the stage for achieving a molecular understanding. In the case of the mating program, I concentrate on papers that explore how the two distinct haploid mating types, or cell types as they are sometimes called, are established and what regulatory strategies and mechanisms endow each type with unique developmental potential. The third differentiation program, meiosis and spore development, is the focus of Chapter 9.

CELL TYPE: GENETICS AS THE ENTRÉE

Wild strains of *Saccharomyces cerevisiae* are homothallic; i.e., when a haploid spore germinates and grows into a colony, most—if not all—of the cells in that colony are diploid as a result of a mating process described below (more on homothallism later). The development of yeast as a genetic organism required the isolation of heterothallic variants in which the haploid state was stable, and the investigator could therefore control the life cycle. Lindegren achieved this in the 1940s (Lindegren and Lindegren 1943) and demonstrated that there were two distinct mating types, dubbed **a** or α. Over the years, it became apparent that the **a** and α phenotypes are complex. Each mating type secretes a unique pheromone, and it was inferred that each cell type must produce a unique receptor. For example, **a** cells must synthesize a receptor that enables them to respond to α factor, the pheromone secreted by α cells. Each cell type also produces a unique agglutinin protein; interaction of the **a** and α agglutinins promotes adhesion of the mating pair. Finally, **a** cells secrete an activity, initially termed Barrier, that is now known to be a protease that degrades α factor. Despite this complexity, the **a** and α cellular phenotypes were shown to be determined by a single locus, *MAT*, which could be mapped to a discrete location on chromosome III of the yeast genome: **a** cells have the *MAT***a** allele, α cells have the *MAT*α allele, and **a**/α diploid cells have both alleles.

How does a single genetic locus determine these complex phenotypes? At the extreme, one can imagine two models. In the Structural Gene Model, the *MAT* alleles themselves are highly complex, each harboring the structural genes for the many proteins produced uniquely by **a** or α cells. In the Regulatory Gene Model, the *MAT* alleles are much less complex, encoding one or a small number of transcription regulatory proteins, which in turn govern the transcription of the structural genes whose expression is restricted to **a** or α cells. Stated another way, the structural genes for each of the cell-type-specific proteins are present in all three cell types: **a**, α, and **a**/α cells. Whether the structural genes are expressed is determined by the genotype at the *MAT* locus.

The pioneering work of MacKay and Manney (1974a,b) provided the first hint that the Regulatory Gene Model applied, and also provided the mutants that would ultimately lead to an elegant model for how the *MAT* alleles govern cell type. MacKay and Manney isolated mating-defective mutants, and then carried out genetic analysis of these mutants. Although the mutants are mating defective, they do mate at low frequencies ($\sim 10^{-6}$), and it is possible to identify "rare mating" events by demanding complementation of nutritional deficiencies present in the strains being mated. For example, one haploid strain might require histidine and lysine, the other arginine and tryptophan. The rare diploid that is formed will be able to grow on unsupplemented minimal medium and thereby be identified easily. Mutations deficient in mating defined *STE* (sterile) genes. A few of the mutations, isolated in an α background, appeared to be alleles of *MAT*α. However, most mutations were unlinked to the mating-type locus. Among the mutations unlinked to the *MAT* locus were some that affected mating only by α cells; i.e., *MAT*α cells harboring such a mutation (e.g., *ste3*) were mating defective, whereas *MAT***a** cells harboring the same mutation were mating proficient. A reasonable hypothesis is that such α-specific *STE* genes might encode one of the α-specific proteins described above, e.g., the α-factor pheromone or the receptor for **a** factor. Other mutations identified an analogous set of **a**-specific *STE* genes. From these analyses, MacKay and Manney outlined a model that proposed that the *MAT* locus was regulatory and the unlinked *STE* genes were targets of the *MAT*-encoded regulatory activities.

Herskowitz and colleagues began with the rough outline offered by MacKay and Manney and generated a satisfying and pleasingly simple hypothesis for the control of cell type by the *MAT* locus, a hypothesis that evoked three regulatory proteins—two encoded by *MAT*α and one by *MAT***a** (**Strathern et al. 1981**). The hypothesis emerged from the masterly application of the standard, but powerful, tools of genetic analysis: close examination of the phenotype, complementation analysis, and double-mutant analysis. MacKay and Manney had demonstrated that one mutant contained a lesion at *MAT*α. They speculated that a second mutant did as well because diploids formed by mating it to **a** cells were deficient in sporulation. One explanation for this apparent dominant sporulation defect is that the mutation disables a function of *MAT*α required for sporulation. Herskowitz and colleagues confirmed this supposition by tetraploid genetic analysis and also identified two other mutants from the MacKay and Manney collection that mapped to *MAT*α. Most important was the observation that the *mat*α mutants fell into two distinct phenotypic classes. Two of the *mat*α mutations prevented expression of any known α-specific trait. The other two conferred an unusual phenotype: They expressed at least one **a**-specific trait, the Barrier (hereafter called Bar) protease mentioned above, and they had the ability to mate inefficiently as **a** cells. All four mutations were recessive to wild-type *MAT*α. Complementation analysis suggested that the two mutations that prevented expression of α-specific activities formed one complementation group (referred to as *MAT*α*1*) and the two mutations that conferred expression of the Bar protease formed a second complementation group (referred to as *MAT*α*2*).

The observation of complementation by mutations in the genes, now referred to as *MAT*α*1* and *MAT*α*2*, required additional experimental support. After all, a diploid that

tion to the expression of **a**-specific genes as well. Deletion analysis of the control region from the α-specific *STE3* gene had identified a 43-bp element that was sufficient to confer α-specific expression (Jarvis et al. 1988). Comparison of this 43-bp sequence to the sequence from other α-specific genes revealed similar sequences in their presumptive upstream control regions and pointed to a 26-bp sequence as the likely α-specific control element. The consensus sequence that was developed had two notable features. On one side was a 16-bp element that had vestiges of palindromic character. Adjacent to this P element, as it was called, was a 10-bp Q element. Ultimately, synthetic versions of these elements were constructed and tested for their ability to function as upstream activation sequences (UASs). The QP element from *STE3* was able to function as an α-specific UAS. The Q element alone had no UAS activity, nor did the P element as it exists at *STE3*. Unexpectedly, however, a perfect palindromic version of the P element [P(PAL)] had UAS activity. This UAS activity was not limited to the α cell type; it could also drive expression in **a** and **a**/α cells (summarized in Table 1). These UAS studies suggested that α-specific expression required two components, α1 and a second factor present in all three cell types. The second factor could drive transcription from symmetrical P elements, but not from the asymmetrical P elements that actually existed at α-specific genes. Transcription from these asymmetrical elements required the adjacent Q sequence and α1 protein.

Contemporaneously with the studies that defined the nature of the *cis*-acting element that confers α-specific transcription, DNA-binding studies were carried out using α1 protein expressed in *Escherichia coli*. The findings mirrored those obtained for the UAS elements just summarized (Table 1). α1 prepared from *E. coli* could not bind to QP(*STE3*), nor could proteins present in an extract of yeast **a** cells. However, when α1 and the **a** cell extract were mixed, a protein-DNA complex could be detected on QP(*STE3*). No complex could be detected on QP(*STE3*), regardless of the cell type that served as a source of the extract, whereas a complex was formed on P(PAL) using any extract. Moreover, P(PAL) DNA could compete for formation of a protein DNA complex on QP(*STE3*). The implication is that a general factor, one present in all three cell types, is capable of activating transcription when bound to P(PAL) elements and can bind to the P elements present at α-specific genes only through cooperative interactions with α1 protein, interactions that also require the adjacent Q element (Fig. 2).

The model developed above suggests that α1-specific transcription is achieved by the α1-dependent recruitment of a general transcription factor to sites where the general factor cannot bind alone. It is possible, however, that α1 does more than simply recruit this general factor; it has been suggested that α1 induces a conformational change in the gen-

TABLE 1. Summary of UAS activity and protein-DNA complex formation

	UAS activity		Protein-DNA complex formation		
	α cell	**a** cell	α1 alone	Mcm1 alone	α1 and Mcm1
QP(*STE3*)	+	–	–	–	+
Q	–	–	–	–	–
P(*STE3*)	–	–	–	–	–
P(PAL)	+	+	–	+	+[a]

[a] A complex is formed in P(PAL), but it does not contain α1.

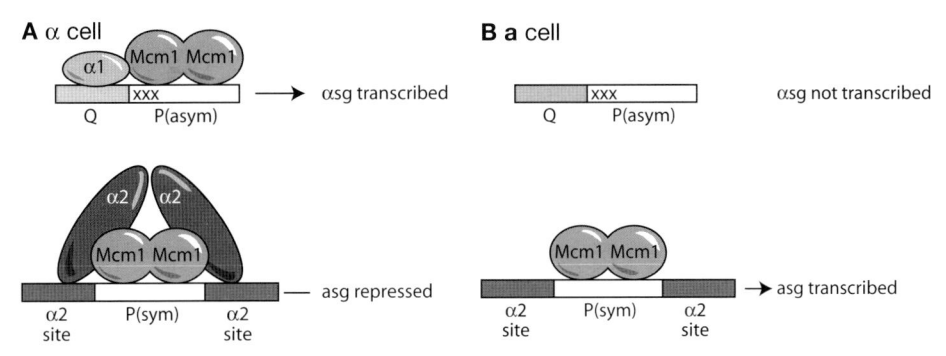

FIGURE 2. (*A*) Molecular model for cell-type-specific transcription in α cells. The DNA-binding sites for proteins encoded by the mating-type locus are shown for the upstream control regions of α-specific and **a**-specific genes. α1 binds cooperatively with Mcm1 to the QP (asym) sequence that functions as the upstream activation sequence (UAS) for α*sg*. QP(asym) is a generic name for the QP sequences at α-specific genes; the asymmetric nature of the P sequences is denoted XXX. α2 binds cooperatively with Mcm1 to operator sites present in the upstream control regions of **a***sg*. The P sequence of the operator is highly symmetric [P(sym)]. The α2•Mcm1 complex brings about repression of the gene set by recruiting Ssn6 and Tup1. (*B*) Molecular model for cell-type-specific transcription in **a** cells. The absence of α1 protein precludes the interdependent binding of α1 and Mcm1 to QP(asym) sites and hence α-specific genes are not transcribed. In contrast, Mcm1 binds well to the P(sym) sequence of **a***sg* and contributes to their transcription. Repression does not occur because α2 is not present to bind and recruit Ssn6 and Tup1.

eral factor to make it competent to activate transcription. Two observations support this view. First, some QP elements have been shown by in vivo footprinting in **a** cells to be bound by protein, presumably the general factor, yet no transcription of these genes occurs (G. Ammerer, pers. comm.). Second, the protease sensitivity of the general factor (which is identified as Mcm1; see below) is different when it is present in a protein-DNA complex that includes α1 than when it alone is bound to DNA (Tan and Richmond 1990).

What is the identity of this general factor, and what other roles might it have? Here, the story turns to α2-mediated repression of **a**-specific genes. Johnson and Herskowitz (1985) had shown that α2 binds to a 32-bp operator upstream of the **a**-specific *STE6* gene. Johnson and colleagues then went on to show that this 32-bp operator is compound in nature. α2 can bind alone to the ends of this operator sequence, but a different protein was found to bind to the center (Keleher et al. 1988). This second protein is present in all three cell types and, like α2, can bind alone to the operator. Together, however, the two proteins bind cooperatively with a 50-fold increase in affinity. Moreover, mutational studies show that the binding of both proteins is required to bring about α2-mediated repression (Fig. 2A). The center of the operator is remarkably similar to the P sequences defined above in the discussion of α-specific control elements, suggesting that the same general factor may be working with α1 and α2. Indeed, work from my lab, the Johnson lab, the Tye lab, and the Ammerer lab has subsequently shown that this general factor is the Mcm1 protein (Jarvis et al. 1989; Keleher et al. 1989; Ammerer 1990), a defining member of the so-called "MADS" family of proteins, and also includes the mammalian serum response factor and two homeotic proteins from *Arabidopsis*. Thus, an intriguing combinatorial strategy is at play in the determination of yeast cell type. Mcm1 interacts cooperatively with α1 to bring about transcription activation of α-specific genes, and Mcm1 interacts cooperatively with α2 to bring about repression of **a**-specific genes. How α1•Mcm1 brings about transcription activation is not understood, but is an intriguing problem for future research especially given that neither α1 nor Mcm1 have the standard hallmarks of transcription activation proteins. How α2•Mcm1 brings about transcription repression is the topic of the next paper in this chapter.

CELL TYPE: LESSONS FROM PLEIOTROPY

α2-mediated repression provided a context for a substantial breakthrough in understanding general repression mechanisms. The yeast literature from the 1970s and 1980s contains a number of examples where the same two genes were identified repeatedly in mutant isolation schemes designed to probe very different biological problems. Today, these genes are commonly called *SSN6* and *TUP1*, but each has many synonyms reflecting the different settings in which mutant versions of the gene have been found. Mutations in these genes influence carbon source utilization, the ability to undergo meiosis and sporulation, the ability to take up extracellular thymidylic acid, and many other physiological processes. However, the most fascinating phenotype—at least from the perspective of our story—is that mutations in these genes also confer an **a**-specific *STE* phenotype. What can account for such puzzling pleiotropy, such a dizzying array of phenotypes? Taking a cue from the **a**-specific *STE* phenotype, the Carlson and Johnson labs speculated that these proteins might be involved in α2-mediated repression (**Keleher et al. 1992**). Indeed, they showed that Ssn6 could bind to α2 protein and that this interaction was required for α2-mediated repression. Moreover, they were able to create a situation in which Ssn6•Tup1 could bring about repression in the absence of α2. A fusion of Ssn6 to LexA, a bacterial DNA-binding protein, conferred repression on a reporter construct when a LexA DNA-binding site was placed adjacent to the UAS element. Moreover, in this setting, repression required Tup1 as well. Thus, these findings take Ssn6•Tup1 out of the relatively small world of α2-mediated repression and suggest that Ssn6•Tup1 might serve as a general repressor. In this view, Ssn6•Tup1 is delivered to target genes via a protein-protein interaction with site-specific DNA-binding proteins, such as α2. This general idea has been verified. For example, Ssn6•Tup1 has been shown to interact with Mig1, a repressor of glucose-sensitive genes, with Rox1, a repressor of oxygen-sensitive genes, and with **a**1•α2, a repressor of *RME1*. In turn, this appreciation of the general role of Ssn6•Tup1 has motivated studies on the mechanism of its action. In some cases, repression is brought about by interactions between Ssn6•Tup1 and histone proteins or histone deacetylases, which lock a nucleosome over the TATA box of the target gene. In other cases, repression involves interactions between Ssn6•Tup1 and subunits of the RNA polymerase II mediator complex.

MOBILE MATING-TYPE INFORMATION: THE CASSETTE HYPOTHESIS

The strains of yeast used commonly in the laboratory have a stable cell type: An **a** cell produces exclusively daughter **a** cells, an α cell exclusively daughter α cells, and so on. Wild yeast cells, however, are homothallic. A spore isolated from a tetrad has a particular mating type, either **a** or α, but after a few cell division cycles, some of the cells now have the opposite mating type. Cells within this microcolony mate to form **a**/α diploids, and, hence, by the time the colony is visible, most of the cells are diploid. (In fact, there are well-described rules for the pattern of mating-type switching, rules that describe which cells have the potential to switch, at what frequency they will switch, and so on [Strathern and Herskowitz 1979].) How do yeast cells achieve this remarkable switch in mating type? Lindegren and Lindegren (1943) identified a variant in which this switching process did not occur: For example, a spore that was born an **a** cell produced a clone composed exclusively of **a** cells. This variant identified the *HO* gene. Wild strains carry the wild-type *HO* allele and are homothallic. The variant strains carry *ho*, have a stable mating type, and are referred to as heterothallic. When the homothallic strain studied by Lindegren was put through meiosis, tetrads were dissected, and the spores were allowed to grow into colonies, all four spores consistently exhibited the nonmating, sporulation-proficient phenotype of **a**/α diploids. In the 1970s, two other genes required for mating-type switching were identified through the study of other wild *Saccharomyces* strains (Santa Maria and Vidal 1970; Naumov and Tolstorukov 1973). (At the time, they were considered different species, but in the current view, they would be considered simply strains of *S. cerevisiae*.) These switching-defective *Saccharomyces* strains showed different segregation patterns. One strain segregated two spores that gave rise to nonmating sporulation-proficient diploids and two spores that gave rise to colonies with the α mating phenotype. These latter segregants could be shown to contain the *HO* allele, implying that they were defective at a different locus, one required for switching from α to **a**. The other strain had the complementary phenotype; i.e., two segregants in a tetrad gave rise to colonies that had the nonmating, sporulation-proficient phenotype and two spores gave rise to colonies that had the **a** mating phenotype. Again, these segregants could be shown to contain the *HO* allele and were therefore inferred to be defective at a locus that was required to switch from **a** to α. The nomenclature that was devised to symbolize these new loci is virtually impenetrable. I refer to them simply as *HM* loci and have provided a cheat sheet (Table 2) that connects this old nomenclature to the modern nomenclature, so that the presented paper can be read and interpreted. In an Herculean mapping effort—Herculean both because of the difficulty of the assay and because of the paucity of genetic markers—Oshima and Takano (1971) mapped these two *HM* loci to opposite arms of chromosome III, near each telomere. Recall that the *MAT* locus is also on chromosome III, on the right arm as it is normally drawn.

Herskowitz and colleagues used the *HM* mapping result in formulating the cassette hypothesis for mating-type switching, or interconversion. They also used a second piece of information: the properties of a strain containing a large deletion on chromosome III, which had been created by D. Hawthorne (1963). Hawthorne had carried out forced mat-

TABLE 2. Cheat sheet connecting old and new nomenclature

Nomenclature		Mating-type information
old	current	
*HM*α	*HMR***a**	**a**
*hm*α	*HMR*α	α
*HM***a**	*HML*α	α
*hm***a**	*HML***a**	**a**

ings between two different α strains and studied the properties of the diploids that were created. One of these diploids behaved phenotypically as if it were an **a**/α diploid; i.e., it had a nonmating phenotype and was sporulation-proficient. When put through meiosis and sporulation, the tetrads produced by this diploid always yielded two live spores that had the α mating phenotype and two dead spores. These dead spores were shown to have the **a** mating phenotype by placing them next to α cells and observing zygotes form. (The "dead" spores were able to mate before they actually died.) Hawthorne showed that the dead spores contained a deletion, which extended from the mating-type locus rightward past *THR4*, but not as far as *MAL2*. The extensive nature of the deletion presumably accounts for the lethality of the haploid strain carrying this chromosome. From the perspective of the cassette hypothesis, however, the interesting aspect of the so-called "Hawthorne's deletion" was the paradox that it created. *MAT***a** and *MAT*α, as has already been discussed, are codominant alleles, implying that both have a function in establishing the **a**/α phenotype. Hawthorne's deletion, however, appeared to convert an α cell to an **a** cell by deletion. The paradox is: How can a deletion be codominant?

Herskowitz and colleagues used these two pieces of information, coupled with a good dose of inspiration, to develop the cassette hypothesis (Fig. 3) (Hicks et al. 1977). According to this hypothesis, mating-type information resides not only at the *MAT* locus (as *MAT***a** or *MAT*α alleles), where it can be expressed, but also at the *HM* loci, where the information is not capable of being expressed. In most strains of *S. cerevisiae*, the *HM* locus on the left arm of chromosome III contains α information and is therefore called *HML*α. The locus on the right arm of chromosome III contains **a** information and is referred to as *HMR***a**. During mating-type interconversion, for example, when a *MAT*α cell is going to switch to a *MAT***a** cell, the information at the *MAT* locus is in some way excised and destroyed and replaced with a copy of the information from (in this case) the *HMR***a** locus. This hypothesis accounts for the switching properties of the variant *Saccharomyces* strains that identified the *HM* loci. For example, the strain that cannot switch from α to **a** has a defective *HMR* locus. It does not have **a** information at the HMR locus (it actually has α information; the *HMR*α allele). The hypothesis also resolves the paradox raised by Hawthorne's deletion: The deletion extends from the *MAT* locus to *HMR***a** and, in essence, fuses the *MAT* locus to *HMR*.

The selected paper describes an elegant experiment that was an important test of this hypothesis and follows from the ability to carry out pedigree analysis in which the switching properties of a given cell can be followed over several generations (**Hicks and Herskowitz 1977**). This pedigree analysis led to the establishment of rules for mating-type interconversion. Imagine a spore that contains *MAT*α and *HO*. This spore germinates and produces a daughter cell via mitosis. After the first cell division cycle, both the spore and the daughter are α. By micromanipulation, the spore and daughter are separated and allowed to go through another cell division cycle. In 80% of the second cell division cycles, the spore and the second daughter switch mating type and are now **a**. The first daughter and its daughter never switch—they retain the α state. Pedigree analysis can be extended for a number of generations, simply by separating the cells via micromanipulation. Once the spore has undergone a switching event, it can switch as often as every subsequent cell division. A newborn daughter cell never switches in its first cell division cycle but, once it has undergone a cycle, it can begin switching, again as often as every cell division cycle.

Cassette Hypothesis

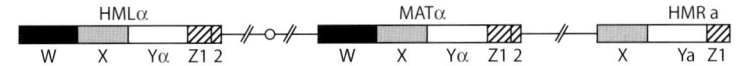

```
      HMLα                         MATα                        HMR a
   ■■■ ▓▓ □ ▧▧  //—o—//       ■■■ ▓▓ □ ▧▧   //—//        ▓▓ □ ▧
   W   X  Yα Z1 2              W   X  Yα Z1 2              X  Ya Z1
```

Controlling Element Hypothesis

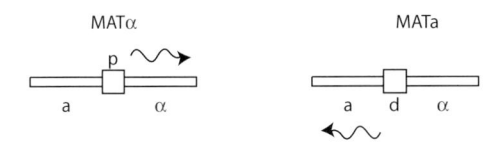

```
        MATα                              MATa
              p ↝
      ═══□═══                          ═══□═══
       a    α                           a  d  α
                                         ↜
```

FIGURE 3. The cassette and controlling element models for mating-type interconversion. (*Top*) The cassette model. The structures of *HML*α, *MAT*α, and *HMR***a** are shown. The cassettes are divided into five regions—W, X, Yα, Z1, and Z2—which represent regions of identity among the cassettes based on heteroduplex mapping of cloned DNA. Both *MAT* and *HML* have the W and Z2 regions, whereas *HMR* does not. The Y region is nonhomologous between **a** and α information, and includes the promoter, transcription start sites for the divergently transcribed *MAT* transcripts. During mating-type interconversion, the information at *MAT* is excised and replaced with a copy of the information from either *HML* or *HMR*. (*Bottom*) Controlling element model. The *MAT* locus contains both **a** and α information. Which information is expressed depends on the orientation of a controlling element, here denoted by an open box. The letter "P" indicates the orientation of the element. In this model, mating-type interconversion results from the inversion of the controlling element, a process governed by the *HO* and the *HM* loci.

Herskowitz and colleagues realized that the ability of the cell to switch repeatedly provided an opportunity to test the notion that *HML*α indeed contained α information, which was used when an **a** cell switched to α. They reasoned that if a spore containing both a *mat*α*1* mutation and the *HO* allele was allowed to go through a series of mating-type interconversion events, first switching to **a** using *HMR***a** as the source of information and then back to α using *HML*α, the original *mat*α*1* mutation could be "healed." In a competing model, inspired by examples of mobile genetic elements such as flippable promoters in bacterial systems (Shapiro 1983), the *MAT* locus might contain both **a** and α information, and some sort of controlling element whose orientation determines which information is expressed (Fig. 3). In this model, mating-type switching would involve a reorientation of the controlling element, and this reorientation would be controlled by the *HM* loci. In this model, then, a *mat*α*1* mutation would not be healed by switching events. The spore would switch to **a** by inversion of the controlling element, but in the next round of switching, the controlling element would once again drive expression of the mutant *mat*α1 allele. In the paper provided in this section, healing was observed, supporting the cassette hypothesis. Subsequently, the Herskowitz lab (Kushner et al. 1979) and Klar and Fogel (1979) performed "wounding" experiments. They isolated mutations that mapped to *HM* loci. When these mutant versions of the *HM* loci were present in strains undergoing mating-type switching, the new allele created at *MAT* was defective. Together, these healing and wounding experiments provided strong genetic support for the cassette hypothesis. The model was ultimately proven when the *MAT* locus and the *HM* loci were cloned, demonstrating that the *HM* loci indeed contained the same **a** and α information that could reside at the *MAT* locus (Hicks et al. 1979).

What insights has the cassette hypothesis afforded in the ensuing years? One line of investigation has sought to understand the actual mechanism of mating-type interconversion. *HO* encodes a site-specific endonuclease that makes a double-strand break within the *MAT* locus (Kostriken et al. 1983), and this double-strand break then initiates a repair process using one of the *HM* loci as a template. A second line of investigation has sought to understand the rules of switching, which were briefly outlined above. Only cells that have previously been mothers can undergo switching, and switching events always generate two cells with changed mating-type information, the latter observation implying that switching occurs in the G_1 phase of the cell cycle before DNA replication. The cell cycle dependence of mating-type switching is due to G_1-specific transcription of *HO*, which has one of the largest upstream control regions known for yeast genes (Nasmyth 1985a,b). The effort to understand this G_1-specific transcription led to the identification of the SBF and MBF transcription factors and the Swi/Snf chromatin-remodeling complex. The effort to understand the mother-daughter asymmetry switching potential led to the identification of the first example of localized mRNA in yeast. Ash1 mRNA is specifically localized to the daughter cell and specifies a repressor of *HO* transcription. Finally, the effort to understand why the *HM* loci are silent, rather than expressed, led to the identification of the *SIR* genes and the roles of their products in establishing regions of heterochromatin in the yeast genome.

FILAMENTATION: HUNTING FOR FOOD

In 1992, the Fink lab discovered that yeast cells, and the colonies that they produced, could adopt a mode of growth different from the vegetative style familiar to yeast researchers (**Gimeno et al. 1992**). Instead of forming ovoid buds that separate cleanly from the mother cell, under conditions of nutrient limitation, yeast cells grew as filaments in which the daughter cells remain adhered to the mother cell (Fig. 4). A report of this phenomenon had appeared earlier (Guilliermond 1920), so the Fink lab discovery might be considered a rediscovery. In any event, the important part of the Fink lab contribution is not the discovery of the phenomenon, but rather the careful description of the physiological events that characterize the phenomenon and the investigation of the molecular mechanisms that underlie these physiological events.

The phenomenon was first discovered in a particular diploid strain (Σ1978b) as a response to nitrogen limitation. Three gross morphological changes characterize the switch from vegetative growth to filamentous growth. First, upon nutrient limitation, buds that were produced were considerably elongated, compared to the ovoid cells produced under vegetative conditions. Second, the budding pattern of the cells changed. Diploid cells normally bud in a bipolar fashion, in which either of the two poles of the ovoid mother cell is equally likely to be the site of emergence of a new bud. In the filamentous mode, cells bud in a unipolar fashion: A new bud always emerges at the pole opposite the birth scar of the mother cell. Third, the cells show increased tendency to adhere to each other. This property, coupled with the change in budding pattern, results in a chain of cells that remain connected. The change to filamentous growth in diploid cells has been termed "pseudohyphal" development.

Haploid cells undergo a similar switch from vegetative to filamentous growth upon nutrient limitation. As for diploid filamentation, the cells become elongated, there is a change in budding pattern, and the daughter cells tend to adhere to one another. Subtle dif-

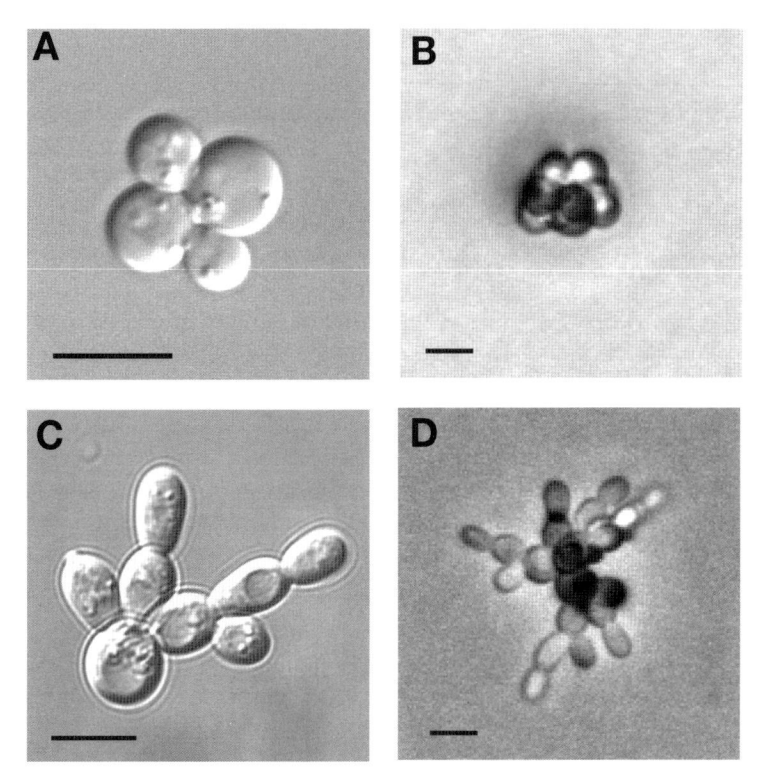

FIGURE 4. Comparison of yeast form and filamentous form growth. Microcolonies of cells growing in the yeast form (*A* and *B*) or filamentous form (*C* and *D*) are shown. Bar, 10 μm.

ferences distinguish the diploid and haploid phenomena, however. First, glucose limitation, rather than nitrogen limitation, is thought to be the trigger in the case of haploid filamentation, although diploid cells have been seen to show features of filamentous growth following glucose limitation, so this distinction may not be real. Second, haploid cells—but not diploid cells—gain the ability to invade the agar substrate.

The developmental switch from vegetative to filamentous form is a fascinating phenomenon, and it presents several intriguing molecular puzzles that remain to be solved. Two of these puzzles center on signal transduction pathways that become activated upon nutrient limitation and that are required for the developmental transition. At least three pathways are known to be required: the Ras/PKA pathway, the Snf1 glucose-sensing pathway, and a pathway that includes elements of the pheromone response pathway. One question, then, is how these distinct signals are integrated to give the overall physiological response. Recent studies show that one point of integration is the transcriptional control region for the gene *FLO11*. This gene encodes a surface glycoprotein, adhesin, that is required for filamentous growth. The control region for this gene is unusually large and contains binding sites for transcription factors known to be targets of these distinct pathways. Presumably, there are other points of integration as well. The second signal transduction puzzle concerns the participation of some components from the pheromone pathway, in particular, several of the protein kinases from the MAP kinase cascade (Roberts and Fink 1994). Given this sharing of components between the pheromone and filamentous pathway, how is the specificity of pathway signaling achieved? Other challenges for the future include understanding how these (or other) signal transduction pathways interface with the bud-site selection machinery to confer unipolar budding, and understanding how these pathways interface with the cell cycle and cell polarity machinery to cause a change in cell shape.

STUDY QUESTIONS

1. Imagine that you have identified (by EMSA [electrophoresis mobility shift assay]) a protein-DNA complex that forms on a UAS that you are interested in. Further imagine that you know that Yfp is required for formation of the complex (no complex is detected if extracts are prepared from *yfp* mutants). What sorts of experiments, short of purifying Yfp, could be done to determine whether Yfp is actually part of the complex that you have detected?

2. According to the α1-α2 hypothesis, α1 is a positive regulator of αsg. What is the genetic argument that it is a positive (in a formal sense) rather than a negative regulator? Need α1 be a direct positive regulator of αsg, as we now know to be the case? On the basis of the genetic evidence that Herskowitz et al. had at hand (Strathern et al. 1981), could they eliminate the possibility that α1 was a negative regulator of a repressor (call it Rep1) of αsg? What would be the phenotype of a *MAT*α *rep1* strain? A *mat*α*1 rep1* strain? A *MAT***a** *rep1* strain?

3. Diagram the pedigree of cell divisions for the healing experiment and for a wounding experiment.

4. What might the role of the *HM* loci be in the controlling element model of mating-type switching? The integration and excision of bacteriophage λ may be useful to think about.

REFERENCES

Ammerer G. 1990. Identification, purification, and cloning of a polypeptide (PRTF/GRM) that binds to mating-specific promoter elements in yeast. *Genes Dev.* **4:** 299–312.

Bender A. and Sprague G.F., Jr. 1987. MATα1 protein, a yeast transcription activator, binds synergistically with a second protein to a set of cell-type-specific genes. *Cell* **50:** 681–691.

Gimeno C.J., Ljungdahl P.O., Styles C.A., and Fink G.R. 1992. Unipolar cell divisions in the yeast *S. cerevisiae* lead to filamentous growth: Regulation by starvation and RAS. *Cell* **68:** 1077–1090.

Goutte C. and Johnson A.D. 1988. **a**1 protein alters the DNA binding specificity of α2 repressor. *Cell* **52:** 875–882.

Guilliermond A. 1920. *The yeasts*. John Wiley and Sons, New York.

Hawthorne D.C. 1963. A deletion in yeast and its bearing on the structure of the mating type locus. *Genetics* **48:** 1727–1729.

Hicks J.B. and Herskowitz I. 1977. Interconversion of yeast mating types. II. Restoration of mating ability to sterile mutants in homothallic and heterothallic strains. *Genetics* **85:** 373–393.

Hicks J.B., Strathern J.N., and Herskowitz I. 1977. The cassette model of mating-type interconversion. In *DNA insertion elements, plasmids and episomes* (ed. A.I. Bukhari et al.), pp. 457–462. Cold Spring Harbor Laboratory, Cold Spring Harbor, New York.

Hicks J., Strathern J.N., and Klar A.J. 1979. Transposable mating type genes in *Saccharomyces cerevisiae*. *Nature* **282:** 478–483.

Jarvis E.E., Clark K.L., and Sprague G.F., Jr. 1989. The yeast transcription activator PRTF, a homolog of the mammalian serum response factor, is encoded by the *MCM1* gene. *Genes Dev.* **3:** 936–945.

Jarvis E.E., Hagen D.C., and Sprague G.F., Jr. 1988. Identification of a DNA segment that is necessary and sufficient for α-specific gene control in *Saccharomyces cerevisiae*: Implications for regulation of α-specific and **a**-specific genes. *Mol. Cell. Biol.* **8:** 309–320.

Johnson A.D. and Herskowitz I. 1985. A repressor (MATα2 product) and its operator control expression of a set of cell type specific genes in yeast. *Cell* **42:** 237–241.

Kassir Y. and Simchen G. 1976. Regulation of mating and meiosis in yeast by the mating-type region. *Genetics* **82:** 187–206.

Keleher C.A., Goutte C., and Johnson A.D. 1988. The yeast-cell-type-specific repressor α2 acts cooperatively with a non-cell-type-specific protein. *Cell* **53:** 927–936.

Keleher C.A., Passmore S., and Johnson A.D. 1989. Yeast repressor α2 binds to its operator cooperatively with yeast protein Mcm1. *Mol. Cell. Biol.* **9:** 5228–5230.

Keleher C.A., Redd M.J., Schultz J., Carlson M., and Johnson A.D. 1992. Ssn6-Tup1 is a general repressor of transcription in yeast. *Cell* **68:** 709–719.

Klar A.J.S. and Fogel S. 1979. Activation of mating type genes by transposition in *Saccharomyces cerevisiae*. *Proc. Natl. Acad. Sci.* **76:** 4539–4543.

Klar A.J.S., Strathern J.N., Broach J.R., and Hicks J.B. 1981. Regulation of transcription in expressed and unexpressed mating type cassettes of yeast. *Nature* **289:** 239–244.

Kostriken R., Strathern J.N., Klar A.J.S., Hicks J.B., and Heffron F. 1983. A site-specific endonuclease essential for mating-type switching in *Saccharomyces cerevisiae*. *Cell* **35:** 167–174.

Kushner P.J., Blair L.C., and Herskowitz I. 1979. Control of yeast cell types by mobile genes—A test. *Proc. Natl. Acad. Sci.* **76:** 5264–5268.

Lindegren C.C. and Lindegren G. 1943. A new method for hybridizing yeast. *Proc. Natl. Acad. Sci.* **29:** 306–308.

MacKay V.L. and Manney T.R. 1974a. Mutations affecting sexual conjugation and related processes in *Saccharomyces cerevisiae*. I. Isolation and phenotypic characterization of nonmating mutants. *Genetics* **76:** 255–271.

———. 1974b. Mutations affecting sexual conjugation and related processes in *Saccharomyces cerevisiae*. II. Genetic analysis of nonmating mutants. *Genetics* **76:** 273–288.

Nasmyth K.A. 1985a. At least 1400 base pairs of 5´-flanking DNA is required for the correct expression of the *HO* gene in yeast. *Cell* **42:** 213–223.

———. 1985b. A repetitive DNA sequence that confers cell-cycle START (*CDC28*)-dependent transcription of the *HO* gene in yeast. *Cell* **42:** 225–235.

Nasmyth K.A., Tatchell K., Hall B.D., Astell C., and Smith M. 1981. A position effect in the control of transcription at yeast mating type loci. *Nature* **289:** 244–250.

Naumov G.I. and Tolstorukov I.I. 1973. Comparative genetics of yeast. X. Re-identification of mutators of mating types in *Saccharomyces*. *Genetika* **9:** 82–91.

Oshima Y. and Takano I. 1971. Mating types in *Saccharomyces*: Their convertibility and homothallism. *Genetics* **67:** 327–335.

Roberts R.L. and Fink G.R. 1994. Elements of a single MAP kinase cascade in *Saccharomyces cerevisiae* mediate two developmental programs in the same cell type: Mating and invasive growth. *Genes Dev.* **15:** 2974–2975.

Santa Maria J. and Vidal D. 1970. Segregación anormal del "mating type" en *Saccharomyces*. *Inst. Nac. Invest. Agron. Conf.* **30:** 1–21.

Shapiro J.A., Ed. 1983. *Mobile genetic elements*. Academic Press, New York, New York.

Sprague G.F., Jr., Jenson R., and Herskowitz I. 1983. Control of yeast cell type by the mating type locus: Positive regulation of the α-specific *STE3* gene by the *MATα1* product. *Cell* **32:** 409–415.

Strathern J.N. and Herskowitz I. 1979. Asymmetry and directionality in production of new cell types during clonal growth: The switching pattern of homothallic yeast. *Cell* **17:** 371–381.

Strathern J., Hicks J., and Herskowitz I. 1981. Control of cell type in yeast by the mating type locus: The α1-α2 hypothesis. *J. Mol. Biol.* **147:** 357–372.

Tan S. and Richmond T.J. 1990. DNA binding-induced conformational change of the yeast transcriptional activator PRTF. *Cell* **62:** 367–377.

Tatchell K., Nasmyth K.A., Hall B.D., Astell C., and Smith M. 1981. In vitro mutation analysis of the mating-type locus in yeast. *Cell* **27:** 25.

Wilson K.L. and Herskowitz I. 1984. Negative regulation of *STE6* gene expression by the α2 product of *Saccharomyces cerevisiae*. *Mol. Cell. Biol.* **4:** 2420–2427.

9

Meiosis and Spore Development

Rochelle Easton Esposito
The University of Chicago
Chicago, Illinois 60637

Identifying genes essential for meiosis and spore development, 160
Esposito R.E., Frink N., Bernstein P., and Esposito M.S. 1972. The genetic control of sporulation in *Saccharomyces*. II. Dominance and complementation of mutants of meiosis and spore formation. *Mol. Gen. Genet.* **114:** 241–248.

The mitotic/meiotic decision and initiation of meiosis, 165
Kassir Y., Granot D., and Simchen G. 1988. *IME1*, a positive regulator gene of meiosis in *S. cerevisiae*. *Cell* **52:** 853–862.

Role of the synaptonemal complex, 167
Sym M., Engebrecht J.A., and Roeder G.S. 1993. ZIP1 is a synaptonemal complex protein required for meiotic chromosome synapsis. *Cell* **72:** 365–378.

How homologs separate in MI, 171
Watanabe Y. and Nurse P. 1999. Cohesin Rec8 is required for reductional chromosome segregation at meiosis. *Nature* **400:** 461–464.

Regulation of meiosis and checkpoint controls, 178
Hepworth S.R., Friesen H., and Segall J. 1998. *NDT80* and the meiotic recombination checkpoint regulate expression of middle sporulation-specific genes in *Saccharomyces cerevisiae*. *Mol. Cell. Biol.* **10:** 5750–5761.

Note: The landmark papers listed above are those discussed in this chapter. Each landmark paper is preceded by the name of the section (with starting page number) where the paper is first discussed in detail.

INTRODUCTION

What Does Meiosis Accomplish?

Discovered more than a century ago (van Beneden 1883; Boveri 1887; Hertwig 1890), meiosis continues to fascinate biologists because of its fundamental importance in sexual reproduction. It plays a vital role in promoting genetic diversity, ridding populations of recessive lethals, and maintaining constancy in chromosome numbers from one generation to the next. This is accomplished by three critical changes in chromosome behavior compared to mito-

Note: Boldfaced references in the text denote landmark papers that are on the accompanying CD.

sis: (1) a dramatic increase in recombination between parental genomes, (2) sister-centromere cohesion and coorientation coupled to separation and independent assortment of parental chromosomes, and (3) alteration of the cell division cycle to include two successive rounds of M phase without an intervening S phase. These unique "modifications of mitosis" enable first segregation of homologous chromosomes at meiosis I (reductional division), followed by separation of previously replicated sister chromatids at meiosis II (equational division). Further coupling of these specialized nuclear divisions with morphogenic differentiation events to form mature functional gametes (ova, sperm, spores, pollen) results in meiotic products containing rearranged genomes with half the chromosome number of the original cell.

Context of Meiotic Studies

Understanding the genetic basis of both mitosis and meiosis has been vigorously pursued with increased availability of sophisticated molecular, genetic, and cytological tools to explore these processes. The pioneering analysis of conditional mutations in T4 morphogenesis (Edgar et al. 1964), demonstrating the feasibility of dissecting complex developmental processes by genetics, provided a powerful model for these studies. Within this framework, unraveling the mechanisms driving meiosis has been of interest from two distinct perspectives: One can be thought of as determining the "genetics of genetics," i.e., how recombination and homolog segregation occur, the two events of major genetic consequences in meiosis. The other, viewed in the broader context of cell differentiation, aims to uncover the strategies ensuring the orderly progression of individual events of chromosome behavior into a successful developmental pathway. The switch from mitosis to meiosis raises many interesting questions about the mechanisms controlling cell division and differentiation. For example: What causes cells to initiate meiosis? Is there a single "start" stage (as in mitosis) committing cells to meiosis and spore development? How are meiosis-specific functions kept off in mitosis and activated in meiosis? How does the cell cycle machinery interact with meiosis-specific controls to change the division pattern? How are high recombination levels achieved and what is the role of the synaptonemal complex in this process? How are centromere behavior and anaphase separation at meiosis I and meiosis II regulated? What mechanisms specify the orderly progression of meiosis and how are the meiotic divisions coordinated with spore development?

Five Seminal Papers

This chapter describes five seminal yeast papers (mostly in budding yeast) that advanced the field and set the stage for finding answers to many of these questions. The main topics focus on (1) detection of genes essential to meiosis, (2) control systems altering the nuclear division pattern, especially centromere and sister chromatid behavior, and (3) regulatory mechanisms coordinating the transcriptional program with signal transduction pathways to exquisitely order gene expression. The important area of genetic recombination, reviewed separately by L. Symington in Chapter 2, is covered only in the context of synaptonemal complex function to avoid redundancy. Finally, the difficult task of choosing a limited number of landmark papers leaves out many publications worthy of this designation. These contributions and related studies in fission yeast are recognized in the text.

Special Advantages of Yeast for Analysis of Meiosis

Budding and fission yeast have been foremost among the organisms examined to determine the mechanisms governing meiosis because of (1) their unicellular character allowing recovery of large populations of single cells at similar stages of development, (2) the ability to initiate and to interrupt meiosis by simple change of nutritional conditions, (3) their genetic and molecular manipulability, and (4) most significantly, the ease of recovery of recessive and dominant mutations affecting the process. Indeed, more is currently known about meiosis in these yeasts than any other eukaryotic system. The information derived from these studies has provided groundbreaking paradigms for understanding meiotic development across species. In yeast, meiosis is one of several cell fates that occur in response to external nutritional cues. Depending on cell type and carbon and nitrogen availability, vegetative cells either divide by mitosis, arrest in G_1 and mate, undergo filamentous "foraging" growth, or proceed to G_0, a resting stage where G_1-arrested cells cease general transcription and translation and accumulate stress response proteins and complex carbohydrates. In budding yeast, meiosis is typically triggered when **a**/α cells deprived of glucose and an essential nutrient (nitrogen, sulfur, or phosphorus) enter G_0 in the presence of an oxidizable carbon source (e.g., acetate).

Major Developmental Landmarks

Premeiotic DNA synthesis, recombination, and segregation during meiosis I (MI) and meiosis II (MII) in yeast are generally similar to those of higher cells except that spindle assembly and chromosome separation occur within an intact nuclear membrane. Morphogenesis of at least two nucleus-associated structures has important roles in coupling specific landmarks. First, the synaptonemal complex (SC) in budding yeast and other systems where it exists, although not required to initiate recombination, facilitates completion of most (80–90%) reciprocal exchange, formation of chiasmata, and proper microtubule attachment of homologs during MI segregation. Second, embedded in the nuclear membrane, spindle pole bodies (SPBs) have a unique dual function in meiotic development. In addition to their well-known role as microtubule-organizing centers nucleating spindle assembly at each M phase, they also initiate and direct prospore membrane formation, thereby coordinating spore differentiation with the meiotic divisions. Early in MII, as spindles develop, the outer plaques of the four SPBs (at the MII spindle poles) enlarge and, via interaction with the secretory pathway, recruit vesicles that fuse initiating prospore membrane synthesis. Then, with perfect harmony as MII segregation progresses pulling chromatids to opposite poles, prospore membranes grow bidirectionally from each SPB tracking along the nuclear membrane, ultimately encapsulating four meiotic products into prospores. These further mature by acquiring additional protective carbohydrate and protein layers, enabling them to survive a variety of environmental stresses such as desiccation and heat.

Progress on Understanding Meiosis: An Overview

Early analysis of meiosis and spore development suggested that 100–200 genes specifically control the process (for review, see Esposito and Klapholz 1981). A little more than 15 years ago, no meiosis-specific genes were cloned and virtually nothing was known about their gene products or regulation. Since then, an explosion of genetic studies and whole-genome analyses have identified more than 900 "meiotically regulated" genes in at least seven expression classes (Chu et al. 1998; Primig et al. 2000). Among these, about 100 are exclusively transcribed in meiosis across species and are "meiosis-specific" (Schlecht and Primig 2003). Most of these are essential for the process (Rabitsch et al. 2001; Briza et al. 2002; Enyenihi and Saunders 2003) in remarkably close agreement with earlier genetic estimates. Many other loci expressed during vegetative growth, either up-regulated in meiosis or expressed constitutively, are also crucial for meiotic development, with deletion of some genes actually improving its efficiency (Deutschbauer et al. 2002). All told, approximately 10% of the protein-coding genes in the genome are essential for the process. Key loci governing meiotic transcription, mRNA processing and stability, and protein modification, localization, and degradation are now defined, and many gene functions are related to specific morphological events (for review, see Kupiec et al. 1997). As a result of these studies, the basic architecture of the gene expression program and recombination, segregation, and spore development pathways is becoming clear. These findings are providing a wealth of new information about the genetic control of meiosis in yeast and other organisms. So, how did we get to this point in such a relatively short time? Much of the early progress in the field was spurred by advances described in the next sections, which had a critical role in transforming yeast into an attractive and tractable system for meiotic analysis.

IDENTIFYING GENES ESSENTIAL FOR MEIOSIS AND SPORE DEVELOPMENT

Recovery of Recessive Mutants Defective in Ascus Production

An early crucial breakthrough occurred with the development of procedures for the systematic recovery of recessive mutations affecting meiosis and spore formation. At the time, how to design a comprehensive screen for such mutants was not obvious as meiosis was known only to occur in diploid cells and cells of higher ploidy. Since most mutations are recessive, mutagenesis of either diploid or haploid cells was considered impractical. In 1967, M.S. Esposito and R.E. Esposito devised a novel scheme to overcome this based on the behavior of homothallic strains bearing the D (diploidization) gene (Winge and Roberts 1949). A few years earlier, this gene (now known as HO) was found to promote mating-type switching of both **a** and α spores (Hawthorne 1963). After a couple of divisions, single spores give rise to both **a** and α progeny that mate to produce diploids homozygous for all genes except the mating-type locus (see Chapter 8). In a first report, whole asci from a D gene strain were mutagenized with UV to a level where most surviving colonies resulted from the growth of a single spore (Esposito and Esposito 1969). Survivors competent for mating-type switching, respiration, and mitotic growth were sporulated again and examined by light microscopy for reduced ascospore or ascus formation at diagnostic temperatures. Those (~75/1000) sporulating at least three standard deviations from the wild-type mean at one or more temperatures and having a consistent phenotype through several cycles of growth and sporulation were analyzed further. Outcrossing of putative mutants, complementation testing of alleles segregating in single Mendelian fashion with the clearest conditional phenotypes, and cytological analysis of arrest points were described in a second publication (**Esposito et al. 1972**). This landmark paper defined the first 11 sporulation-defective (*spo*) genes affecting various stages of meiosis in budding yeast (summarized in Esposito and Esposito 1975). Some were later shown to have specific defects in meiotic entry (*spo8/ime4*), initiation of recombination (*spo11*), SPB duplication, spindle assembly, and spore formation (*spo1*), and prospore membrane synthesis (*spo3/ssp1, spo4, spo5*). This study provided the first estimate of the number of genes (50–100) in the yeast genome specifically required for gametogenesis (based on allelic identity in crosses to the original collection). A subsequent mutant hunt (mutagenizing random spores with ethylmethyl sulfonate [EMS]) screening for inviable as well as reduced gamete production revised the estimate to 100–200 genes (cited in Esposito and Klapholz 1981). This analysis showed the feasibility of genetically dissecting a complex process such as meiosis in a simple model organism and had a major impact in attracting other investigators into the field.

Bypassing Normal Controls of Meiotic Initiation

The homothallic system was subsequently used to obtain recessive mutants, bypassing nutritional control of meiosis forming asci in the presence of nitrogen (Dawes 1975). A related approach recovered mutants overcoming **a**/α mating-type control of the process, allowing **a**/**a** or α/α cells to sporulate (Hopper and Hall 1975). This led to the identification of a major *r*epressor of *me*iosis, *RME1* (Hopper and Hall 1975; Kassir and Simchen 1976), down-regulated by mating-type gene products (Mitchell and Herskowitz 1986). These were among the first uses of the "bypass rationale" to study development in eukaryotes. This method detects mutations in negative controls enabling the process to go forward (not necessarily at wild-type levels) where normally it does not. Such mutations are unlikely to be found in simple loss-of-function screens unless the gene also has a positive role (e.g., negatively regulates a downstream negative regulator). This rationale also paved the way for identification of positive meiotic regulators. For example, selection for loci whose overexpression overrides the negative regulation of *RME1* led to the landmark recovery of *IME1*, a key *i*nducer of *me*iosis (Kassir et al. 1988), and later *IME2*, encoding an important kinase controlling meiotic progression (Smith and Mitchell 1989). The above screens detected alleles altering ascus production needed throughout the process. Since then, other approaches were designed to recover defects in specific landmarks. Overall, the criteria used include (1) reduction or gain in ascus formation, (2) reduction or gain in the number of spores per ascus, (3) spore inviability, (4) specific loss of recombination and/or proper segregation, and (5) altered gene expression. Figure 1 indicates the sporulation behavior of some of the mutants recovered in these schemes.

Single Division and Diploid versus Haploid Meiosis

Natural variants producing two-spored asci with diploid spores (Grewal and Miller 1972) led to the invention of other novel tools to unravel the genetic control of meiosis. Their analysis resulted in the recovery of two recessive mutations, *spo12* and *spo13*, each causing a dyad phenotype (Klapholz and Esposito 1980a). Cells homozygous for either one undergo a single meiotic division where sister chromatids separate mostly equationally on the MI spindle (Klapholz and Esposito 1980b). SPBs at the two poles become modified as in MII (Moens 1974), and prospore membrane growth encapsulates the two diploid spores from the single (mitotic-like) division (Fig. 1, row 3). The regular separation of sisters in the

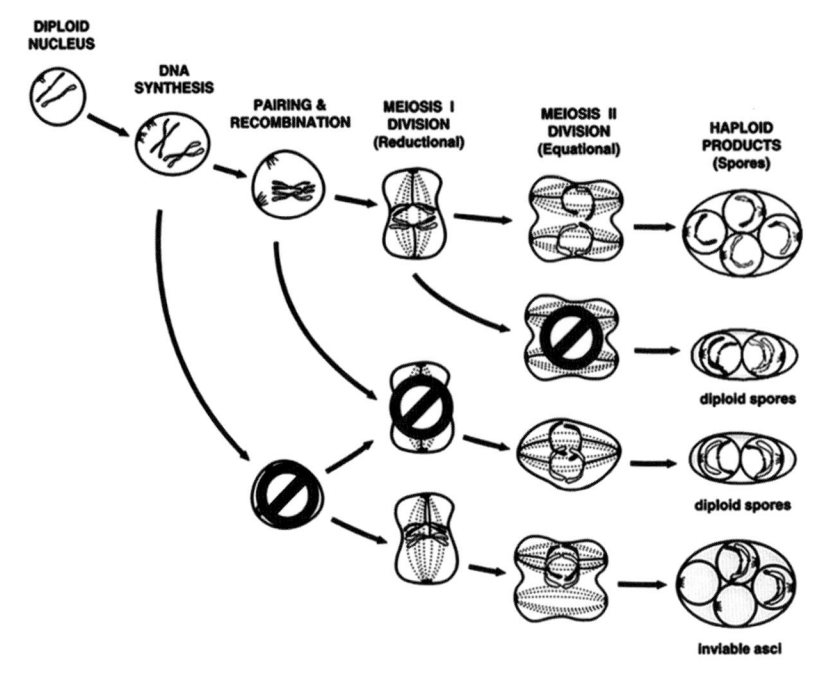

FIGURE 1. Schematic of meiotic chromosome behavior in wild type and mutants yielding (*top*) tetrads in the wild type; (*2nd row*) dyads with diploid spores due to failure to complete MII equational division and packaging of MI products into single spores; (*3rd row*) dyads with diploid spores due to failure of MI reductional division with mostly equational separation of sister chromatids on an MI spindle with modified SPBs; (*4th row*) inviable aneuploid spores due to failure of crossing over. The *arrow* between rows 3 and 4 shows that certain Rec⁻ mutants (e.g., *spo11*, lacking double-strand breaks and yielding inviable spores from random MI homolog segregation without chiasmata) can be rescued by coupling to a mutant (*spo13*) executing a single largely equational division, which does not depend on chiasmata to occur properly. This schematic is not meant to imply the timing of when events occur in mutants but rather what happens to chromosomes (see text for details).

mutants implies a lack or eventual loss of sister-centromere cohesion and/or coorientation in individual homologs during their divisions. Depending on genetic background, a variable level of aberrant and reductional segregation also occurs, indicating that other modifiers can affect this process (Klapholz and Esposito 1980b; Klapholz et al. 1985; Hollingsworth and Byers 1989; Hugerat and Simchen 1993). The *spo13* mutation has the special property of allowing its unique division to be completed with unpaired (univalent) chromosomes which otherwise missegregate during meiosis. This was shown in diploid recombination-deficient (Rec⁻) mutants failing to initiate exchange and lacking chiasmata to hold homologs together (Malone and Esposito 1981; Malone 1983) and in haploids containing only one member of each chromosome (Wagstaff et al. 1982). Such strains with normal sister-centromere cohesion and coorientation typically produce inviable meiotic products because of random homolog segregation to one pole or the other at MI (Klapholz and Esposito 1982). When sisters separate at MII, genomes arise with extra or missing chromosomes (e.g., Fig. 1, row 4). Depending on the severity of aneuploidy and gene dosage imbalance, they fail to be packaged or form immature and/or mostly inviable spores. In contrast, *spo13* Rec⁻ diploid and haploid cells yield a high level of viable meiotic products because sister chromatids from unpaired homologs can separate (divide equationally) during the single division (Fig. 1, arrow between rows 3 and 4), bypassing the lethal effects of random homolog segregation when sisters stay together at MI. This rescue behavior, which occurs only marginally in *spo12*, defective later in MI (after recombined homologs attach to the MI spindle), led to new convenient assays to identify and order gene functions in both recombination and segregation pathways (for details, see Esposito et al. 1991).

Haploid meiosis is especially useful in detecting and analyzing recessive alleles in cells that can be directly tested for sporulation competency. For example, *spo13* haploids expressing both mating types (e.g., bearing an **a**/α disome, a mutation allowing silent mating-type information to be expressed or a plasmid with the opposite mating-type allele) can enter meiosis and progress through S phase, recombination, and sister separation on the MI spindle, forming dyads with two viable haploid spores. This led to the recovery of one of the first SC components, Hop1, specifically required for exchange between homologs, e.g., assayed in a disome (Hollingsworth and Byers 1989), and key regulators, including Ndt80, Sin3, and Rpd3, of genes expressed in mid meiosis (Hepworth et al. 1998). Coupled with a fluorescence spore test (Fig. 2f), it also defined *DIT* loci required for spore maturation (Briza et al. 1990), new alleles of *spo13* (Rutkowski and Esposito 2000), and other genes required for meiosis and spore formation.

ASSAYING SPECIFIC MEIOTIC LANDMARKS

Cytology of Meiosis

Development of methods to better fix and examine yeast cells by electron microscopy provided another crucial advance in understanding the genetic control of meiosis. Pioneering studies by Moens and co-workers showed that extensive enzymatic digestion of the cell wall, although failing to preserve cytoplasmic structures, allowed optimal visualization of nuclear components. Through careful serial sectioning and reconstruction of whole yeast cells, these authors reported the first detailed electron microscopic description of meiotic

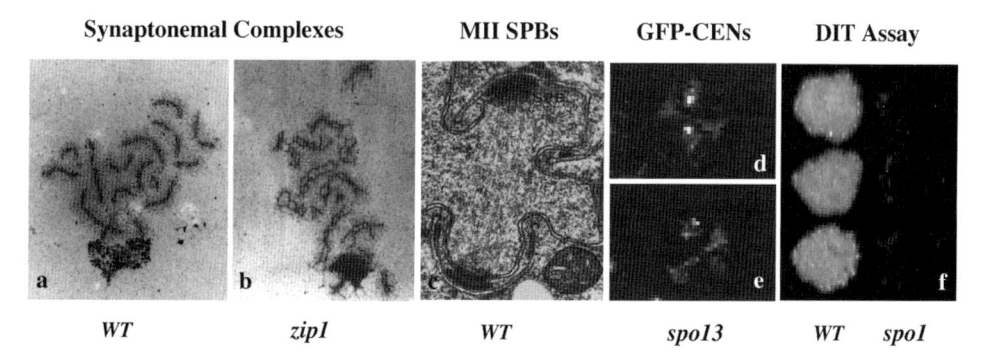

FIGURE 2. (*From left*) Electron micrographs of wild-type (*a*) and *zip1* (*b*) nuclear spreads of SCs (courtesy G.S. Roeder); (*c*) MII-modified SPBs and prospore membrane growth (courtesy S. Klapholz). Dyads from *spo13* haploid meiosis with a single chromosome III centromere (from a disomic pair) marked with the *tet* operator and GFP-tet repressor showing equational (*d*) and reductional (*e*) segregation in Rec⁻ and Rec⁺ strains, respectively (courtesy L. Rutkowski). DIT (dityrosine) spore fluorescence assay (*f*) in wild type and *spo1* (courtesy G. Tevzadze).

development (Moens 1971; Moens and Rapport 1971a, b). They found that the nuclear membrane remains intact throughout the process and described yeast SCs, meiotic spindle assembly, and SPB modification associated with prospore wall membrane growth (Fig. 2c). These observations set the stage for more detailed analysis of the morphogenic events in meiosis. For example, key studies by the Byers group provided a precise chromosome number from the meiotic SC karyotype of diploid and tetraploid budding yeast (Byers and Goetsch 1975). Adaptation of nuclear spreading procedures (developed by M. Moses in higher systems) later allowed convenient visualization of SCs by electron microscopy without serial sectioning (see Fig. 2a,b) and by light microscopy (Goetsch and Byers 1982; Dresser and Giroux 1988). This led to the identification of SC components by immunofluorescence localization in a variety of mutants and a comprehensive view of the stages involved in SC assembly (for review, see Zickler and Kleckner 1998; Bishop and Zickler 2004; Page and Hawley 2004). Cytogenetic analysis also showed definitively that SPB modification is required for prospore membrane synthesis. This was shown in a mutant yielding asci with two haploid spores initiating only at one modified SPB at each pole of the MII spindles (Davidow et al. 1980). Later, recovery of proteins localizing to modified SPBs and growing prospore membranes stimulated a bourgeoning new field defining factors and signal pathways regulating the morphological transformation of meiotic products into functional gametes (for review, see Engebrecht 2003; Moreno-Borchart and Knop 2003).

Return to Growth

Another useful tool was the return-to-growth protocol providing complementary information on the timing and order of events. This procedure interrupts cells during meiosis by returning them to a rich media where sporulation usually does not occur and events in progress generally, but not always, cease. The switch in culture conditions (such as mutant analysis) allows inferences to be made about the order and dependency of specific landmarks based on whether they can be uncoupled during return to growth. The method was initially used to assay commitment to meiosis, the stage when the decision is made to proceed with spore formation even in growth media (Ganesan 1958). It was also employed to assay the timing of commitment to gene conversion (Sherman and Roman 1963), reciprocal recombination (R.E. Esposito and M.S. Esposito 1974), and the meiotic divisions, initially by plating cells on selective growth media to detect specific recombinants and expression of recessive drug resistance markers reflecting homolog segregation. Real-time assays directly measuring recombined DNA molecules and centromere segregation during meiosis now show that the decision to commit to and the actual execution of an event generally occur at the same time (Borts et al. 1986). One exception is commitment to reciprocal exchange where a decision about the distribution of events is made before SC assembly, whereas their completion occurs afterward. Another is spore formation where commitment is detected when binucleate cells appear at MI completion (perhaps when SPBs are modified), well before mature spores are actually seen (see below).

Recombination, Segregation, and Spore Formation Assays

The emergence of newer molecular assays for detecting specific stages in recombination, segregation, and spore formation further fueled rapid growth of the field. For example, assays detecting double-strand breaks (DSBs) and sister and interhomolog exchange using polymorphic markers and two-dimensional gels (see Chapter 2) led to a more sophisticated understanding of the roles of specific genes in forming recombination intermediates. The advent of centromere tags using multiple copies of bacterial regulatory elements integrated near centromeres (e.g., *lacZ* or *tet* operator sequences) and green fluorescent protein (GFP)-tagged repressors binding to these sequences (Fig. 2d,e) had a similar effect on deciphering the genetic basis of meiotic segregation (for review, see Nasmyth 2002). Likewise, the dityrosine plate assay, allowing convenient and sensitive detection of the natural fluorescence of dityrosine in the spore wall (Fig. 2d), proved especially useful for quickly assessing sporulation competency and detecting mutants in spore development and other events (Briza et al. 1986, 1990). Finally, analyses of the precise timing and order of specific landmarks have been greatly improved by the use of strains (such as SK1) that sporulate rapidly, synchronously, and to a relatively high level (Kane and Roth 1974).

THE MITOTIC/MEIOTIC DECISION AND INITIATION OF MEIOSIS

Cell-type Control of Entry into Meiosis

Initiation of meiosis, as noted earlier, depends on both cell type and nutritional cues. Proteins responding to these cues often act as developmental regulatory switches inducing meiosis while at the same time repressing mitosis, thereby reinforcing the choice of cell fate. The presence of both mating-type alleles rather than diploidy per se was discovered early on to be critical for initiating meiotic development since **a**/**a** and α/α cells are unable to sporulate (Roman and Sands 1953). These diploids fail to initiate premeiotic DNA synthesis and subsequent events (Roth and Lusnak 1970). In contrast, haploid cells expressing both mating-type alleles enter meiosis and progress to spore formation, although they rarely mature because of aneuploidy due to unequal partitioning of single chromosomes at MI (Roth and Lusnak 1970; Roth and Fogel 1971; Wagstaff et al. 1982). The mating type **a**1 and α2 proteins are specifically required to initiate meiosis (Strathern et al. 1981). They form a negative regulator that directly represses the *RME1* gene (Covitz et al. 1991). Rme1 is one example of a regulatory switch that makes mitosis and meiosis incompatible. It is a dual-role transcription factor promoting mitosis by inducing the mitotic G_1 cyclin *CLN2* (Toone et al. 1995), while at the same time negatively regulating meiosis by repressing the *IME1* positive regulator of early meiotic expression (**Kassir et al. 1988**). This theme occurs repeatedly for nutritional controls of meiotic initiation, which converge with cell-type signals to regulate *IME1*.

Nutritional Control of Initiation

Uncovering the mechanisms in nutritional regulation of meiosis has been difficult given the multiplicity of signal pathways participating in this process. Besides proper cell type, three key nutritional conditions are crucial for entry into meiosis: (1) starvation for an essential nutrient such as nitrogen, sulfur, or phosphorous (Freese et al. 1982), (2) glucose deprivation, and (3) availability of an oxidizable (nonfermentable) carbon source (for review, see Esposito and Klapholz 1981). The first two cause cells to enter a resting state known as G_0 (Pinon 1977). Respiratory-competent cells leave G_0 when exposed to an oxidizable carbon source embarking on meiosis from the G_1 stage of the mitotic cell cycle (M.S. Esposito and R.E. Esposito 1974; Hartwell 1974; Pinon 1977). The nutritional signals are transmitted to Ime1 by a combination of transcriptional and posttranslational mechanisms involving at least three signal switches that simultaneously promote mitosis while blocking meiosis and vice versa. First, nitrogen activates G_1 cyclins, which repress transcription of *IME1* primarily via Cln2 (Purnapatre et al. 2002). Phosphorylation of the Ime1 protein by cyclin-dependent kinase (Cdk-Cln) further prevents residual protein from entering the nucleus and executing its function (Colomina et al. 1999). Thus, when Clns are expressed, mitosis is favored; in contrast, when they are repressed during nitrogen starvation, meiosis is favored (Gallego et al. 1997). Second, glucose negatively regulates a kinase (Snf1) that participates in regulating Ime1 expression (Honigberg and Lee 1998). Relief of Cln-mediated nitrogen repression causes a change from low to moderate Ime1 expression, rising to higher levels needed for meiotic initiation when glucose repression of Snf kinase is also lifted (Purnapatre et al. 2002). Third, glucose also acts through stimulation of the Ras-cyclic AMP signaling pathway. This pathway increases both GTP pools and PKA (protein kinase A) activities that promote mitosis and repress meiosis (Matsumoto et al. 1983; Varma et al. 1985). This pathway acts in part through PKA phosphorylation of two DNA-binding proteins, Sok2 and the transcriptional activator complex Msn2-Msn4, which comprise another mitotic/meiotic switch (Shenhar and Kassir 2001). When PKA activity is high, Sok2^P functions as both an activator of mitosis and repressor of *IME1*, while Msn2-Msn4^P is sequestered from the nucleus (Gorner et al. 1998). When PKA activity is low, unphosphorylated Sok2 enables Msn2-Msn4 to bind to the *IME1* promoter inducing its transcription (Y. Kassir, unpubl.). Interestingly, complete deletion of Cln3 (Colomina et al. 2003) or Sok2 (M. Primig, unpubl.) allows more rapid meiotic initiation, providing potential new experimental tools for improving meiotic synchrony. Finally, Tor signaling required for growth in response to nitrogen has important roles in meiosis not yet fully understood. It prevents cells from entering G_0 (Pedruzzi et al. 2003), thereby inhibiting meiotic initiation, but then is subsequently required for Ime1 nuclear localization (Colomina et al. 2003), progression through pachytene, and spore packaging (Zheng and Schreiber 1997). Other kinases (Mck; Neigeborn and Mitchell 1991) and pathways (*RIM 1, 8, 9,* and *13*; Su and Mitchell 1993) also respond to nutritional cues. How they interact with the systems described above remains to be determined.

Recovery and Function of Ime1

The identification and analysis of *IME1* provided a major breakthrough in determining how meiotic development begins (Kassir et al. 1988). *IME1*, an essential meiosis-specific gene, was isolated by selecting for plasmids (from a high-copy yeast library), allowing a marked diploid expressing the *RME1* repressor of meiosis to sporulate. The diploid had a mutation in *MAT***a** and functional mating-type information only at *MAT*α, thereby permitting *RME1* expression. One plasmid containing the *IME1* gene had properties consistent with a positive regulator of meiosis acting downstream from both *RME1* and nutritional controls. For example, (1) overexpression allowed **a/a** and α/α cells to sporulate, overriding the negative effect of Rme1; (2) this occurred even in rich media in the presence of glucose and nitrogen, indicating it bypasses both these controls, whereas absence of Rme1 alone does not; and (3) disruption of the gene resulted in a Spo$^-$ phenotype, even in *rme1* cells, implying it is likely a target rather than a negative regulator of *RME1*. The gene was induced at low levels insufficient for sporulation in **a**/α acetate grown cells (with nitrogen), indicating independent effects of cell type and nutritional controls on its expression. Detailed analysis of its promoter reflects the complexity of its regulation. Distinct elements respond to cell type/Rme1 regulation, glucose/Sok2 repression, and acetate induction (Sagee et al. 1998). Remarkably, *IME1* expression alone is sufficient to induce meiotic development, underscoring its crucial role in positively regulating the process. Indeed, it is now known that Ime1 promotes early meiotic transcription in a manner not yet fully understood via interaction with Ume6, a DNA-binding protein that binds to early gene promoters and regulates their expression (see below).

Ime2 and Initiation of Premeiotic DNA Synthesis

Once *IME1* is transcribed at appropriate levels and early genes are induced, progression to premeiotic S phase (an early developmental landmark) is similar in many respects to mitosis. Mitotic S phase is regulated by phosphorylation and subsequent degradation of Sic1. This protein is a key inhibitor of the cyclin-dependent kinases Cdc28-Clb5,6 essential for initiation of both mitotic and meiotic S phase (Stuart and Wittenberg 1998). During vegetative growth, Sic1 phosphorylation is mediated by Cdc28-Cln1,2, but in meiosis, where G_1 cyclins decrease during nitrogen starvation, it is dependent instead on another kinase, Ime2. This important kinase controls progression of several stages of meiosis, together with Cdc28. *IME2* is an early meiotic gene induced by Ime1. It was identified by screening for plasmids in a high-copy library allowing **a**/**a** diploids to bypass negative regulation of *RME1* overexpression, permitting them to enter meiosis and undergo genetic recombination on sporulation plates (Smith and Mitchell 1989). It was later shown to be a kinase (Yoshida et al. 1990; Kominami et al. 1993) essential for premeiotic S via Sic1 phosphorylation (Dirick et al. 1998). It also phosphorylates Ime1 leading to its degradation (Guttmann-Raviv et al. 2001). Ime2 is an unstable protein (Guttmann-Raviv et al. 2001; Bolte et al. 2003) with glucose at moderate levels directly inhibiting Ime2 function by stimulating its proteolysis (Purnapatre et al. 2005). Interestingly, the requirement for Clb5,6 in initiating S phase implies Cdc28 is further needed for its progression, but initial examination of Cdc28 mutants failed to detect arrest at this stage because of their leaky phenotype (Shuster and Byers 1989). The role of Cdc28 in premeiotic S phase has now been confirmed using conditional mutants sensitive to inhibitors (Benjamin et al. 2003). It also becomes more critical for S phase when Ime2 is deleted, suggesting that Ime2 and Cdc28 have partially redundant functions at this stage (Guttmann-Raviv et al. 2001). After premeiotic S phase, meiotic cells enter prophase where recombination occurs and the SC, an elaborate chromosomal structure, assembles.

ROLE OF THE SYNAPTONEMAL COMPLEX

SCs, Chiasmata, and Accurate Segregation

The SC is composed of more than a dozen proteins amassing between homologs in prophase of MI (for review, see Page and Hawley 2004). Its role in meiosis has aroused keen interest since its discovery nearly 50 years ago (Moses 1956). SC architecture, conserved across species, consists of two lateral elements (LEs) approximately 1000 Å apart, each derived from a single axial core/element (AE) shared between sister chromatids of one homolog. Transverse filaments run perpendicular across the central region/element (CE), giving SCs a ladder-like appearance. Densely staining bodies, recombination nodules, associate with the structure. Their frequency and distribution in mid prophase (pachytene) correspond to sites of reciprocal recombination between nonsister chromatids that later form chiasmata (cross connections) between homologs (Carpenter 1975, 1987). In nearly all organisms where they are found, SCs are required for chiasmata (von Wettstein et al. 1984), which in turn are critical for accurate and regular homolog segregation. Elegant manipulation studies of chromosomes on spindles demonstrate that chiasmata promote proper homolog attachment to polar microtubules carrying them to opposite spindle poles at anaphase I (Nicklas and Koch 1969). They provide an opposing force to that generated by microtubules pulling homologs in opposite directions. A biochemical signal responding to tension arising from these opposing forces is believed to stabilize kinetochore-microtubule attachments required for correct segregation (Fig. 3, right). Without tension (in Rec⁻ strains) homologs randomly attach to microtubules from the same or opposite poles, leading to MI nondisjunction and aneuploid products as described earlier (Fig. 1, row 4). Although one chiasmata per bivalent is often sufficient to ensure proper segregation, in most organisms approximately three chiasmata per bivalent occur regardless of DNA content, which generally increases in higher organisms. This suggests that reciprocal exchange frequencies in meiosis are selected mainly to direct proper segregation, not specifically to enhance diversity through increased recombination. However, recombination alone does not produce chiasmata; nonsister crossover chromatids must also remain associated with their original homolog via sister-chromatid arm cohesion. Once such cohesion is removed, recombined nonsister arms can untangle from their sisters, resolving chiasmata and allowing homologs to detach from each other. Tension-mediated microtubule attachment also operates for sister chromatids during MII and in mitosis. In these cases, where sister chromatids and not homologs separate, recombination is dispensable and sister cohesion alone suffices to provide the opposing force to that produced by microtubules pulling chromatids to opposite poles.

SCs and Initiation of Recombination

Because of their intimate association with chiasmata, SCs were classically thought to have a key role in aligning homologs for genetic recombination. However, in budding yeast, this

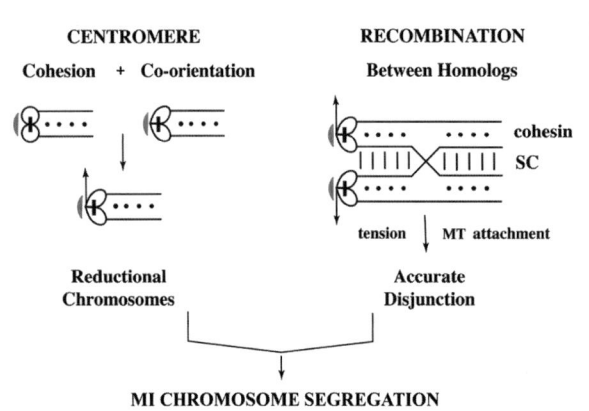

FIGURE 3. Requirements for MI segregation of homologs. (*Left*) State of the centromere: Rec8 cohesin along sister-chromatid arms (*black dots*) and at sister centromeres (*black vertical line in white circles*), protected by shugoshin (*gray semicircle*) with monopolar orientation (*side-by-side ovals*) leading to "reductional homologs" in which sisters segregate together. (*Right*) Recombination and chiasmata in bivalent with synaptonemal complex. Tension is produced between a pair of recombined homologs containing chiasmata when monopolar-oriented centromeres attach to microtubules from opposite poles.

is not the case. It is now clear that DSBs, gene conversion, and even decisions about the distribution of reciprocal exchanges (i.e., interference) are all made prior to SC assembly and are actually required to initiate synapsis, not the other way around (for review, see Bishop and Zickler 2004; Page and Hawley 2004). This conclusion is based on the time of appearance of various Rec intermediates and behavior of mutants defective in SC formation. For example, an early indication that initiation of recombination precedes SC formation came from analysis of *MER2*, a high-copy suppressor of *mer1*, deficient in gene conversion, reciprocal exchange, and SCs. Simultaneous restoration of gene conversion and SCs, but not reciprocal events, in suppressed strains raised the possibility that gene conversion occurs prior to and is a precondition for SC formation, not the reverse (Engebrecht et al. 1990). This study also supported the view that reciprocal exchange leading to chiasmata in the context of the SC directs proper disjunction, whereas Rec events (10%) in the absence of SCs do not (see also Rockmill and Roeder 1990). Overall, these findings favor the conclusion that SCs are required for completion of reciprocal exchange, chiasmata, and chromosome disjunction, rather than for initiating exchange. Consistent with this idea are the findings that DSBs mediated by Spo11 (Keeney et al. 1997), which promote most meiotic exchange, (1) appear prior to SCs (Padmore et al. 1991), (2) are detected in their absence (de Massy et al. 1994), and (3) are proportional to and essential for formation of synaptic initiation complexes (SICs) upon which SC assembly depends (Fung et al. 2004; Henderson and Keeney 2004). SICs, in turn, arise before early recombination intermediates called single end invasions detected during SC assembly (Borner et al. 2004). Interestingly, in budding yeast, no null DSB mutants with normal SC levels exist, although such mutants occur in flies and worms, suggesting in these systems homology search mechanisms leading to SC assembly are independent of DSBs (for review, see Page and Hawley 2004). Once recombination begins, processing of single end invasion recombination intermediates to double Holliday junctions, and their resolution into reciprocal crossovers and chiasmata are believed to occur largely in the context of the SC.

SC Components

Although the manner in which recombination intermediates mature into chiasmata in the SC remains unclear, a good deal is known about specific SC components and SC assembly during prophase (for review, see Zickler and Kleckner 1998, 1999). For example, *Leptotene* chromosomes contain short AE segments, and their ends are often associated with the nuclear membrane. DNA breaks, detected at this time, are thought to result in the formation of DNA bridges and axial associations (AAs) facilitating subsequent AE alignment. In *zygotene*, AEs extend and move closer, and SC assembly initiates. By *pachytene*, when crossovers are completed, homologs are fully synapsed. Chiasmata become visible in *diplotene* when SC proteins start to disassemble and desynapsis begins. In *diakinesis*, homologs attach to microtubules and are now ready for metaphase I. The first SC components identified by genetic screens, Hop1 (Hollingsworth and Byers 1989; Hollingsworth et al. 1990) and Red1 (Rockmill and Roeder 1988; Smith and Roeder 1997), both localize to AEs and interact with each other. A recent model, with considerable supporting evidence, suggests that in conjunction with a kinase (Mek1), they promote exchange events preferentially between homologs rather than sisters (Wan et al. 2004). Hop1, a DNA-binding protein thought to bind to GC-rich regions where DSBs will form, recruits Red1, a phosphoprotein, which in turn associates with Mek1 (Bailis and Roeder 1998). This kinase when activated is believed to phosphorylate substrates in the DSB recombinational repair pathway involved in strand exchange between homologs via Dmc1 (Bishop et al. 1992). Besides these proteins, cohesins and condensins present in AEs are also essential for proper localization of other SC components needed to complete recombination, such as Zip1, described further below (Klein et al. 1999).

Zip1, SC Assembly, and SC Function

Zip1 is probably the most extensively studied of all SC proteins. Its discovery by the Roeder laboratory (**Sym et al. 1993**) was another landmark that helped clarify the role of the SC in meiosis. Zip1 was recovered in a screen for mutants producing inviable spores and defined as an SC component needed for reciprocal exchange. It was proposed to be a CE component that *zippered* AEs together because it localizes only to synapsed and not unsynapsed AEs. Later careful studies of the distances between AEs in *zip1* deletion and addition mutations showed that Zip1 forms the transverse filament of the SC (Sym and Roeder 1995; Tung and Roeder 1998) and is a founding member of a family of such proteins present in flies, worms, and mammalian cells. Zip1 contains an internal domain that dimerizes, causing carboxy-terminal ends to associate with AEs/LEs and amino-terminal globular regions with each other in the CE. Three critical clues about the role of the SC emerged from studies of Zip1. The first was that AEs are spaced further apart in the absence of Zip1 than in fully synapsed SCs, but come together in close proximity in a regular fashion at special sites designated AAs (Fig. 2b). On the basis of their frequency and nonrandom distribution, AAs were proposed to be sites destined for reciprocal exchange. Later studies confirmed this view and demonstrated that AAs depend on DSBs, strand transfer proteins, and a helicase specifically affecting reciprocal crossing over but not gene conversion. The second clue was that Zip1 is recruited to AAs in early zygotene by two proteins (Zip2 and Zip3) before forming transverse filaments, as part of a SIC, along with other proteins (Msh4 and Msh5) needed for completion of reciprocal exchange (Agarwal and Roeder 2000). SIC complexes of these now called ZYMM proteins form before single end invasion intermediates are detected when SCs assemble (Borner et al. 2004). They are required to process DSBs specifically into reciprocal crossovers and are also associated with late recombination nodules long viewed as containing the enzymatic machinery for exchange (Fung et al. 2004). The final crucial clue was that loss of Zip1 eliminates crossover interference reflecting the decision about where crossovers will occur (Sym and Roeder 1994). Significantly, axial associations and SICs display interference even without Zip1, implying that the crossover distribution decision is made before SC assembly which depends on Zip1 (Fung et al. 2004). This provides conclusive evidence against the view that SCs are a precondition for initiating reciprocal recombination, a long held belief. The absence of interference in Zip1 mutants is thought to be due to failure to complete exchange events already destined to occur at particular positions, rather than to any role of SCs in establishing sites for crossing over. On the basis of these findings, it is now generally accepted that initiation of reciprocal exchange itself promotes SC formation, and that SCs per se have little if any role in governing the distribution of recombination events. The primary role of the SC thus appears to be in maturation of reciprocal crossover intermediates into chiasmata, which are necessary for accurate and regular homolog disjunction.

This description of events in budding yeast is not completely applicable to organisms lacking SCs, such as fission yeast, which exhibit gene conversion and reciprocal exchange but have no synaptonemal complexes or interference. In this case LE fragments, which intriguingly contain protein homologs of Red1, Hop1, and Mek1 (Lorenz et al. 2004), may have an analogous role in linking homologous chromosomes. Finally, the formation of SCs between nonhomologous chromosomes (von Wettstein et al. 1984; Loidl et al. 1991), occurrence of mutants with presumably no breaks and some synapsis (Loidl et al. 1994), and exchange events independent of SCs (for review, see Mao-Draayer et al. 1996) are not yet understood. What is clear is that SCs and reciprocal recombination, where they exist, are critical for directing proper MI segregation. Moreover, manipulation studies moving MI chromosomes to MII spindles unambiguously show that their unique segregation behavior is contained entirely within chromosomes themselves as they retain their reductional segregation properties even on heterologous spindles (Paliulis and Nicklas 2000). The section below describes other chromosomal factors involved in this process.

HOW HOMOLOGS SEPARATE IN MI

Regulation of Sister Centromere Cohesion

Although not entirely separate processes, it is useful to think about the segregation of MI homologs in terms of two chromosomal parameters: (1) the state of the centromere, i.e., centromere cohesion and coorientation necessary for sisters to move together in so-called *reductional* behavior (Fig. 3, left), and (2) the state of chromosome arms, i.e., arm cohesion and nonsister chromatid interactions allowing a pair of homologs to separate *accurately* and *regularly* from each other (Fig. 3, right). Chromosome arm interactions via recombination and SCs were described in detail above. In recent years, several complementary advances have occurred in understanding the state of the MI centromere (for review, see Nasmyth 2003). One is the discovery of the cohesin complex in mitosis responsible for keeping sister centromeres and chromatid arms together until anaphase, when a key cohesin (Scc1) is cleaved releasing them for separation (see also Chapter 6). This led to the identification of an important Scc1 meiosis-specific homolog, called Rec8, first defined in *Schizosaccharomyces pombe*, where it was shown to be required for sister centromere and arm cohesion as well as monopolar centromere orientation. Conceptualizing the role of Rec8 in meiosis, a landmark paper by **Watanabe and Nurse (1999)** highlighted below had a major impact in understanding meiotic chromosome behavior and uncovering other activities in this process.

REC8, initially detected in a screen for Rec-deficient fission yeast (Ponticelli and Smith 1989), is expressed in meiosis beginning in premeiotic S phase (Lin et al. 1992; Mata et al. 2002). Besides a drastic reduction in meiotic recombination, loss of *REC8* causes disruption of the lateral elements in meiotic chromosomes and precocious separation of sister chromatids (Molnar et al. 1995). It was thought to be a member of the cohesin family (Michaelis et al. 1997) on the basis of partial sequence analysis (Lin et al. 1992) and later shown to be so by the Nurse laboratory, who independently cloned and sequenced the gene (Watanabe and Nurse 1999). They reported its homology with *S. pombe* and human

RAD21 (mitosis-specific homologs of *SCC1*) and its ability to complement lethality of a *rad21Δ* mutant when expressed in mitosis. Western analysis showed that it replaces Scc1 and is phosphorylated at later stages of the process. Precocious sister-chromatid separation at MI (monitored by a green fluorescent protein [GFP] centromere tag) confirmed the protein's importance for reductional segregation. Coupling *rec8Δ* and *cdc2* (which leads to MII failure) moreover causes cells to undergo a single MI division with equational rather than reductional chromatid separation, similar to the budding yeast *spo13* mutant. As expected, Rec8 localized to centromeric and centromere-adjacent regions, as well as to chromosome arms beginning in S phase. But, most significantly, although Rec8 disappeared from chromosome arms at anaphase I, it remained tightly associated with centromere regions until anaphase II. The regulated degradation of Rec8, first at arms in MI and then at centromeres in MII, provided a key insight into the control of chromosome behavior in the two successive meiotic divisions (see Fig. 4). It was proposed that Rec8 plays two critical roles in promoting reductional division during MI in *S. pombe* in (1) maintaining cohesion at sister-centromeric regions through anaphase I until anaphase II when sisters separate and (2) facilitating monopolar orientation of sister centromeres during MI so that they attach to microtubules, moving them to the same spindle pole. Rec8 is now known to partner with other proteins in different chromosomal domains to execute these dual roles in fission yeast (Kitajima et al. 2003). For example, it associates with Rec11 at centromere-adjacent regions and chromosome arms to control centromere cohesion and recombination and with Psc3 at centromeres to promote monopolar orientation. In budding yeast, Rec8 shows a similar pattern of localization with retention at centromeric sites through anaphase I until anaphase

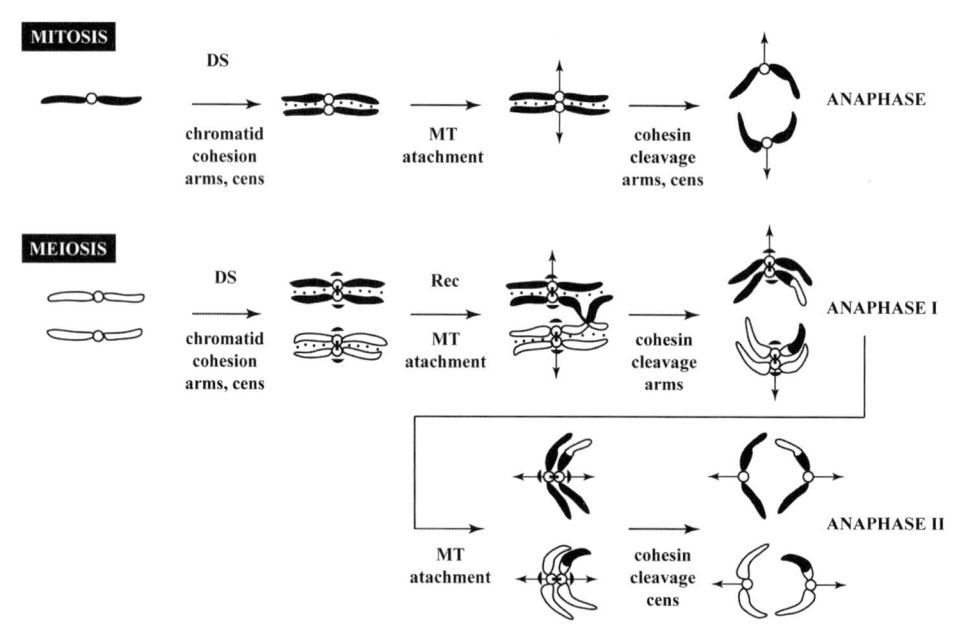

FIGURE 4. Schematic of regulated cohesin cleavage of Scc1 during mitosis and Rec8 in meiosis. In meiosis, Rec8 is found along sister-chromatid arms (*dotted lines*) and at sister centromeres (*black dot between white circles*) where it is protected by shugoshin (*black semicircles*) until anaphase II. Rec8 is cleaved first between chromatid arms at anaphase I and then at centromeres at anaphase II, allowing sister to separate once shugoshin is degraded.

II, when sisters finally separate (Klein et al. 1999; Buonomo et al. 2000). But centromere orientation in this yeast appears to be controlled by an independent complex called monopolin, described below.

An important question about Rec8 behavior in both yeasts is what protects it from degradation at centromeric regions while it is being cleaved by the separase endoprotease (Buonomo et al. 2000) in chromosome arms at anaphase I? After intense pursuit, a "protector" protein called shugoshin (Japanese for "guardian spirit") was found in fission yeast with meiosis-specific and mitotically expressed homologs (Kitajima et al. 2004; Rabitsch et al. 2004). Shugoshin is part of a family defined by *Drosophila* Mei-S332, a centromeric protein regulating cohesion in higher cells. Sgo1, a meiosis-specific protein, was first recovered by screening for loci causing lethality when coexpressed with Rec8 in mitosis (which has little affect alone). Under these conditions, centromeric Rec8 persists, and high levels of reductional sister separation occur during vegetative growth (Kitajima et al. 2004). In meiosis, Sgo1 protects centromeric Rec8 from cleavage until anaphase II, when both proteins are degraded. Surprisingly, in contrast to the loss of Rec8, absence of Sgo1 alone has little or no effect on MI sister segregation, causing their precocious separation and random segregation only in MII. This suggests that Sgo1 has another activity besides Rec8 protection that is partially compensated for by other proteins in MI, not present in MII. It has been proposed that this other activity involves maintenance of centromere orientation. A single shugoshin homolog in budding yeast, which is expressed in both mitosis and meiosis, has similar meiotic behavior (Katis et al. 2004; Kitajima et al. 2004; Marston et al. 2004). In fission yeast, Sgo localization to centromeric regions requires a kinase (Bub1) involved in G_2/M checkpoint control, suggesting that phosphorylation may be necessary for its stabilization and/or localization. Sgo proteins disappear from anaphase II centromeres following degradation by the anaphase-promoting complex (APC). This finding prompted the notion that the budding yeast *SPO13* gene may indirectly control centromere cohesion/coorientation by functioning as a Sgo1 activator (Kitajima et al. 2004) perhaps by preventing its degradation by APC (Katis et al. 2004). This is compatible with observations that Spo13 protects Rec8 from cleavage when both are expressed in mitosis (Schonn et al. 2002) and that it causes mitotic arrest at the metaphase/anaphase transition when overexpressed in mitosis (McCarroll and Esposito 1994). Significantly, prolonging expression of Sgo1 has little effect on meiosis, suggesting that although necessary to protect Rec8, it alone is not sufficient and other proteins are likely involved. At least two other proteins are now known to participate in this process in budding yeast (Marston et al. 2004).

Sister Kinetochore Coorientation and Monopolin

The centromere is the site where spindle microtubules attach to chromosomes (see Chapter 3). But, microtubules do not attach directly to the DNA. They associate with a protein complex containing more than 50 subunits that assembles at centromeres forming a kinetochore. Each sister chromatid typically has one kinetochore-microtubule attachment site that binds to a single microtubule as in budding yeast, or several microtubules as in fission yeast and higher cells (see Chapter 11). During mitosis and in MII, sister kinetochores are bipolar-oriented in what is thought to be a "back-to-back" configuration enabling them to attach to microtubules emanating from opposite spindle poles so that they separate during division. In MI, sister kinetochores are in a monopolar or "side-by-side" orientation, causing both sisters to attach to a microtubule from the same pole so that sisters segregate together (for review, see Hauf and Watanabe 2004). What controls the transition from bipolar to monopolar orientation? The discovery of the monopolin complex in budding yeast, which regulates sister-centromere orientation independently of Rec8, was another important advance in answering this question (Toth et al. 2000). The first member of this complex, Mam1, was identified among deletions of meiotically regulated genes causing precocious sister-centromere separation, assayed by a GFP centromere tag (Toth et al. 2000). In its absence, sister centromeres appear to separate, but chromatids do not fully disjoin until MII because of persistent cohesion. At MI, there is a short metaphase spindle, and anaphase elongation is delayed even though securin (which inhibits separase cohesin cleavage) has been degraded, a condition that normally triggers anaphase. No real MI division occurs, only a single MII division in which sisters move randomly into the four meiotic products. Significantly, when both Mam1 and Sgo1 are lacking, there is no delay in elongation and two meiotic divisions occur, supporting the view that Sgo1 protects Rec8 from cleavage at centromeres. Although the double mutant yields a large fraction of cells exhibiting a precocious equational segregation, many others undergo reductional segregation, so some retention of cohesion occurs even without Sgo1. This, like the lack of phenotype when Sgo1 expression is extended, suggests other components also act in maintaining cohesion. With respect to monopolin, two other subunits are now known, Lrs4 and Csm1 (both of which are also involved in rDNA silencing), but only Mam1 is meiosis-specific. All three proteins are present in the nucleolus and relocate to the nucleus (Rabitsch et al. 2003), dependent on the Cdc5 polo kinase (Lee and Amon 2003), where they assemble into monopolin. What causes the change from monopolar back to bipolar orientation in late anaphase I? Is this due simply to removal of the monopolin complex or to an active reorientation process or both? That an active process is likely involved is suggested by the fact that mitotic bipolar orientation requires the presence of the Aurora B kinase (Cheeseman et al. 2002). Absence of this protein, which localizes to centromeres, results in reductional sister separation. It is thought to detect tension between sisters due to cohesion and bipolar attachment and to act in releasing incorrectly attached kinetochores from microtubules. Mad2, a spindle checkpoint function, is also believed to have a role in facilitating bipolar orientation (Schonn et al. 2003). Finally, Spo13, in addition to protecting Rec8 and Sgo1 from degradation, is also implicated in monopolin recruitment to kinetochores as well as cohesion maintenance (Schonn et al. 2002). Spo13 degradation is therefore also likely crucial for the transition to MII bipolar centromere orientation.

EXIT FROM MI, MII, AND SPORE DEVELOPMENT

M-phase Exit and MII

During mitosis, the transitions from the metaphase to anaphase (M/A) and anaphase to exit (A/Exit) are controlled by APC ubiquitin-mediated proteolysis of first securin (releasing separase for cohesin cleavage, thereby driving anaphase) and then G_2 Clb cyclins (to reset the cycle), respectively. Sequential APC activity at these stages is controlled by distinct regulators associating with APC at M/A and A/Exit (see Chapter 6), which also act in meiosis. During the M/A transition, the APC regulator controlling M-phase exit (Cdh1), is inactivated by phosphorylation, ensuring that cells do not execute this stage prematurely. Subsequent function by a critical phosphatase, Cdc14, activates the exit pathway and is under elaborate control by more than a dozen genes in the FEAR (Cdc Fourteen Early Anaphase Release) and MEN (Mitotic Exit Network) systems, which also act in meiosis (Stegmeier et al. 2002). These systems function sequentially to release Cdc14, sequestered in an inactive form in the nucleolus (for review, see Cohen-Fix 2003; D'Amours and Amon 2004). Spindle disassembly, thought to occur as a result of Clb degradation and Cdk down-regulation after anaphase, requires FEAR, whose components include Spo12 and Slk19 among others (Marston et al. 2003). Mutants in these genes (and *CDC14*) defective late in MI fail to disassemble the MI spindle. However, instead of arresting at anaphase I, they bypass this stage and proceed to MII SPB modification and sister separation on the single MI spindle. The resulting single division yields a mix of equational and reductional segregation depending on strain background (Klapholz and Esposito 1980b). These findings demonstrate that although FEAR-mediated disassembly is clearly not required for meiotic progression, it is essential for two-division meiosis and for ensuring that MI and MII segregation occur on separate spindles (D'Amours and Amon 2004). Interestingly, *spo12* and *slk19*, defective late in MI (i.e., after the point at which paired, recombined homologs are required for proper attachment to the MI spindle), unlike *spo13*, thus cannot rescue Rec⁻ mutants into viable dyads. This implies that the Spo13 function in sister-centromere cohesion/coorientation occurs at or before chromosome attachment so that in its absence, bioriented sister centromeres can proceed to attach and segregate from one another. This difference in the ability of early and late MI segregation functions to rescue Rec⁻ mutants thus provides a useful genetic tool to order genes in the chromosome segregation pathway (for review, see Esposito et al. 1991) and to identify other genes acting in centromere orientation early in MI (Rabitsch et al. 2003).

Spore Development

As in other eukaryotes, the segregation of chromatin at MII is exquisitely coordinated with gamete development. The steps involved in terminal differentiation are currently being rapidly uncovered (for review, see Smits et al. 2001; Coluccio et al. 2004) with SPBs having a central role in the process. During MI, SPBs duplicate, separate to opposite poles of the nuclear membrane, and initiate spindle formation. Following SPB duplication at MII, they undergo a unique meiotic modification essential for spore formation, limiting their ability to associate with cytoplasmic microtubules while promoting meiosis-specific vesicle recruitment for prospore wall synthesis. Specifically, an SPB component that binds to tubulin is lost while several new critical proteins are added. These include four proteins (Ady4, Mpc50, Spo21, and Spo74) expressed in mid-meiosis during the MI/MII transition (Knop and Strasser 2000; Bajgier et al. 2001). Their recruitment is stimulated by a nutritional signal involving acetate utilization (Nickas et al. 2004). Once SPBs with modified outer plaques (MOPs) separate to opposite poles, MII spindle formation and prospore synthesis initiate simultaneously. A specialized branch of the secretory pathway is activated (Neiman 1998) recruiting cytoplasmic vesicles (5–10 nm) behind each modified outer plaque where they fuse to form a flattened sac closely associated with the nuclear membrane. Each spore wall sac then grows bidirectionally along the outer surface of the nuclear membrane. Development of a leading-edge protein (LEP) coat at the tip of the prospore membrane sac requires additional proteins, including Spo14 (Honigberg et al. 1992; Rudge et al. 1998), Ady3 (Neiman et al. 2000), Ssp1/Spo3, and Don1 (Moreno-Borchart et al. 2001). At least four septins are also involved, two of which (Spr3 and Spr28) are specifically expressed in meiosis. They form ring-like structures around each SPB and extend under the membrane as the prospore wall grows (De Virgilio et al. 1996; Fares et al. 1996). Once prospore wall closure occurs, the spore wall matures by sequential addition of two inner layers composed of mannan and glucan, and two outer layers of chitosan (Briza et al. 1988) and dityrosine (Briza et al. 1986), respectively. Further details of spore wall assembly are elegantly described in Coluccio et al. (2004).

A number of protein kinases, phospholipases, and phosphatases are also essential for spore formation, implying the existence of signaling pathways coordinating these events. For example, two meiosis-specific kinases Smk1 (Krisak et al. 1994) and Sps1 (Friesen et al. 1994) and other regulatory kinases—Mps1, involved in spindle checkpoint control (Straight et al. 2000), Cak1, a regulator of Cdc28 (Wagner et al. 1997), and Tor growth control proteins (Zheng and Schreiber 1997)—are all needed. In addition, Spo1, a meiosis-specific phospholipase B (Tevzadze et al. 2000), and Spo14, a phospholipase D (Rose et al. 1995), which act on phosphatidylinositol, a well-known signaling molecule (for review, see Engebrecht 2003), are required for MI and/or MII and spore formation. Finally, phosphatases Tep1 (Heymont et al. 2000) and Glc7 (Tachikawa et al. 2001), which colocalize with septins and regulate their assembly, also have a role. How these signaling molecules interface with meiotic division controls and those regulating spore development is an important area for future work.

COMMITMENT TO MEIOSIS AND DEPENDENCY RELATIONSHIPS OF MEIOTIC EVENTS

Commitment

Commitment to meiosis remains an enigmatic phenomenon. When cells are interrupted in meiotic development and returned to growth they initially retain the ability to revert to mitosis. However, about the time that binucleate cells appear, reflecting that MI is occurring, they lose this ability. Instead, they complete both meiotic divisions and form spores that germinate before reentering the mitotic cell cycle. This transition defines commitment to meiosis. However, as meiosis progresses, cells commit to genetic exchange and the meiotic divisions separately (Sherman and Roman 1963). For example, when diploid cells return to vegetative growth, uncommitted cells exhibit increasing levels of exchange, reaching full intergenic map distances without becoming committed to chromosome segregation or haploidization (R.E. Esposito and M.S. Esposito 1974; Esposito et al. 1974). Some differences are known to occur in the way strand exchange is completed (Zenvirth et al. 1997), but the pathway(s) by which meiotic cells reenter the mitotic cell cycle remains largely undefined. Remarkably, cells are uncommitted to the meiotic divisions even after initiation of the transcriptional program and SC assembly. This may, in part, reflect the fact that early meiotic messages are highly unstable and rapidly disappear during growth, whereas those made after commitment are relatively long-lived (Surosky and Esposito 1992). Most surprisingly, however, it is now clear that commitment to meiotic development is not irreversible as once thought. Cells arrested after MI by high temperature or mutations which block progression to MII (*spo14*) or to spores (*spo3*) retain the capacity to directly revert to mitosis by a default mechanism when terminal differentiation is blocked (Honigberg et al. 1992). On the basis of these findings, it has been proposed that commitment to meiosis results from a competitive process between going forward in meiosis and returning to mitosis, favoring meiotic events once commitment occurs (Honigberg and Esposito 1994). Consistent with this idea is the finding that the budding process is delayed in cells that return to mitosis after MI (in temperature-arrested wild-type cells or mutants blocked in further progression) compared to cells that return earlier before MI. This suggests that commitment to meiosis may result from a programmed event(s) between recombination and MI that transiently delays budding once the meiotic divisions begin so that cells returned to growth first complete meiosis before initiating cell division.

Dependency of Events

A principal insight gained from analysis of meiotic mutants in yeast is that there is a remarkable flexibility in the cell's ability to omit certain stages and still produce viable products (Figs. 1 and 4). The temporal order of events is not controlled by a sequential dependency pathway with a single START stage committing cells to successive landmarks as in mitosis. Instead, they become committed to meiotic landmarks independently without strictly depending on the initiation of a prior event for their progression. In some cases, loss of specific gene functions result in arrest of meiotic development. As in mitosis, such functions may define crucial steps whose absence causes surveillance systems to terminate the process. Figure 5 summarizes the phenotypes and interdependence of a subset of known meiotic mutants. These relationships have led to several important conclusions about the coordination of meiotic events.

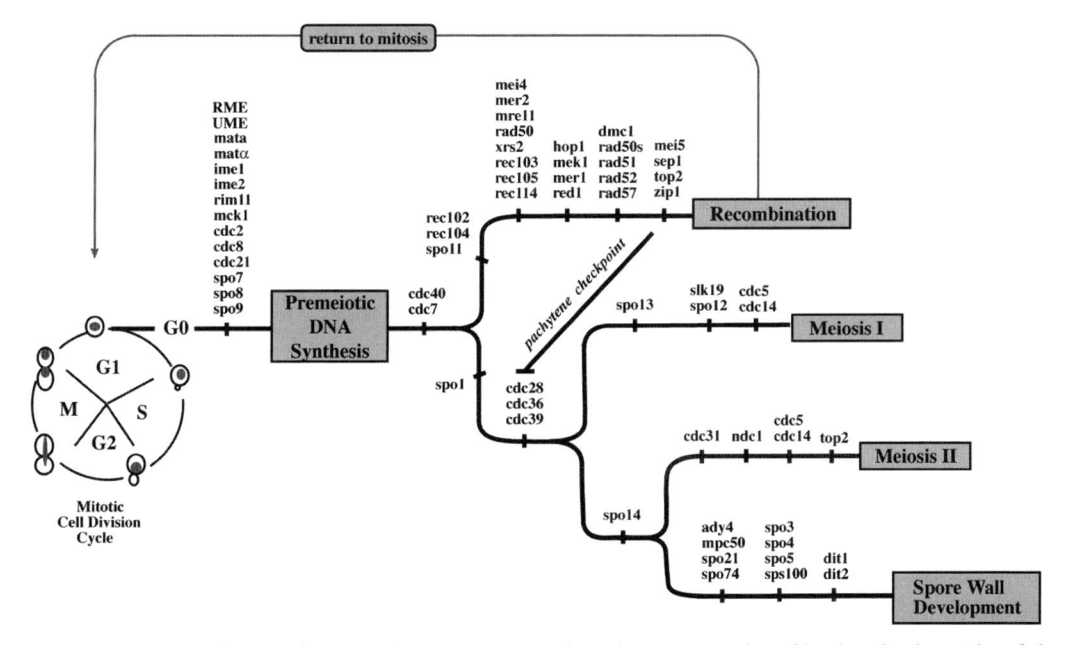

FIGURE 5. Dependency scheme with return to growth pathway. Examples of landmarks that either fail to occur or are defective in some well-studied mutants illustrate various branches leading to recombination; MI, MII, and spore formation can occur independently of one another. See text for details.

1. After meiotic entry, premeiotic DNA synthesis is essential for all subsequent landmarks as all mutants blocking DNA synthesis are unable to progress further in the process (M.S. Esposito and R.E. Esposito 1974).

2. Events up to and including recombination initiation are neither essential for nor commit cells to MI separation of homologs. Cells can still return to mitosis and separate sister chromatids directly after meiotic recombination as shown in return-to-growth studies (R.E. Esposito and M.S. Esposito 1974). In addition, recombination initiation mutants (e.g., *rad50* and *spo11*) can undergo spindle assembly, cosegregation of sisters at MI, sister separation at MII, and spore formation, although immature or inviable spores form due to random segregation of unpaired homologs at MI (Klapholz and Esposito 1982). Once recombination starts and DSBs are made, checkpoint controls arrest development until the DSBs are repaired (for review, see Bailis and Roeder 2000). However, without broken chromosomes, recombination per se is not required for subsequent progression.

3. Prior separation of homologs at MI is neither essential for nor commits cells to subsequent separation of sisters as equational division can occur in the absence of reductional division (Klapholz and Esposito 1980b) and vice versa (Schild and Byers 1980). MI spindle disassembly is also dispensable for subsequent sister separation as this fails to occur in *slk19* and *spo12*, which nonetheless undergo equational division (Marston et al. 2003).

4. MII completion is not required to initiate spore formation as mutants can form spores after MI (e.g., *cdc5*, *cdc14*, *cdc28* at semipermissive temperatures) (Schild and Byers 1980). Indeed, MI homolog separation is also not a precondition since cells can form spores after a single equational division (e.g., in *spo12*, *spo13*) (Klapholz and Esposito 1980b). SPB modification appears to be all that is required for prospore membrane synthesis (Davidow et al. 1980). Although SPB modification and spore formation typically occur after SPB duplication at MII (just before equational separation), it can occur precociously after SPB duplication at MI, in the absence of sister-centromere cohesion/coorientation (in *spo12* and spo13) (Moens 1974), suggesting it may be negatively regulated by the state of the MI centromere.

These findings raise an important question. If the primary landmarks of meiosis can occur independently of initiation of prior events, then how is the temporal order of events determined and how are they integrated with cell division cycle controls? Work from a number of labs has shown that although progression through the meiotic cell cycle is driven by much of the same machinery as mitosis, it also requires unique products whose regulation is governed by a meiotic transcriptional cascade. These meiosis-specific components replace and/or modify critical mitotic components to alter the pattern of cell division and control meiotic progression.

REGULATION OF MEIOSIS AND CHECKPOINT CONTROLS

The Meiotic Transcriptional Program

Both entry into meiosis and subsequent progression are driven by a highly regulated transcriptional program activated by signal pathways responding to nutritional and cell-type cues as described earlier (for review, see Vershon and Pierce 2000; Kassir et al. 2003). The transcriptional program (see Fig. 6) was initially uncovered by analyzing steady-state messsage levels of known genes specifically required for the process and using them as genetic tools to identify regulators governing their expression. The first genes analyzed defined the very early (Kassir et al. 1988), early (Atcheson et al. 1987; Wang et al. 1987; Smith and Mitchell 1989), and middle (Malavasic and Elder 1990) expression classes. Complementary reverse genetic approaches recovering relatively abundant meiosis-specific messages detected middle, mid-late, and late classes (Clancy et al. 1983; Percival-Smith and Segall 1984). More than a decade after these pioneering efforts, a dramatic leap forward occurred with application of DNA microarray analysis to budding yeast meiosis (Chu et al. 1998; Primig et al. 2000). These studies detected nearly 900 developmentally regulated core transcripts in two strains commonly used in meiotic analyses. Most (~95%) are expressed only in **a**/α and not in **a**/**a** or α/α nutritionally deprived cells and are specific to meiosis rather than starvation per se, with approximately 500 of these having a very high correlation coefficient (.8) in their pattern of expression in the two strains (Primig et al. 2000). Strikingly, only a subset of transcripts (<20%) are essential for the process (Rabitsch et al. 2001). The remaining may encode redundant functions, may be fortuitously expressed by common regulatory elements with meiotic genes, or are indirect targets of nonspecific regulators expressed during the process. Interestingly, most early messages have extremely short half-lives (~3 min) compared to middle and late classes (~15 min) (Surosky and Esposito 1992) and are found in lower abundancy (Primig et al. 2000). Although regulation of transcription and not message stability is the major mechanism responsible for differential expression classes, the rapid degradation of certain transcripts after induction ceases may be significant for meiotic progression.

Several proteins are presently known to promote the transcriptional program once cells enter meiosis: (1) the Ime1-Ume6 complex, whose formation is a precondition for the induction of most, if not all, early genes (Bowdish et al. 1995; Steber and Esposito 1995; Rubin-Bejerano et al. 1996), (2) Ndt80, required for expression of many middle and mid/late loci (Chu and Herskowitz 1998; Hepworth et al. 1998), and (3) Abf1, a general constitutive transcription factor that stimulates induction in all classes (Prinz et al. 1995; Gailus-Durner et al. 1996; Ozsarac et al. 1997). No specific transcription factors for late gene expression are known. Promoter elements needed for meiotic expression include URS1 (upstream repression sequence 1) in early genes (Buckingham et al. 1990), MSE (mid-sporulation element) in middle genes (Hepworth et al. 1995), and UAS (upstream activation sequence) elements in all classes (Vershon et al. 1992; Bowdish and Mitchell 1993). Other key regulators also have essential roles in this process. For example, nine *UME* loci negatively controlling early meiotic transcription in mitosis were found by selecting for mutations causing "*u*nscheduled *m*eiotic *e*xpression" during growth (Strich et al. 1989). Surprisingly, the derepressed expression of early genes in these mutants has minimal effect on growth in glucose medium. Several are part of a critical early regulatory switch controlling both repression and induction of most early genes. Ume6, a central component of this switch, binds to URS1 elements upstream of early genes (Strich et al. 1994). During mitosis, it keeps early expression off by associating with at least two chromatin-remodeling complexes preventing recruitment of RNA polymerase to the transcriptional initiation site: one, a histone deacetylase complex Sin3/Rpd3 (Kadosh and Struhl 1997) recovered as Ume4/Ume7, and the other, an ATPase, Isw2 (Kadosh and Struhl 1997; Goldmark et al. 2000). During meiosis, early expression is turned on by interaction between Ume6 and

Ime1, which is proposed to provide a meiosis-specific transcriptional activation domain (Smith et al. 1993; Mandel et al. 1994). Significantly, a general activation domain (e.g., GAL4) fused to Ume6 cannot substitute for Ime1 as long as the Ume6-binding site for the deacetylase complex is present. However, it can partially replace Ime1 when the Ume6-binding site for the deacetylase complex is mutated so that it is nonfunctional (Washburn and Esposito 2001). This implies that initiation of early gene expression is a two-step process in which Ime1 both relieves deacetylase corepressor function and promotes activation. Recent studies indicating that Rim11-mediated phosphorylation of Ime1 alleviates deacetylase repression (Rubin-Bejerano et al. 2004) support this view. Rim15 also appears to play a role in this process (independent of Ime1), as it is required to transiently eject the deacetylase from the Ume6/URS1 complex during meiosis (Prueli et al. 2004).

Interaction between Ume6 and Ime1 and subsequent activation of meiotic transcription requires nutritional signaling, which in part depends on Rim11 and Rim15 kinases (Bowdish et al. 1994; Rubin-Bejerano et al. 1996; Vidan and Mitchell 1997). These genes were identified by screening for mutants in Regulators of *IME2* expression, which detected an allele of *UME6* and other genes mediating nutritional regulation (Su and Mitchell 1993). Meiotic gene activation also requires other chromatin-remodeling factors including histone acetylase (Burgess et al. 1999) and RSC proteins (Yukawa et al. 1999). mRNA methylation is additionally thought to have a role in transcript accumulation during sporulation (Clancy et al. 2002). The precise mechanism regulating the switch from repression to

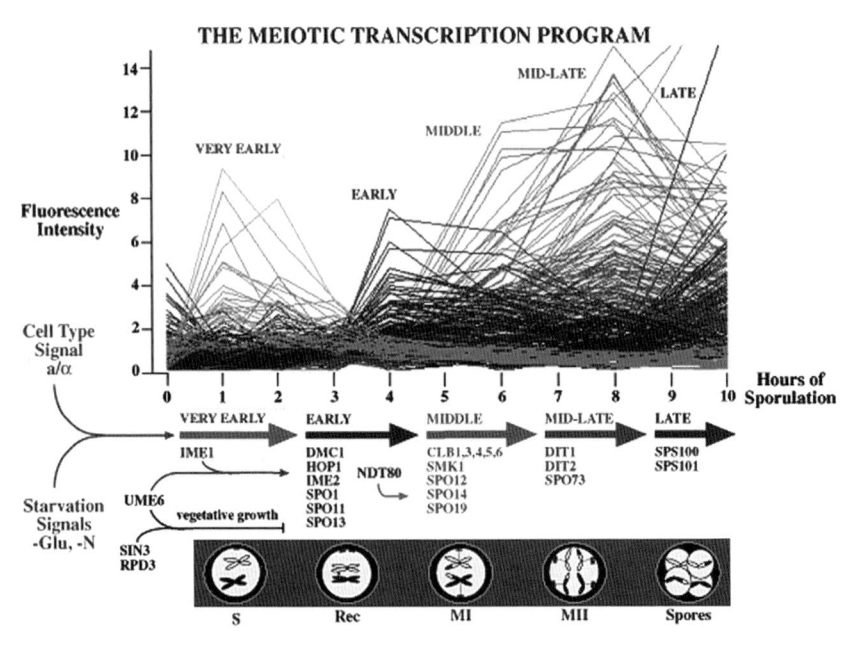

FIGURE 6. (*Top*) Waves of transcription of different expression classes during sporulation of strain SK1, detected by DNA microarray analysis using Affymetrix GeneChips; only lower abundancy transcripts are shown for comparison (data from Primig et al. 2000). (*Middle*) Schematic of cell type and nutritional signals leading to Ume6-Ime1 induction of early meiotic genes, followed by Ndt80 activation of middle and mid-late genes with examples of prototype genes in these classes. (*Bottom*) Stages of meiosis occurring cotemporally with various expression classes.

activation at early promoters is still under study. However, DNA microarray analysis now indicates that approximately 80 genes in the yeast genome are regulated by the Ume6/URS1 complex (Williams et al. 2002). About half are early meiotic genes and the rest act primarily in carbon and nitrogen metabolism during starvation, suggesting that Ume6 is a global regulator coupling the starvation response to subsequent induction of early meiotic events. Among the other Ume proteins required for repression, Ume1 is part of the Sin3-Rpd3 repression complex (Mallory and Strich 2003), whereas Ume2, Ume3, and Ume5 are components of the RNA polymerase II holoenzyme, i.e., Srb9, Srb11, and Srb10 (Liao et al. 1995). Interestingly, Ume2 and Ume5 (a Ser/Thr kinase) promote rapid turnover of early meiotic mRNAs (Surosky et al. 1994), whereas Ume3 and Ume5 form a nonessential cyclin-C-dependent kinase (Cdk) that phosphorylates RNA polymerase II to negatively regulate its function. Ume3, unlike other cyclins, does not vary during the mitotic cell cycle, but it is specifically degraded prior to MI by ubiquitin-mediated proteolysis and is required for efficient MI (Cooper et al. 1997; Cooper and Strich 2002).

Ndt80, the Pachytene Checkpoint, and the Transition to Middle Gene Expression

Once early genes are expressed, the next major transition to middle gene induction is mediated by Ndt80. Like Ime1, Ndt80 is regulated by both transcriptional and posttranslational mechanisms. It is induced as part of a distinct pre-middle class at the end of pachytene (Hepworth et al. 1998) and binds to MSE promoter elements (Hepworth et al. 1995; Ozsarac et al. 1997) in many, but not all, middle genes, including itself (Chu and Herskowitz 1998). During vegetative growth and early in meiosis, middle genes are kept off by a repressor, Sum1, which also binds to MSEs (Xie et al. 1999). The Ndt80 promoter has both URS1 and MSE sites, which account for its unique expression between early and mid genes (Pak and Segall 2002a). Early in meiosis, Sum1 binding to the MSE prevents Ime1 activation at the URS1 site. Later, Sum1 becomes destabilized (Lindgren et al. 2000) and low-level expression occurs, permitting Ndt80 autoregulation and induction of other middle genes. Ime2 also phosphorylates and contributes to full activity of Ndt80 but is not essential for its function (Sopko et al. 2002; Shubassi et al. 2003). Importantly, this leads to higher levels of expression of genes encoding G_2 cyclins and thus Cdc28-Clb activity, which together with Ndt80 acts in a feedback loop increasing Ime2 kinase levels needed for MI and MII (Benjamin et al. 2003).

The early-to-middle expression transition is controlled, in part, by the pachytene checkpoint (for review, see Bailis and Roeder 2000), which keeps middle genes off until recombination is completed. Pachytene arrest was first described in yeast by the Byers group, who studied the behavior of Cdc28 mutants in meiosis (Shuster and Byers 1989). Conditional mutants in *Cdc28* enter meiosis, execute premeiotic S phase, and arrest at pachytene with duplicated but unseparated SPBs and full-length SCs. Mutation of *NDT80*, a gene that was initially found in a screen for general Spo⁻ mutants, led to a similar phenotype, prompting the suggestion that it might be needed for pachytene exit by regulating Cdc28 activity (Xu et al. 1995). Rec⁻ mutants defective in completing exchange, such as *dmc1* (Bishop et al. 1992) and *zip1* (Sym et al. 1993), also arrested at this crucial point, dependent on DNA-damage-sensing checkpoint functions such as Rad17, Rad24, and Mec1 (Lydall et al. 1996). A landmark paper in understanding exit from pachytene occurred with the identification of Ndt80, in the Segall laboratory, as a regulator of middle gene expression (**Hepworth et al. 1998**). This report provided clear evidence that the pachytene checkpoint regulates progression to the meiotic divisions by controlling Ndt80 induction of middle gene expression. Ndt80 was found in this case by selecting for mutants failing to express a middle gene reporter during haploid meiosis while still expressing early genes. Induction of Ndt80 and subsequent middle expression was shown to be required for entry into the meiotic divisions and not the reverse, since mutations in G_2 cyclins (*CLB1*, *CLB3*, and *CLB4*) needed for the divisions did not block middle expression. Significantly, Ndt80 activity and middle gene expression were blocked in a Rec⁻ mutant that provoked pachytene arrest (*dmc1*), but restored by mutations in the DNA-damage checkpoint genes. This finding provided the critical link between completion of recombination, lifting of checkpoint controls, and induction of middle gene expression. More extensive genome-wide DNA microarray analysis soon after defined a large class of middle genes under Ndt80 control (Chu et al. 1998), among them the G_2 cyclins (Chu and Herskowitz 1998) needed to initiate the meiotic divisions. The finding that pachytene arrest could be bypassed by mutations that enhanced middle expression and thus Clb expression implies that the checkpoint ultimately acts to prevent Cdc28-Clb activity needed for progression of the meiotic divisions (Pak and Segall 2002b). It does this in multiple ways by (1) stabilizing Sum1 to repress Ndt80 (Lindgren et al. 2000), (2) inhibiting activation of any protein made by preventing Ndt80 phosphorylation (Tung et al. 2000), and (3) promoting Swe1 phosphorylation of Cdc28 to block any existing Cdc28-Clb function (Leu and Roeder 1999). The pachytene checkpoint system provoked by unrepaired DNA breaks is probably not the only pathway regulating this crucial transition. Evidence exists that other parallel pathways transiently stop entry into M phase until proper recombination (A. Hochwagen and A. Amon, unpubl.) and centromere cohesion/coorientation (McCarroll and Esposito 1994) are completed.

The Spindle Checkpoint and Subsequent Progression

Once cells enter M phase, another surveillance system—the spindle checkpoint pathway (mediated by Mad proteins)—monitors chromosome-microtubule attachment and tension, arresting cells at metaphase when the segregation machinery is defective. The Mad system, extensively studied in mitosis, blocks anaphase by interacting with Cdc20, an activator of APC needed to degrade securin and release separase-triggering anaphase (see Chapter 6). In meiosis, the Mad system also senses the lack of tension (e.g., in a *spo11* Rec⁻ mutants) and delays destruction of securin (Schonn et al. 2000). Mad2 has also been shown to have a role in proper attachment of homologs providing time for reorientation of misaligned chromosomes (Schonn et al. 2003). Finally, the spindle checkpoint appears to be responsible for the single division phenotype of *spo13* mutants, since checkpoint mutants permit two divisions to occur (producing highly aneuploid, mostly inviable products). In this case, metaphase delay triggered by the absence of Spo13 allows spores to form before cells can enter MII (Schonn et al. 2002). In mitosis, as noted earlier, sequential APC degradation of securin and then of Clb2 (mediated by a Cdc20-like regulator, Cdh1) regulates first anaphase and then exit from M phase, respectively. Interestingly, in meiosis, a new developmental-specific Cdc20 family member (Ama1) mediates Clb1 degradation at MI exit and is needed for late gene expression and spore assembly as well (Cooper et al. 2000). The precise regulatory mechanisms, however, coordinating progression through the divisions with later transcription and spore development remain to be determined.

CONCLUDING REMARKS

The extensive knowledge of meiosis in yeast is providing an important model for understanding this process in other systems. As a result of genome-wide DNA microarray analysis of meiotically regulated genes and deletion studies, we now know for the first time nearly all of the genes required for meiosis in a simple eukaryote. Similar studies are being pursued in a variety of organisms including flies, worms, plants, and mammals (see GermOnline Database http://www.germonline.org) (Primig et al. 2003). Comparative genomics of the behavior of essential meiotic genes in these systems will ultimately lead to a more comprehensive view not only of critical genetic controls of this important process, but also of its evolution across species.

ACKNOWLEDGMENTS

Special thanks to Y. Kassir, M. Primig, J. Segall, and G. Tevzadze for valuable discussions and helpful comments on this chapter and to J. Stortz for assistance with the figures. The many contributions of former members of my lab to my understanding of meiosis are gratefully acknowledged. Finally, sincere apologies to colleagues whose work I was not able to cite. The writing of this chapter was supported by National Institutes of Health grant GM-29182.

STUDY QUESTIONS

1. What are the key differences in chromosome behavior between mitosis and meiosis?

2. What is haploid meiosis? What are the requirements for this process and why is it useful?

3. How has the "bypass rationale" for obtaining mutants been useful in the study of meiosis?

4. What are the key factors that promote entry into meiotic development? What are the roles of RME, IME, and UME genes in this process?

5. Why are centromere cohesion and coorientation important in meiosis? How does the regulation of the meiosis-specific Rec8 cohesin affect the behavior of chromosomes in the two divisions?

6. How does the pachytene checkpoint regulate meiotic progression?

7. Discuss how a mutant early in the Rec pathway can be used to order early and late MI segregation functions and how a mutant early in the MI segregation pathway can be used to order early and late Rec functions.

8. Discuss the overall dependence of events in meiosis and key controls regulating meiotic progression.

temperature-sensitive sporulation-deficient mutants. *Genetics* **61:** 79–89.
———. 1974. Genes controlling meiosis and spore formation in yeast. *Genetics* **78:** 215–225.
———. 1975. Mutants of meiosis and ascospore formation. *Methods Cell Biol.* **11:** 303–326.
Esposito R.E. and Esposito M.S. 1974. Genetic recombination and commitment to meiosis in *Saccharomyces*. *Proc. Natl. Acad. Sci.* **71:** 3172–3176.
Esposito R.E. and Klapholz S. 1981. Meiosis and ascospore development. In *The molecular biology of the yeast* Saccharomyces: *Life cycle and inheritance* (ed. J.N. Strathern et al.), pp. 211–287. Cold Spring Harbor Laboratory Press, Cold Spring Harbor, New York.
Esposito R.E., Dresser M., and Breitenbach M. 1991. Identifying sporulation genes, visualizing synaptonemal complexes and large-scale spore and spore-wall purification. *Methods Enzymol.* **194:** 110–131.
Esposito R.E., Plotkin D.J., and Esposito M.S. 1974. The relationship between genetic recombination and commitment to chromosomal segregation in meiosis. In *Mechanisms of recombination* (ed. R.F. Grell), pp. 277–285. Plenum Press, New York.
Esposito R.E., Frink N., Bernstein P., and Esposito M.S. 1972. The genetic control of sporulation in *Saccharomyces*. II. Dominance and complementation of mutants of meiosis and spore formation. *Mol. Gen. Genet.* **114:** 241–248.
Fares H., Goetsch L., and Pringle J.R. 1996. Identification of a developmentally regulated septin and involvement of the septins in spore formation in *Saccharomyces cerevisiae*. *J. Cell Biol.* **132:** 399–411.
Freese E.B., Chu M.I., and Freese E. 1982. Initiation of yeast sporulation of partial carbon, nitrogen, or phosphate deprivation. *J. Bacteriol.* **149:** 840–851.
Friesen H., Lunz R., Doyle S., and Segall J. 1994. Mutation of the SPS1-encoded protein kinase of *Saccharomyces cerevisiae* leads to defects in transcription and morphology during spore formation. *Genes Dev.* **8:** 2162–2175.
Fung J.C., Rockmill B., Odell M., and Roeder R.G. 2004. Imposition of crossover interference through the nonrandom distribution of synapsis initiation complexes. *Cell* **116:** 795–802.
Gailus-Durner V., Xie J., Chintamaneni C., and Vershon A.K. 1996. Participation of the yeast activator Abf1 in meiosis-specific expression of the *HOP1* gene. *Mol. Cell. Biol.* **16:** 2777–2786.
Gallego C., Gari E., Colomina N., Herrero E., and Aldea M. 1997. The Cln3 cyclin is down-regulated by translational repression and degradation during G1 arrest caused by nitrogen deprivation in budding yeast. *EMBO J.* **16:** 7196–7206.
Ganesan A.T., Holter H., and Roberts R. 1958. Some observations on sporulation in *Saccharomyces*. *C.R. Lab. Carlsberg* **13:** 1–6.
Goetsch L. and Byers B. 1982. Meiotic cytology of *Saccharomyces cerevisiae* in protoplast lysates. *Mol. Gen. Genet.* **187:** 54–60.
Goldmark J.P., Fazzio T.G., Estep P.W., Church G.M., and Tsukiyama T. 2000. The Isw2 chromatin remodeling complex represses early meiotic genes upon recruitment by Ume6p. *Cell* **103:** 423–433.
Gorner W., Durchschlag E., Martinez-Pastor M.T., Estruch F., Ammerer G., Hamilton B., Ruis H., and Schuller C. 1998. Nuclear localization of the C2H2 zinc finger protein Msn2p is regulated by stress and protein kinase A activity. *Genes Dev.* **12:** 586–597.
Grewal N.S. and Miller J.J. 1972. Formation of asci with two diploid spores by diploid cells of *Saccharomyces cerevisiae*. *Can. J. Microbiol.* **18:** 1897–1905.
Guttmann-Raviv N., Boger-Nadjar E., Edri I., and Kassir Y. 2001. Cdc28 and Ime2 possess redundant functions in promoting entry into premeiotic DNA replication in *Saccharomyces cerevisiae*. *Genetics* **159:** 1547–1558.
Hartwell L. 1974. *Saccharomyces cerevisiae* cell cycle. *Bacteriol. Rev.* **38:** 164–198.
Hauf S. and Watanabe Y. 2004. Kinetochore orientation in mitosis and meiosis. *Cell* **119:** 317–327.
Hawthorne D.C. 1963. Directed mutation of the mating type allele as an explanation of homothallism in yeast. *Proc. Int. Congr. Genet.* **1:** 34–35 (Abstr.).
Henderson K.A. and Keeney S. 2004. Tying synaptonemal complex initiation to the formation and programmed repair of DNA double-strand breaks. *Proc. Natl. Acad. Sci.* **101:** 4519–4524.
Hepworth S.R., Ebisuzaki L.K., and Segall J. 1995. A 15 base-pair element activates the *SPS4* gene midway through sporulation in *Saccharomyces cerevisiae*. *Mol. Cell. Biol.* **15:** 3934–3944.
Hepworth S.R., Friesen H., and Segall J. 1998. *NDT80* and the meiotic recombination checkpoint regulate expression of middle sporulation-specific genes in *Saccharomyces cerevisiae*. *Mol. Cell. Biol.* **10:** 5750–5761.
Hertwig O. 1890. Vergleich der Ei- und Samenbildung bei Nematoden: Eine Grundlage für celluläre

Streitfragen. *Arch. Mikrosc. Anat.* **36:** 1–138.
Heymont J., Berenfeld L., Collins J., Kaganovich A., Maynes B., Moulin A., Ratskovskaya I., Poon P.P., Johnston G.C., Kamenetsky M., DeSilva J., Sun H., Petsko G.A., and Engebrecht J. 2000. TEP1, the yeast homolog of the human tumor suppressor gene PTEN/MMAC1/TEP1, is linked to the phosphatidylinositol pathway and plays a role in the developmental process of sporulation. *Proc. Natl. Acad. Sci.* **97:** 12672–12677.
Hollingsworth N.M. and Byers B. 1989. *HOP1:* A yeast meiotic pairing gene. *Genetics* **121:** 445–462.
Hollingsworth N.M., Goetsch L., and Byers B. 1990. The *HOP1* gene encodes a meiosis-specific component of yeast chromosomes. *Cell* **61:** 73–84.
Honigberg S.M. and Esposito R.E. 1994. Reversal of cell determination in yeast meiosis: Post-commitment arrest allows return to mitotic growth. *Proc. Natl. Acad. Sci.* **91:** 6559–6563.
Honigberg S.M. and Lee R.H. 1998. Snf1 kinase connects nutritional pathways controlling meiosis in *Saccharomyces cerevisiae*. *Mol. Cell. Biol.* **18:** 4548–4555.
Honigberg S.M., Conicella C., and Esposito R.E. 1992. Commitment to meiosis in *Saccharomyces cerevisiae:* Involvement of the *SPO14* gene. *Genetics* **130:** 703–716.
Hopper A.K. and Hall B.D. 1975. Mating type and sporulation in yeast. I. Mutations which alter mating-type control over sporulation. *Genetics* **80:** 41–59.
Hugerat Y. and Simchen F. 1993. Mixed segregation and recombination of chromosomes and YACs during single division meiosis in *spo13* strains of *Saccharomyces cerevisiae*. *Genetics* **155:** 297–308.
Kadosh D. and Struhl K. 1997. Repression by Ume6 involves recruitment of a complex containing Sin3 corepressor and Rpd3 histone deacetylase to target promoters. *Cell* **89:** 365–371.
Kane S. and Roth R. 1974. Carbohydrate metabolism during ascospore development in yeast. *J. Bacteriol.* **118:** 8–14.
Kassir Y. and Simchen G. 1976. Regulation of mating and meiosis in yeast by the mating-type region. *Genetics* **82:** 187–206.
Kassir Y., Granot D., and Simchen G. 1988. IME1, a positive regulator gene of meiosis in *S. cerevisiae*. *Cell* **52:** 853–862.
Kassir Y., Adir N., Boger-Nadjar E., Raviv N.G., Rubin-Bejerano I., Sagee S., and Shenhar G. 2003. Transcriptional regulation of meiosis in budding yeast. *Int. Rev. Cytol.* **224:** 111–171.
Katis V.L., Galova M., Rabitsch K.P., Gregan J., and Nasmyth K. 2004. Maintenance of cohesin at centromeres after meiosis I in budding yeast requires a kinetochore-associated protein related to mei-S332. *Curr. Biol.* **14:** 560–572.
Keeney S., Girouz C.N., and Kleckner N. 1997. Meiosis-specific double strand breaks are catalyzed by Spo11, a member of a widely conserved protein family. *Cell* **88:** 375–384.
Kitajima T.S., Kawashima S.A., and Watanabe Y. 2004. The conserved kinetochore protein shugoshin protects centromeric cohesion during meiosis. *Nature* **427:** 510–517.
Kitajima T.S., Yokobayashi S., Yamamoto M., and Watanabe Y. 2003. Distinct cohesin complexes organize meiotic chromosome domains. *Science* **300:** 1152–1155.
Klapholz S. and Esposito R.E. 1980a. Isolation of *spo12-1* and *spo13-1* from a natural variant of yeast that undergoes a single meiotic division. *Genetics* **96:** 567–588.
———. 1980b. Recombination and chromosome segregation during the single division meiosis in *spo12-1* and *spo13-1* diploids. *Genetics* **96:** 589–611.
———. 1982. A new mapping method employing a meiotic Rec-mutant of yeast. *Genetics* **100:** 387–412.
Klapholz S., Waddell C.S., and Esposito R.E. 1985. The role of the *SPO11* gene in meiotic recombination in yeast. *Genetics* **110:** 187–216.
Klein F., Mahr P., Galova M., Buonomo S.B., Michaelis C., Nairz K., and Nasmyth L. 1999. A central role for cohesins in sister chromatid cohesin, formation of axial elements and recombination during yeast meiosis. *Cell* **98:** 91–103.
Knop M. and Strasser K. 2000. Role of the spindle pole body of yeast in mediating assembly of the prospore membrane during meiosis. *EMBO J.* **19:** 3657–3667.
Kominami K.Y., Sakata Y., Sakai M., and Yamashita I. 1993. Protein kinase activity associated with the *IME2* gene product, a meiotic inducer in the yeast *Saccharomyces cerevisiae*. *Biosci. Biotechnol. Biochem.* **57:** 1731–1735.
Krisak L., Strich R., Winters R.S., Hall J.P., Mallory M.J., Kreitzer D., Tuan R.S., and Winter E. 1994. SMK1,

a developmentally regulated MAP kinase, is required for spore wall assembly in *Saccharomyces cerevisiae*. *Genes Dev.* **8:** 2151–2161.

Kupiec M., Byers B., Esposito R.E., and Mitchell A.P. 1997. Meiosis and sporulation in *Saccharomyces cerevisiae*. In *The molecular and cellular biology of the yeast* Saccharomyces: *Cell cycle and cell biology* (ed. J.R. Pringle et al.), vol. 3, pp. 889–1036. Cold Spring Harbor Laboratory Press, Cold Spring Harbor, New York.

Lee B.H. and Amon A. 2003. Role of Polo-like kinase Cdc5 in programming meiosis chromosome segregation. *Science* **300:** 482–486.

Leu J.Y. and Roeder G.S. 1999. The pachytene checkpoint in *S. cerevisiae* depends on Swe1-mediated phosphorylation of the cyclin-dependent kinase Cdc28. *Mol. Cell* **4:** 805–814.

Liao S.M., Zhang J., Jeffery D.A., Koleske A.J., Thompson C.M., Chao D.M., Viljoen M., van Vuuren H.J., and Young R.A. 1995. A kinase-cyclin pair in the RNA polymerase II holoenzyme. *Nature* **374:** 193–196.

Lin Y., Larson K.L., Dorr R., and Smith G.R. 1992. Meiotically induced *rec7* and *rec8* genes in *Schizosaccharomyces pombe*. *Genetics* **132:** 75–85.

Lindgren A., Bungard D., Pierce M., Xie J., Vershon A., and Winter E. 2000. The pachytene checkpoint in *Saccharomyces cerevisiae* requires the Sum1 transcriptional repressor. *EMBO J.* **19:** 6489–6497.

Loidl J., Klein F., and Scherthan H. 1994. Homologous pairing is reduced but not abolished in asynaptic mutants of yeast. *J. Cell Biol.* **125:** 1191–1200.

Loidl J., Nairz K., and Klein F. 1991. Meiotic chromosome synapsis in a haploid yeast. *Chromosoma* **100:** 221–228.

Lorenz A., Wells J.A., Pryce D.W., Novatchkova M., Eisenhaber F., McFarlane R.J., and Loidl J. 2004. *S. pombe* meiotic linear elements contain proteins related to synaptonemal complex components. *J. Cell Sci.* **117:** 3343–3351.

Lydall D., Nikolsky Y., Bishop D.K., and Weinert T. 1996. A meiotic recombination checkpoint controlled by mitotic checkpoint genes. *Nature* **383:** 840–843.

Malavasic M.J. and Elder R.T. 1990. Complementary transcripts from two genes necessary for normal meiosis in the yeast *Saccharomyces cerevisiae*. *Mol. Cell. Biol.* **10:** 2809–2819.

Mallory M.J. and Strich R. 2003. Ume1p represses meiotic gene transcription in *Saccharomyces cerevisiae* through interaction with the histone deacetylase Rpd3. *J. Biol. Chem.* **278:** 44727–44734.

Malone R.E. 1983. Multiple mutant analysis of recombination in yeast. *Mol. Gen. Genet.* **189:** 405–412.

Malone R.E. and Esposito R.E. 1981. Recombinationless meiosis in *Saccharomyces cerevisiae*. *Mol. Cell. Biol.* **1:** 891–901.

Mandel S., Robzyk K., and Kassir Y. 1994. *IME1* gene encodes a transcription factor which is required to induce meiosis in *Saccharomyces cerevisiae*. *Dev. Genet.* **15:** 139–147.

Mao-Draayer Y., Galbraith A.M., Pittman D.L., Cool M., and Malone R.E. 1996. Analysis of meiotic recombination pathways in the yeast *Saccharomyces cerevisiae*. *Genetics* **144:** 71–86.

Marston A., Lee B.H., and Amon A. 2003. The Cdc14 phosphatase and the FEAR network control meiotic spindle disassembly and chromosome segregation. *Dev. Cell* **4:** 711–728.

Marston A., Tham W.H., Shah H., and Amon A. 2004. A genome-wide screen identifies genes required for centromeric cohesion. *Science* **303:** 1367–1370.

Mata J., Lyne R., Burns G., and Bahler J. 2002. The transcriptional program of meiosis and sporulation in fission yeast. *Nat. Genet.* **32:** 143–147.

Matsumoto K., Uno I., and Ishikawa T. 1983. Initiation of meiosis in yeast mutants defective in adenylate cyclase and cyclic AMP-dependent protein kinase. *Cell* **32:** 417–423.

McCarroll R.M. and Esposito R.E. 1994. *SPO13* negatively regulates the progression of mitotic and meiotic nuclear division in *Saccharomyces cerevisiae*. *Genetics* **138:** 47–60.

Michaelis C., Ciosk R., and Nasmyth K. 1997. Cohesins: Chromosomal proteins that prevent premature separation of sister chromatids. *Cell* **91:** 35–45.

Mitchell A.P. and Herskowitz I. 1986. Activation of meiosis and sporulation by repression of the RME1 product in yeast. *Nature* **319:** 738–742.

Moens P.B. 1971. Fine structure of ascospore development in the yeast *Saccharomyces cerevisiae*. *Can. J. Microbiol.* **17:** 507–510.

———. 1974. Modification of sporulation in yeast strains with two-spored asci (*Saccharomyces,* Ascomycetes). *J. Cell Sci.* **16:** 519–527.

Moens P. and Rapport E. 1971a. Synaptic structures in the nuclei of sporulating yeast, *Saccharomyces cerevisiae* (Hansen). *J. Cell Sci.* **9:** 665–677.
———. 1971b. Spindles, spindle plaques, and meiosis in the yeast *Saccharomyces cerevisiae* (Hansen). *J. Cell Biol.* **50:** 344–361.
Molnar M., Bahler J., Sipiczki M., and Kohli J. 1995. The *rec8* gene of *S. pombe* is involved in linear element formation, chromosome pairing and sister chromatid cohesion in meiosis. *Genetics* **141:** 61–73.
Moreno-Borchart A.C. and Knop M. 2003. Prospore membrane formation: How budding yeast gets shaped in meiosis. *Microbiol. Res.* **158:** 83–90.
Moreno-Borchart A.C., Strasser K., Finkbeiner M.G., Shevchenko A., Shevchenko A., and Knop M. 2001. Prospore membrane formation linked to the leading edge protein (LEP) coat assembly. *EMBO J.* **20:** 6946–6957.
Moses M.J. 1956. Chromosomes structures in crayfish spermatocytes. *J. Biophys. Biochem. Cytol.* **2:** 215–218.
Nasmyth K. 2002. Segregating sister genomes: The molecular biology of chromosome segregation. *Science* **297:** 559–564.
———. 2003. Un menage a quatre: The molecular biology of chromosome segregation in meiosis. *Cell* **12:** 423–440.
Neigeborn L. and Mitchell A.P. 1991. The yeast *MCK1* gene encodes a protein kinase homolog that activates early meiotic gene expression. *Genes Dev.* **5:** 533–548.
Neiman A.M. 1998. Prospore membrane formation defines a developmentally regulated branch of the secretory pathway in yeast. *J. Cell Biol.* **140:** 29–37.
Neiman A.M., Katz L., and Brennwald P.J. 2000. Identification of domains required for developmentally regulated SNARE function in *Saccharomyces cerevisiae*. *Genetics* **155:** 1643–1655.
Nickas M.E., Diamond A.E., Yang M.J., and Neiman A.M. 2004. Regulation of spindle pole function by an intermediary metabolite. *Mol. Biol. Cell* **15:** 2606–2616.
Nicklas R.B. and Koch C.A. 1969. Chromosome micromanipulation. III. Spindle fiber tension and the reorientation of mal-oriented chromosomes. *J. Cell Biol.* **43:** 40–50.
Ozsarac N., Straffon M.J., Dalton H.E., and Dawes I.W. 1997. Regulation of gene expression during meiosis in *Saccharomyces cerevisiae*: *SPR3* is controlled by both ABFI and a new sporulation control element. *Mol. Cell. Biol.* **17:** 1152–1159.
Padmore R., Cao L., and Kleckner N. 1991. Temporal comparison of recombination and synaptonemal complex formation during meiosis in *S. cerevisiae*. *Cell* **66:** 1239–1256.
Page S.L. and Hawley R.S. 2004. The genetics and molecular biology of the synaptonemal complex. *Annu. Rev. Cell Dev. Biol.* **20:** 525–558.
Pak J. and Segall J. 2002a. Regulation of the premiddle and middle phases of expression of the *NDT80* gene during sporulation of *Saccharomyces cerevisiae*. *Mol. Cell. Biol.* **22:** 6417–6429.
———. 2002b. Role of Ndt80, Sum1 and Swe1 as targets of the meiotic recombination checkpoint that control exit from pachytene and spore formation in *Saccharomyces cerevisiae*. *Mol. Cell. Biol.* **22:** 6430–6440.
Paliulis L.V. and Nicklas R.B. 2000. The reduction of chromosome number in meiosis is determined by properties built into chromosomes. *J. Cell Biol.* **150:** 1223–1231.
Pedruzzi I., Dubouloz F., Cameroni E., Wanke V., Roosen J., Winderickx J., and De Virgilio C. 2003. TOR and PKA signalling pathways converge on the protein kinase Rim15 to control entry into Go. *Mol. Cell* **12:** 1607–1613.
Percival-Smith A. and Segall J. 1984. Isolation of DNS sequences preferentially expressed during sporulation in *Saccharomyces cerevisiae*. *Mol. Cell. Biol.* **4:** 142–150.
Pinon R. 1977. Effects of ammonium ions on sporulation of *Saccharomyces cerevisiae*. *Exp. Cell Res.* **105:** 367–378.
Pnueli L., Edry I., Cohen M., and Kassir Y. 2004. Glucose and nitrogen regulate the switch from histone deacetylation to acetylation for expression of early meiosis-specific genes in budding yeast. *Mol. Cell. Biol.* **24:** 5197–5208.
Ponticelli A.S. and Smith G.R. 1989. Meiotic recombination-deficient mutants of *Schizosaccharomyces pombe*. *Genetics* **123:** 45–54.
Primig M., Williams R.M., Winzeler E.A., Tevzadze G.G., Conway A.R., Hwang S.Y., Davis R.W., and Esposito R.E. 2000. The core meiotic transcriptome in budding yeast. *Nat. Genet.* **26:** 415–423.
Primig M., Wiederkehr C., Basavaraj R., Sarrauste de Menthiere C., Hermida L., Koch R., Schlecht U.,

Dickinson H.G., Fellous M., Grootegoed J.A., Hawley R.S., Jegou B., Maro B., Nicolas A., Orr-Weaver T., Schedl T., Villeneuve A., Wolgemuth D.J., Yamamoto M., Zickler D., Lamb N., and Esposito R.E. 2003. GermOnline, a new cross-species community annotation database on germ-line development and gametogenesis. *Nat. Genet.* **35:** 291–292.

Prinz S., Klein F., Auer H., Schweizer D., and Primig M. 1995. A DNA binding factor (UBF) interacts with a positive regulatory element in the promoters of genes expressed during meiosis and vegetative growth in yeast. *Nucleic Acids Res.* **23:** 3449–3456.

Purnapatre K., Gray M., Piccirillo S., and Honigberg S.M. 2005. Glucose inhibits meiotic DNA replication through SCF Grr1p-dependent destruction of Ime2p kinase. *Mol. Cell. Biol.* **25:** 440–450.

Purnapatre K., Piccirillo S., Schneider B.L., and Honigberg S.M. 2002. The *CLN3/SWI6/CLN2* pathway and *SNF1* act sequentially to regulate meiotic initiation in *Saccharomyces cerevisiae*. *Genes Cells* **7:** 675–691.

Rabitsch K.P., Gregan J., Schleiffer A., Javerzat J.P., Eisenhaber F., and Nasmyth K. 2004. Two fission yeast homologs of *Drosophila* Mei-S332 are required for chromosome segregation during meiosis I and meiosis II. *Curr. Biol.* **14:** 287–301.

Rabitsch K.P., Petronczki M., Javerzat J.P., Genier S., Chwalla B., Schleiffer A., Tanaka T.U., and Nasmyth K. 2003. Kinetochore recruitment of two nucleolar proteins is required for homolog segregation in meiosis I. *Dev. Cell* **4:** 535–548.

Rabitsch K.P., Toth A., Galova M., Schleiffer A., Schaffner G., Aigner E., Rupp C., Penkner A.M., Moreno-Borchart A.C., Primig M., Esposito R.E., Klein F., Knop M., and Nasmyth K. 2001. A screen for genes required for meiosis and spore formation based on whole-genome expression. *Curr. Biol.* **11:** 1001–1009.

Rockmill B. and Roeder G.S. 1988. *RED1*: A yeast gene required for the segregation of chromosomes during the reductional division of meiosis. *Proc. Natl. Acad. Sci.* **85:** 6057–6061.

———. 1990. Meiosis in asynaptic yeast. *Genetics* **126:** 563–574.

Roman H. and Sands S.M. 1953. Heterogeneity of clones of *Saccharomyces* derived from haploid ascospores. *Proc. Natl. Acad. Sci.* **39:** 171–179.

Rose K., Rudge S.A., Frohman M.A., Morris A.J., and Engebrecht J. 1995. Phospholipase D signaling is essential for meiosis. *Proc. Natl. Acad. Sci.* **92:** 12151–12155.

Roth R. and Fogel S. 1971. A selective system for yeast mutants defective in meiotic recombination. *Mol. Gen. Genet.* **112:** 295–305.

Roth R. and Lusnak K. 1970. DNA synthesis during yeast sporulation: Genetic control of an early developmental event. *Science* **168:** 493–494.

Rubin-Bejerano I., Mandel S., Robzyk K., and Kassir Y. 1996. Induction of meiosis in *Saccharomyces cerevisiae* depends on conversion of the transcriptional repressor Ume6 to a positive regulator by its regulated association with the transcriptional activator Ime1. *Mol. Cell. Biol.* **16:** 2518–2526.

Rubin-Bejerano I., Sagee S., Friedman O., Pnueli L., and Kassir Y. 2004. The in vivo activity of Ime1, the key transcriptional activator of meiosis-specific genes in *Saccharomyces cerevisiae*, is inhibited by the cyclic AMP/PKA signal pathway through the glycogen synthase kinase 3-beta homolog Rim11. *Mol. Cell. Biol.* **24:** 6967–6979.

Rudge S.A., Morris A.J., and Engebrecht J. 1998. Relocalization of phospholipase D activity mediates membrane formation during meiosis. *J. Cell Biol.* **140:** 81–90.

Rutkowski L.H. and Esposito R.E. 2000. Recombination can partially substitute for *SPO13* in regulating meiosis I in budding yeast. *Genetics* **155:** 1607–1621.

Sagee S., Sherman A., Shenhar G., Robzyk K., Ben-Doy N., Simchen G., and Kassir Y. 1998. Multiple and distinct activation and repression sequences mediate the regulated transcription of *IME1*, a transcriptional activator of meiosis-specific genes in *Saccharomyces cerevisiae*. *Mol. Cell. Biol.* **18:** 1985–1995.

Schild D. and Byers B. 1980. Diploid spore formation and other meiotic effects of two cell-division-cycle mutations of *Saccharomyces cerevisiae*. *Genetics* **96:** 859–876.

Schlecht U. and Primig M. 2003. Mining meiosis and gametogenesis with DNA microarrays. *Reproduction* **125:** 447–456.

Schonn M.A., McCarroll R., and Murray A.W. 2000. Requirement of the spindle checkpoint for proper chromosome segregation in budding yeast meiosis. *Science* **289:** 300–303.

———. 2002. Spo13 protects meiotic cohesion at centromeres in meiosis I. *Genes Dev.* **16:** 1659–1671.

Schonn M.A., Murray A.L., and Murray A.W. 2003. Spindle checkpoint component Mad2 contributes to bior-

ientation of homologous chromosomes. *Curr. Biol.* **13:** 1979–1984.
Shenhar G. and Kassir Y. 2001. A positive regulator of mitosis, Sok2, functions as a negative regulator of meiosis in *Saccharomyces cerevisiae*. *Mol. Cell. Biol.* **21:** 1603–1612.
Sherman F. and Roman H. 1963. Evidence for two types of allelic recombination in yeast. *Genetics* **48:** 255–261.
Shubassi G., Luca N., Pak J., and Segall J. 2003. Activity of phosphoforms and truncated versions of Ndt80, a checkpoint-regulated sporulation-specific transcription factor of *Saccharomyces cerevisiae*. *Mol. Gen. Genomics* **270:** 324–336.
Shuster E.O. and Byers B. 1989. Pachytene arrest and other meiotic effects of the start mutations in *Saccharomyces cerevisiae*. *Genetics* **123:** 29–43.
Smith A.V. and Roeder G.S. 1997. The yeast Red1 protein localizes to the cores of meiotic chromosomes. *J. Cell Biol.* **136:** 957–967.
Smith H.E. and Mitchell A.P. 1989. A transcriptional cascade governs entry into meiosis in *Saccharomyces cerevisiae*. *Mol. Cell. Biol.* **9:** 2142–2152.
Smith H.E., Driscoll S.E., Sia R.A., Yuan H.E., and Mitchell A.P. 1993. Genetic evidence for transcriptional activation by the yeast *IME1* gene product. *Genetics* **133:** 775–784.
Smits G.J., van den Ende H., and Klis F.M. 2001. Differential regulation of cell wall biogenesis during growth and development in yeast. *Microbiology* **147:** 781–794.
Sopko R., Raithatha S., and Stuart D. 2002. Phosphorylation and maximal activity of the *Saccharomyces cerevisiae* meiosis-specific transcription factor NDT80 is dependent on Ime2. *Mol. Cell. Biol.* **22:** 7024–7040.
Steber C.M. and Esposito R.E. 1995. Ume6 is a central component of a developmental regulatory switch controlling meiosis-specific gene expression. *Proc. Natl. Acad. Sci.* **92:** 12490–12494.
Stegmeier F., Visintin R., and Amon A. 2002. Separase, polo kinase, the kinetochore protein Slk19 and Spo12 function in a network that controls Cdc14 localization during early anaphase. *Cell* **108:** 207–220.
Straight P.D., Giddings T.H., and Winey M. 2000. Mps1p regulates meiotic spindle pole body duplication in addition to having novel roles during sporulation. *Mol. Biol. Cell.* **11:** 3525–3537.
Strathern J., Hicks J., and Herskowitz I. 1981. Control of cell type in yeast by the mating-type locus. The α1-α2 hypothesis. *J. Mol. Biol.* **147:** 357–372.
Strich R., Slater M.R., and Esposito R.E. 1989. Identification of negative regulatory genes that govern the expression of early meiotic genes in yeast. *Proc. Natl. Acad. Sci.* **86:** 10018–10022.
Strich R., Surosky R.T., Steber C., Dubois E., Messenguy F., and Esposito R.E. 1994. *UME6* is a key regulator of nitrogen repression and meiotic development. *Genes Dev.* **8:** 796–810.
Stuart D. and Wittenberg C. 1998. *CLB5* and *CLB6* are required for premeiotic DNA replication and activation of the meiotic S/M checkpoint. *Genes Dev.* **12:** 2698–2710.
Su S.S.Y. and Mitchell A.P. 1993. Identification of functionally related genes that stimulate early meiotic gene expression in yeast. *Genetics* **133:** 67–77.
Surosky R.T. and Esposito R.E. 1992. Early meiotic transcripts are highly unstable in *Saccharomyces cerevisiae*. *Mol. Cell. Biol.* **12:** 3948–3958.
Surosky R.T., Strich R., and Esposito R.E. 1994. The yeast *UME5* gene regulates the stability of meiotic mRNAs in response to glucose. *Mol. Cell. Biol.* **14:** 3446–3458.
Sym M. and Roeder G.S. 1994. Crossover interference is abolished in the absence of a synaptonemal complex protein. *Cell* **79:** 283–292.
———. 1995. Zip1-induced changes in SC structure and polycomplex assembly. *J. Cell Biol.* **128:** 455–466.
Sym M., Engebrecht J., and Roeder G.S. 1993. ZIP1 is a synaptonemal complex protein required for meiotic chromosome synapsis. *Cell* **72:** 365–378.
Tachikawa H., Bloecher A., Tatchell K., and Neiman A.M. 2001. A Gip1p-Glc7p phosphatase complex regulates septin organization and spore wall formation. *J. Cell Biol.* **155:** 797–808.
Tevzadze G.G., Swift H., and Esposito R.E. 2000. Spo1, a phospholipase B homolog, is required for spindle pole body duplication during meiosis in *Saccharomyces cerevisiae*. *Chromosoma* **109:** 72–85.
Toone W.M., Johnson A.L., Banks G.R., Toyn J.H., Stuart D., Wittenberg C., and Johnston L.H. 1995. Rme1, a negative regulator of meiosis, is also a positive activator of G1 cyclin gene expression. *EMBO J.* **14:** 5824–5832.
Toth A., Rabitsch K.P., Galova M., Schleiffer A., Buonomo S.B., and Nasmyth K. 2000. Functional genomics identifies monopolin: A kinetochore protein required for segregation of homologs during meiosis I. *Cell*

103: 1156–1168.
Tung K.S. and Roeder G.S. 1998. Meiotic chromosome morphology and behavior in *zip1* mutants of *Saccharomyces cerevisiae. Genetics* **149:** 817–832.
Tung K.S., Hong E.J., and Roeder G.S. 2000. The pachytene checkpoint prevents accumulation and phosphorylation of the meiosis-specific transcription factor Ndt80. *Proc. Natl. Acad. Sci.* **97:** 12187–12192.
van Beneden E. 1883. Recherches sur la maturation de l'oeuf et la fécondation. *Arch. Biol.* **4:** 610–620.
Varma A., Freese E.B., and Freese E. 1985. Partial deprivation of GTP initiates meiosis and sporulation in *Saccharomyces cerevisiae. Mol. Gen. Genet.* **201:** 1–6.
Vershon A.K. and Pierce M. 2000. Transcription regulation of meiosis in yeast. *Curr. Opin. Cell Biol.* **12:** 334–339.
Vershon A.K., Hollingsworth N.M., and Johnson A.D. 1992. Meiotic induction of the yeast *HOP1* gene is controlled by positive and negative regulatory sites. *Mol. Cell. Biol.* **12:** 3706–3714.
Vidan S. and Mitchell A.P. 1997. Stimulation of yeast meiotic gene expression by the glucose-repressible protein kinase Rim15p. *Mol. Cell. Biol.* **17:** 2688–2697.
von Wettstein D., Rasmussen S.W., and Holm P.B. 1984. The synaptonemal complex in genetic segregation. *Annu. Rev. Genet.* **16:** 331–413.
Wagner M., Pierce M., and Winter E. 1997. The CDK-activating kinase *CAK1* can dosage suppress sporulation defects of smk1 MAP kinase mutants and is required for spore wall morphogenesis in *Saccharomyces cerevisiae. EMBO J.* **16:** 1305–1317.
Wagstaff J.E., Klapholz S., and Esposito R.E. 1982. Meiosis in haploid yeast. *Proc. Natl. Acad. Sci.* **79:** 2986–2990.
Wan L., de los Santos T., Zhang C., Shokat K., and Hollingsworth N.M. 2004. Mek1 kinase activity functions downstream of *RED1* in the regulation of meiotic double strand break repair in budding yeast. *Mol. Biol. Cell* **15:** 11–23.
Wang H.T., Frackman S., Kowalisyn J., Esposito R.E., and Elder R. 1987. Developmental regulation of *SPO13*, a gene required for segregation of homologous chromosomes at meiosis I. *Mol. Cell. Biol.* **7:** 1425–1435.
Washburn B.K. and Esposito R.E. 2001. Identification of the Sin3-binding site in Ume6 defines a two-step process for conversion of Ume6 from a transcriptional repressor to an activator in yeast. *Mol. Cell. Biol.* **21:** 2057–2069.
Watanabe Y. and Nurse P. 1999. Cohesion Rec8 is required for reductional chromosome segregation at meiosis. *Nature* **400:** 461–464.
Williams R.M., Primig M., Washburn B.K., Winzeler E.A., Bellis M., Sarrauste de Menthiere C., Davis R.W., and Esposito R.E. 2002. The Ume6 regulon coordinates metabolic and meiotic gene expression in yeast. *Proc. Natl. Acad. Sci.* **99:** 13431–13436.
Winge Ø. and Roberts C. 1949. A gene for diploidization in yeasts. *C.R. Trav. Lab. Carlsberg Ser. Physiol.* **24:** 341–346.
Xie J., Pierce M., Gailus-Durner V., Wagner M., Winter E., and Vershon A.K. 1999. Sum1 and Hst1 repress middle sporulation-specific gene expression during mitosis in *Saccharomyces cerevisiae. EMBO J.* **18:** 6448–6454.
Xu L., Ajimura M., Padmore R., Klein C., and Kleckner N. 1995. *NDT80*, a meiosis-specific gene required for exit from pachytene in *Saccharomyces cerevisiae. Mol. Cell. Biol.* **15:** 6572–6581.
Yoshida M., Kawaguchi H., Sakata Y., Kominami K., Hirano M., Shima H., Akada R., and Yamashita I. 1990. Initiation of meiosis and sporulation in *Saccharomyces cerevisiae* requires a novel protein kinase homologue. *Mol. Gen. Genet.* **221:** 176–186.
Yukawa M., Katoh S., Miyakawa T., and Tsuchiya E. 1999. Nps1/Sth1p, a component of an essential chromatin-remodeling complex of *Saccharomyces cerevisiae*, is required for the maximal expression of early meiotic genes. *Genes Cells* **4:** 99–110.
Zenvirth D., Loidl J., Klein S., Arbel A., Shemesh R., and Simchen G. 1997. Switching yeast from meiosis to mitosis: Double-strand break repair, recombination and synaptonemal complex. *Genes Cells* **2:** 487–498.
Zheng X.F. and Schreiber S.L. 1997. Target of rapamycin proteins and their kinase activities are required for meiosis. *Proc. Natl. Acad. Sci.* **94:** 3070–3075.
Zickler D. and Kleckner N. 1998. The leptotene-zygotene transition of meiosis. *Annu. Rev. Genet.* **32:** 619–697.
———. 1999. Meiotic chromosomes: Integrating structure and function. *Annu. Rev. Genet.* **33:** 603–754.

10

Signal Transduction

Jeremy Thorner

Department of Molecular and Cell Biology
Divisions of Biochemistry and Molecular Biology and Cell and Developmental Biology
University of California
Berkeley, California 94720-3202

Identification of nonmating mutants defines *STE* genes required for pheromone response, 194

Hartwell L.H. 1980. Mutants of *Saccharomyces cerevisiae* unresponsive to cell division control by polypeptide mating hormone. *J. Cell Biol.* **85:** 811–822.

***STE* genes represent components of a pathway initiated by a G-protein-coupled receptor, 198**

Whiteway M., Hougan L., Dignard D., Thomas D.Y., Bell L., Saari G.C., Grant F.J., O'Hara P., and MacKay V.L. 1989. The *STE4* and *STE18* genes of yeast encode potential β and γ subunits of the mating factor receptor-coupled G protein. *Cell* **56:** 467–477.

Other *STE* gene products define a modular protein kinase cascade, 199

Stevenson B.J., Rhodes N., Errede B., and Sprague G.F., Jr. 1992. Constitutive mutants of the protein kinase STE11 activate the yeast pheromone response pathway in the absence of the G protein. *Genes Dev.* **6:** 1293–1304.

Development of the concept of a scaffold protein, 201

Choi K-Y., Satterberg B., Lyons D.M., and Elion E.A. 1994. Ste5 tethers multiple protein kinases in the MAP kinase cascade required for mating in *S. cerevisiae*. *Cell* **78:** 499–512.

Note: The landmark papers listed above are those discussed in this chapter. Each landmark paper is preceded by the name of the section (with starting page number) where the paper is first discussed in detail.

MANY FUNDAMENTAL PROCESSES IN CELL REGULATION THAT FALL under the rubric of signal transduction mechanisms and are now known to be conserved throughout the evolution of eukaryotes were first uncovered through studies of the biology of *Saccharomyces cerevisiae*. These successes have occurred in areas too numerous to be treated in a single chapter in this volume. For this reason, the focus here is on how examination of the response of haploid yeast cells to their secreted, peptide-mating pheromones was instrumental in showing how G-protein-coupled receptors (GPCRs) and their cognate heterotrimeric G proteins act and in the discovery of mitogen-activated protein kinases (MAPKs) and MAPK cascades.

Note: Boldfaced references in the text denote landmark papers that are on the accompanying CD.

Several additional chapters in this volume provide perspectives on the development of our current understanding of other signaling pathways that control the ability of the cell to adjust appropriately to a diversity of changes in its external environment.

HISTORICAL PERSPECTIVE: OVERVIEW AND BACKGROUND

S. cerevisiae (baker's yeast), a budding ascomycete, exists in three distinct cell types. There are two haploids, dubbed **a** cells and α cells. The third cell type, an **a**/α diploid, is formed by the conjugation or "mating" of an **a** and an α cell, just as any two haploid gametes fuse to form a diploid zygote. The **a**/α diploid cannot mate, but it can undergo meiosis and sporulation (which the haploids cannot), yielding four spores (two **a** and two α). It was inferred almost 50 years ago that haploid yeast cells secrete a diffusible active substance that induces in a haploid cell of the opposite type its acquisition of gamete-like characteristics (Levi 1956). It took nearly another 20 years before it was shown definitively that the agent from α cells responsible for this induction in **a** cells is a peptide of defined sequence (Düntze et al. 1970; Ciejek et al. 1977). Indeed, when the mating response was first reviewed comprehensively from the point of view of its serving as a model for elucidating how extracellular peptides trigger a developmental decision (Thorner 1981), the only component thoroughly described at the molecular level was the mating pheromone (α-factor) secreted by α haploid cells. The pheromone (**a**-factor) secreted by **a** cells that acts on the α cells was not fully characterized and shown to be a lipopeptide for more than another half dozen years (Anderegg et al. 1988).

Remarkably, however, by the time the field was reviewed comprehensively again just a bit over a decade later, the majority of the gene products involved in the processes required for pheromone response and mating were identified, cloned, and biochemically characterized (Sprague and Thorner 1992). Moreover, the strategies that were devised for analyzing the functions of these proteins and for determining the sequence in which they act provided a generally applicable scheme for logically delineating the order of function of the participants in any signaling pathway. Since that time, our level of understanding of these signaling events and the network of interactions among them has reached a very sophisticated level (Dohlman and Thorner 2001). During the course of this continued study, yet again, more features and components have been revealed that are also conserved throughout eukaryotic evolution, and they have pivotal roles in the regulation of all signaling pathways. Examples include the principle that MAPKs are recruited to their activators, substrates, and regulators by binding to high-affinity docking motifs (Bardwell and Thorner 1996; Tanoue and Nishida 2002) and the role of the regulator of G-protein signaling (RGS) family of proteins in down-modulating GPCR-initiated signaling (Dohlman and Thorner 1997). The four papers selected and discussed below were particularly seminal in providing major advances in our knowledge and/or our conceptual understanding and led to very rapid progress; most significantly, they had a major impact on redirecting the thinking of scientists studying signaling in other organisms.

IDENTIFICATION OF NONMATING MUTANTS DEFINES *STE* GENES REQUIRED FOR PHEROMONE RESPONSE

It was already known from pioneering genetic studies of Robert K. Mortimer and Donald C. Hawthorne that the properties of the three different cell types (**a** and α haploids and the

a/α diploid) were dictated by a single genetic locus (*MAT*) near the centromere of chromosome III (Mortimer and Hawthorne 1966). How to go about determining whether or not all of the genes required for the mating process reside at *MAT*? Vivian L. MacKay and Thomas R. Manney took the approach of devising a genetic selection for identifying mating-deficient mutants and using it to isolate a large collection of mating-defective cells (MacKay and Manney 1974a,b). Genetic analysis of these mutants (complementation tests, segregation analysis, mapping) all required crosses, which demanded, of course, that the cells retain at least some residual capacity for mating. In some cases, the strains mated at such a low frequency that it could not be determined with confidence whether the resulting diploid arose from the original mutant haploid or because that mutant haploid picked up an additional rare mutation (suppressor) that allowed it to mate. Despite these drawbacks, MacKay and Manney were able to carry out an analysis of their collection. They found that some of the mutants (obtained from α cells, but not **a** cells) carried mutations at the *MAT* locus itself. However, quite a number of the nonmating ("sterile" or *ste*) mutants in their collection did not map to *MAT*. On the basis of these findings, MacKay and Manney proposed that *MAT* encodes regulatory proteins that control the expression of other genes that specify the products needed for mating- and sporulation-specific functions (MacKay and Manney 1974b). This prescient hypothesis was borne out, with the advent of recombinant DNA technology, by molecular analysis of the DNA and the transcripts from the *MAT***a** and *MAT*α loci (Herskowitz and Oshima 1981; Nasmyth 1982; Sprague et al. 1983).

As far as signaling goes, however, there should be, conceptually speaking, two basic classes of nonmating mutants: those that cannot produce the signal (pheromone), and those that cannot respond to it. The latter are the more interesting class from the point of view of dissecting the gene products required by the cell for transducing its exposure to a peptide signal into appropriate physiological responses. In the mid 1970s, Leland H. Hartwell, in collaboration with Manney, had demonstrated that the presence of the α mating pheromone caused **a** cells to arrest their growth specifically in the G_1 phase of the cell division cycle (Fig. 1) (Bücking-Throm et al. 1973). As described in the first landmark paper discussed here (**Hartwell 1980**), because of his interest in how the action of mating pheromone interdicts cell cycle progression, Hartwell constructed a strain that allowed him to isolate temperature-conditional mutations that caused **a** cells to be insensitive to α-factor-induced cell division arrest. Strikingly, he found that all of the mutants he isolated were unable to mate at the nonpermissive temperature, suggesting that cell division arrest is an essential aspect of the mating process. These mutants could mate at essentially normal frequency at the permissive temperature; hence, they could be readily crossed against α cells and the mutations present analyzed with confidence. After meiosis and sporulation, the mutations could be segregated into both **a** and α cells. By this means, the cell type specificity of each mutation could be determined (i.e., whether it caused a mating defect only in the **a** cells in which it was isolated or also conferred sterility when present in the background of an α cell). Furthermore, having the mutations segregated into both **a** and α cells permitted backcrosses against each of the other mutant classes and against the preexisting *ste* mutants of MacKay and Manney, thereby allowing complementation tests among the mutations.

Hartwell's findings about the pheromone arrest-resistant *ste* mutations permitted some of the first incisive inferences to be made about the rules governing the wiring of this signal transduction pathway. First, none of the genes defined by Hartwell's mutations were

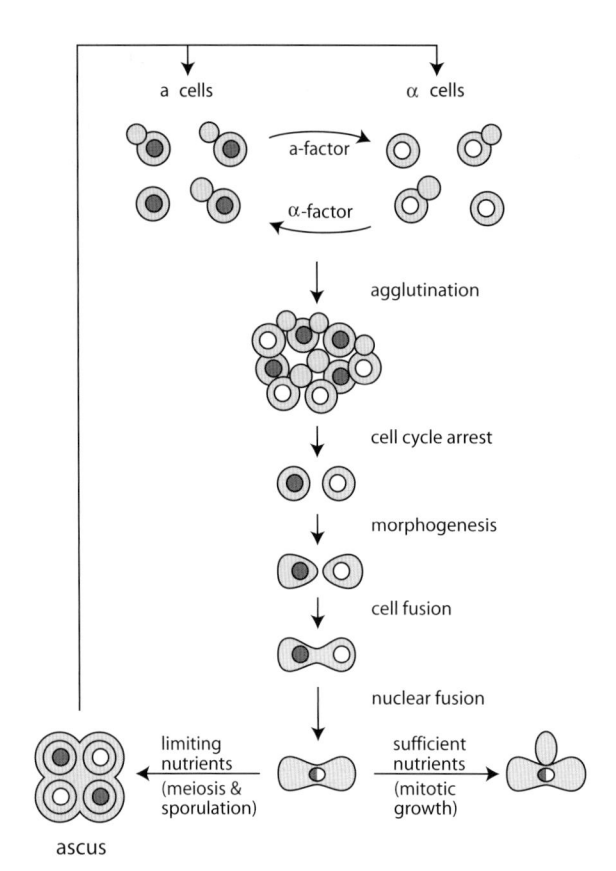

FIGURE 1. Schematic representation of the conjugation process in *S. cerevisiae*. (Redrawn, with permission, from Thorner 1981.)

linked to *MAT*, three of them (*ste2, ste4,* and *ste5*) corresponded to loci previously found by MacKay and Manney, and only one of them (*ste2*) was specific to **a** cells. Likewise, MacKay and Manney had found only one *ste* mutation unlinked to *MAT* that was α-cell-specific (*ste3*). It had already been demonstrated that purified, and even synthetic, α-factor was able to cause G_1 arrest and to trigger all other known **a** cell responses (i.e., no other product of, or contact with, live α cells seemed to be necessary) (Ciejek et al. 1977). Likewise, α cells displayed all known mating-specific responses when exposed to purified **a**-factor (Betz et al. 1977). Therefore, Hartwell proposed that *STE2* encodes the receptor on **a** cells for α-factor, and *STE3* encodes the receptor on α cells for **a**-factor, conclusions that were amply confirmed by subsequent work (Hagen et al. 1986; Blumer et al. 1988). In fact, the deduced primary structures derived from nucleotide sequence analysis of *STE2* and *STE3* (Burkholder and Hartwell 1985; Nakayama et al. 1985) predicted Ste2 and Ste3 to be the second and third seven-transmembrane-spanning segment (7-TMS) proteins identified in any organism (the first being bovine retinal rhodopsin [Nathans and Hogness 1983] and the fourth being the hamster β-adrenergic receptor [Dixon et al. 1986]). Now, thousands of such 7-TMS receptors are known and are responsible for transducing the responses of eukaryotic cells to a plethora of stimuli, including light, odorants, neurotransmitters, peptide hormones, chemokines, opiates, and many other classes of bioactive compounds (Dohlman et al. 1991; Pierce et al. 2002).

The second telling conclusion that Hartwell reached from examining his mutants is that the remainder of the genes required for response must perform functions in both **a** and α cells that are essentially equivalent since mutations in these genes caused both kinds of haploids to be equally mating-defective. Moreover, all of the mutations were recessive, indicative of loss-of-function alleles. Furthermore, the presence of wild-type cells of the same mating type could not restore mating proficiency to any of the mating-defective mutants (nor did the mating-defective mutants vitiate the mating ability of the wild-type cells), indicating that the defects were cell autonomous. Taken together, these characteristics are those expected for mutations that affect gene products involved in propagating an intracellular signal.

Hartwell reasoned further that the response to pheromone and the acquisition of mating competence must involve processes in addition to the imposition of cell division arrest because mating was not restored to any of the mutants if they also carried a temperature-sensitive *cdc28* mutation, which arrests cells in the G_1 phase at the restrictive temperature. Hartwell's group had shown earlier that cells carrying the *cdc28-1* allele held in G_1 at the nonpermissive temperature were still competent to mate, whereas cells held at other cell cycle stages by other *cdc* mutations were not (Reid and Hartwell 1977). *CDC28* and its fission yeast counterpart, *CDC2*, were shown only years later to encode the class of protein kinase (CDK) that is the major cell cycle driver (Hindley and Phear 1984; Lorincz and Reed 1984) (see also Chapter 6). It turns out, fortuitously, that *cdc28-1* cells do not lose viability rapidly at restrictive temperature, but cells carrying other *cdc28* alleles do. In any event, based on the fact that the mating defect of his *ste* mutations was epistatic to the mating proficiency of *cdc28-1* cells, Hartwell concluded that there was more to the induction of mating competence than simply being held in G_1 phase. Indeed, this third deep insight turned out to be correct. It has been shown subsequently that pheromone action sets off multiple independent events (Leberer et al. 1997a; Elion 2000; Dohlman and Thorner 2001; Gulli and Peter 2001; van Drogen and Peter 2001), all of which are required for efficient mating. An especially important event is transcriptional induction of scores of genes (as has now been revealed in great detail using DNA microarray analysis [Roberts et al. 2000] of the entire genome of *S. cerevisiae*, which was the first eukaryote to have its complete genome sequenced [Goffeau et al. 1996]).

Finally, Hartwell observed that his *ste* mutations caused, in some cases, only modest defects in the ability of **a** cells to produce their cognate pheromone (**a**-factor) or to express the secreted endoprotease (*BAR1* gene product) that cleaves and inactivates α-factor (Hicks and Herskowitz 1976; Ciejek and Thorner 1979), but always caused much more profound defects in the properties of **a** cells that seem to be elicited only after exposure to α-factor (induction of agglutinability and imposition of G_1 arrest) (Moore 1983). We now know that the low, constitutive level of expression of many of the genes involved in mating is due to a basal level of stimulation through this pathway that occurs even in the absence of pheromone (Hasson et al. 1994; Siekhaus and Drubin 2003). Nonetheless, despite this ambiguity, the fourth perceptive inference Hartwell made from his findings was that, most likely, the products of his particular set of *STE* genes define components of a pathway required for cells to respond to pheromone, rather than gene products required for cells to generate (synthesize, mature, secrete) an active pheromone signal. Of course, using Hartwell's approach, mutations in redundant genes that encode factors required for mating could not be isolated because both such functions would have to be inactivated, an exceedingly rare event even if mutagens are used to increase the mutation frequency. Moreover, any gene whose product is essential for growth at the restrictive temperature, but is also required for mating, could not have been identified in this study. Indeed, both kinds of gene products—those with significantly overlapping functions (e.g., the MAPKs, Fus3, and Kss1) and those that are necessary for mating, but have other functions that are essential for growth (e.g., the small GTPase, Cdc42)—are required for pheromone response, but were identified by other means.

STE GENES REPRESENT COMPONENTS OF A PATHWAY INITIATED BY A G-PROTEIN-COUPLED RECEPTOR

After the development of recombinant DNA methodology (Morrow et al. 1974) and reproducible procedures for DNA-mediated transformation of *S. cerevisiae* cells (Hinnen et al. 1978), the normal genes corresponding to each of the *ste* mutations could be isolated by complementation using a library of genomic yeast DNA in a suitable vector (Petes et al. 1978; Botstein et al. 1979; MacKay 1983). Because the topology of the proteins predicted by the nucleotide sequences of the *STE2* and *STE3* genes resembled that of demonstrated GPCRs, it seemed likely that they might be coupled to a heterotrimeric guanine-nucleotide-binding (G-) protein. The fact that receptors, like β-adrenergic receptor and rhodopsin, acted via G-proteins, that such G-proteins were composed of three subunits (α, β, and γ), and that exchange of GTP for GDP on the α subunit dissociated the G-protein heterotrimer into its separate Gα and Gβγ constituents and thereby converted Gα into its signaling-competent GTP-bound state, was established in the early to mid 1980s through elegant biochemical studies in animal cells conducted in many laboratories, but largely those from the group of Alfred G. Gilman (1987).

Indeed, working independently, two groups isolated a yeast gene encoding a Gα subunit. A team led by Yoshito Kaziro, Ken-ichi Arai, and Kunihiro Matsumoto applied an antecedent of the candidate gene approach, a tactic that can now be readily applied to any sequenced genome through the use of computational methods (Boguski 1994). In this case, a heterologous probe (cDNA encoding mammalian Gαi) was used to identify and isolate an apparent yeast ortholog, which they dubbed *GPA1* (Nakafuku et al. 1987). They then showed that *GPA1* was expressed only in haploid cells, as expected for a gene specific to pheromone response, and used genetic analysis to implicate *GPA1* in pheromone signaling (Miyajima et al. 1987). Contemporaneously, Janet Kurjan discovered the same gene (Dietzel and Kurjan 1987), which she called *SCG1*, as a dosage suppressor of a mutation (*sst2*) that made haploid cells hypersensitive to the effects of mating pheromone (Chan and Otte 1982). However, *GPA1* did not correspond to any of the *ste* loci pinpointed by Hartwell or by MacKay and Manney. Moreover, both groups showed that rather than blocking pheromone response, the loss of *GPA1* caused cells to behave as if they were responding strongly and constitutively to mating pheromone (Dietzel and Kurjan 1987; Miyajima et al. 1987). These data indicated that Gpa1 is a negative regulator, not a positive effector, of pheromone-initiated signaling. The conclusion that the Gα subunit was dispensable for signaling was heretical because the paradigms established just a few years before for the role of heterotrimeric G-proteins in GPCR-mediated signaling clearly demonstrated that it is the GTP-bound form of the Gα subunit that evokes downstream events, for example, in the stimulation of adenylate cyclase by Gαs (activated by β-adrenergic receptor) (May et al. 1985) or in the stimulation of cGMP phosphodiesterase by Gαt (activated by rhodopsin) (Stryer et al. 1983).

Assuming that the G-protein coupled to Ste2 and Ste3 is heterotrimeric, like its mammalian counterparts, the fact that absence of Gpa1 evoked constitutive signaling akin to that expected for chronic exposure to pheromone suggested that the Gβγ dimer, and not Gα, must be critical for transmitting the signal. The notion that free Gβγ dimers could modulate effectors was strongly inferred from work in animal cells by the late Eva J. Neer and her co-workers (Logothetis et al. 1987); however, her conclusions generally were greeted with considerable skepticism at the time (Birnbaumer 1987). As described in the second landmark paper (**Whiteway et al. 1989**), incontrovertible molecular and genetic evidence was obtained that, in yeast, Gβγ is indeed essential for eliciting a response to pheromone. These findings in yeast tipped the opinion of the G-protein establishment in favor of the idea that Gβγ could modulate downstream effectors and therefore that Gβγ was not simply a passive negative regulator of Gα. Thereafter, many examples of Gβγ-modulated signaling processes in other organisms came to the fore, so that it is now widely accepted that both Gα-GTP and free Gβγ can be effector-stimulating entities (Clapham and Neer 1997).

To initiate the work that led to identification of the *S. cerevisiae* genes encoding the Gβ and Gγ subunits required for the pheromone response (Whiteway et al. 1989), Malcolm S. Whiteway and his co-workers first devised a new selection for isolating pheromone-resistant mutants (Whiteway et al. 1988). They constructed an **a** cell that carried two different kinds of mutations that made them hypersensitive to the effects of α-factor (Chan and Otte 1982) and then introduced a plasmid that forced production of at least some α-factor in these **a** cells. Under these artificially autocrine conditions, only **a** cells that were not subject to the G_1-arrest-inducing activity of α-factor could continue to grow. Seven of the eight pheromone-resistant mutations obtained in this fashion were allelic to the previously described *ste4* locus, but one of the mutations defined a novel locus, *ste18*. The DNAs corresponding to the wild-type *STE4* and *STE18* genes were isolated by virtue of the fact that plasmids containing them could restore mating proficiency to *ste4* and *ste18* cells, respectively. Nucleotide sequence analysis revealed that the predicted *STE4* product was clearly homologous to known mammalian Gβ subunits; likewise, the predicted *STE18* product was related in size and sequence to a known Gγ subunit. Moreover, as expected if their function is necessary for signal propagation in pheromone response, both genes were transcribed only in haploid cells, and cells deleted for either gene were completely insensitive to the presence of exogenously added α factor. Most importantly, absence of either Ste4 or Ste18 prevented the constitutive pheromone-like responses seen in cells lacking Gpa1. The fact that *ste4* and *ste18* mutations were epistatic to *gpa1* mutations provided convincing evidence that free Gβγ is required to initiate signaling in the yeast pheromone response pathway.

OTHER *STE* GENE PRODUCTS DEFINE A MODULAR PROTEIN KINASE CASCADE

Meanwhile, other investigators used plasmid complementation to clone the DNAs corresponding to other *STE* genes. Beverly Errede and co-workers isolated both *STE7* and *STE11* and found that they encode predicted protein kinases (Teague et al. 1986; Rhodes et al. 1990). Stanley Fields found that the *STE12* product was required for the expression of haploid-specific genes, including those induced by pheromone (Fields and Herskowitz 1985), and showed that Ste12 is a DNA-binding transcription factor (Dolan et al. 1989). Two other genes implicated in pheromone response, *KSS1* (Courchesne et al. 1989) and *FUS3* (Elion et al. 1990), were shown to encode highly related protein kinases that were found to be the prototypes for the so-called microtubule-associated protein-2 kinases (also now known as mitogen-activated protein kinases), or MAPKs, of animal cells (Boulton et al. 1990). Neither *fus3* nor *kss1* mutations were among the original *ste* loci because only a *fus3 kss1* double mutant is strongly mating-defective (Elion et al. 1991; Ma et al. 1995). However, given all of these gene products, how could one go about determining their physiological roles and their order of function?

One useful genetic method for delineating the interrelationships among the steps in a pathway is epistasis analysis of double mutants in which an activated allele of one gene is combined with a null allele of another. If the activated gene product functions upstream of the inactivated gene product, it cannot overcome the block and the pathway is not stimulated; conversely, if the activated gene product functions downstream from the inactivated gene product, there is no impediment to stimulation of the pathway. This approach was pioneered for the pheromone response pathway by Duane D. Jenness who isolated activated *STE4* alleles that promoted G_1 arrest and mating in the absence of pheromone and demonstrated that all other known *ste* mutations prevented the effects of these activating mutations (Blinder et al. 1989). This epistasis analysis confirmed that Gβγ functions very early in this signaling pathway.

The third landmark paper discussed here (**Stevenson et al. 1992**) describes a particularly elegant application of such an epistasis approach. This study revealed the order of function of two of the protein kinases, Ste7 and Ste11, involved in pheromone response. Other studies demonstrated that phosphorylation and activation of both Fus3 and Kss1 required all of the other *STE* gene products, except Ste12 (Gartner et al. 1992; Errede et al. 1993; Ma et al. 1995). These findings, combined with the results described here by George F. Sprague, Jr. and his collaborators, and those reported contemporaneously by Roger D. Kornberg and his co-workers (Cairns et al. 1992), were pivotal in establishing the order of function of these three classes of protein kinases.

Sprague and his co-workers reasoned that they could determine which *STE* gene products functioned downstream from Gβγ by isolating mutations that restored mating in cells lacking Ste4. They identified dominant mutations in *STE11* that were able to promote detectable mating of *ste4Δ* and *ste5Δ* cells, suggesting that the *STE5* gene product, like Gβγ, functions earlier in the pheromone response pathway than Ste11. However, the activated *STE11* alleles could not bypass the mating defect of *ste7Δ* or *ste12Δ* cells. These observations placed the action of Ste11 upstream of the function of Ste7. In confirmation of this conclusion, they found that the activated *STE11* mutations caused hyperphosphorylation of Ste7 in the absence of pheromone treatment, a modification of Ste7 that is normally observed in wild-type cells only after they are exposed to pheromone. On the basis of these results, it was proposed that Ste11 and Ste7 represent components of a protein kinase cascade necessary for activation of the transcription factor, Ste12, and that Ste11 acts upstream of (and, most likely, directly upon) Ste7.

These findings in yeast initiated a firestorm of excitement among researchers studying receptor-activated protein kinases in animal cells. Biochemical results and gene cloning

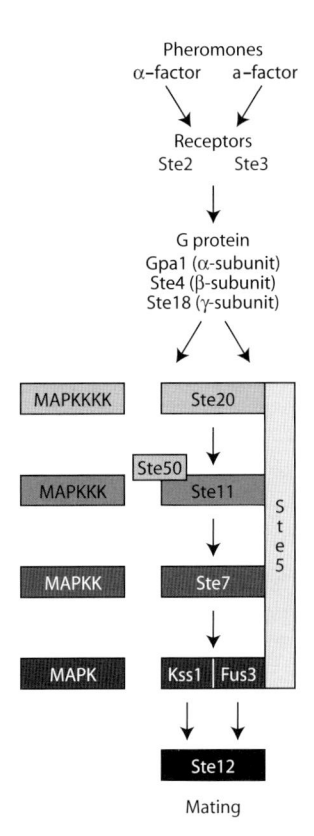

FIGURE 2. The pheromone response cascade. The schematic drawing represents the hierarchy of the different elements of the cascade, known as the MAPK module, with Ste5 as a scaffold protein. (Redrawn and modified from Elion 2000.)

using a variety of vertebrate organisms rapidly confirmed that there were enzymes highly homologous to Ste11 and Ste7, which functioned in the same order, and that homologs of Ste7 were the direct activators of MAPKs/ERKs (Crews and Erikson 1992; Seger at al. 1992; Haystead et al. 1993; Kosako et al. 1993; Lange-Carter et al. 1993). This three-tiered arrangement of protein kinases—MAPK kinase kinase (MEKK), MAPK kinase (MEK), and MAPK (or ERK)—is now known, of course, as a MAPK module (Fig. 2) (Neiman et al. 1993; Zhou et al. 1993; Blumer and Johnson 1994; Pearson et al. 2001).

DEVELOPMENT OF THE CONCEPT OF A SCAFFOLD PROTEIN

A novel gene required for mating, *STE20*, which encodes the first PAK (p21-activated protein kinase) to be identified in any organism (Lim et al. 1996), was isolated by two independent approaches. Overexpression of *STE20* was able to promote the mating of a cell expressing a dominant-negative *ste4* (Gβ) allele (Leberer et al. 1992). Similarly, high-level expression of what turned out to be an amino-terminally truncated derivative of Ste20 was able to activate the pheromone response pathway in the absence of pheromone (Ramer and Davis 1993). Epistasis analysis using all of the then known *ste* mutations indicated that Ste20 functions at the same stage as (or after) Ste4, but upstream of all of the other known *STE* gene products, including Ste5. The *STE5* gene had been cloned early on by complementation of the mating defect of an *ste5* mutant and was shown to be expressed in a haploid-cell-specific manner (Brake et al. 1981; MacKay 1983); but its predicted primary structure deduced from its nucleotide sequence was uniformative about its role in the mechanism of pheromone signaling. Regardless, a mutationally activated *STE5* allele (Hasson et al. 1994) was able to suppress the mating defect of *ste2, ste4*, and *ste18* mutations, as well as of an *ste20* mutation (Leberer et al. 1992), further reinforcing the idea that Ste5 acts downstream from Ste20. Consistent with this view, the *STE5* gene was also isolated as a dosage suppressor of an *ste20* mutant (Leberer et al. 1993). However, confounding this simple interpretation, *STE5* was also cloned as a dosage suppressor of a *cdc25^{ts}* mutation (Perlman et al. 1993). It had already been shown by that time that Cdc25 was the guanine nucleotide exchange factor that promotes GTP-for-GDP exchange in yeast Ras proteins (Powers et al. 1989; Jones et al. 1991).

In any event, these findings placed Ste5 close to the start of the pheromone signaling pathway and suggested that Ste5, Ste20, and Gβγ (Ste4-Ste18) represent some very important, but complicated, nexus required to initiate signaling in response to occupancy of pheromone receptors. To try to sort out these interrelationships, and to determine if any of these interactions are direct, three groups independently applied a newly devised, at the time, genetic method for identifying gene products that physically associate—the yeast two-hybrid screen (Fields and Song 1989; Chien et al. 1991). In the fourth and final landmark paper discussed in this section (**Choi et al. 1994**), Elaine A. Elion and her co-workers demonstrated using the two-hybrid technique that Ste11 (MEKK), Ste7 (MEK), and Fus3 (MAPK) each associate with a discrete region of Ste5 and that none of these associations depend on the presence of the other proteins. They also examined pairwise interactions between all of the same components and found that the strongest interactions were between Ste5 and Ste11. Strikingly, however, Ste11 did not associate detectably with Ste7, despite the overwhelming cumulative evidence that Ste11 is responsible for phosphorylating and activating Ste7 (Cairns et al. 1992; Stevenson et al. 1992; Zhou et al. 1993; Neiman and Herskowitz 1994). This result provided the first inkling that by binding Ste11 and Ste7, Ste5 might serve as a bridge to promote the interaction between these two protein kinases. They also demonstrated that all four proteins (Ste11, Ste7, Ste5, and Fus3) can be copurified in a complex and even cosediment, to a degree, in a glycerol gradient. Most importantly, they showed that these associations are functionally important for the activation of Fus3 in vivo. Essentially contemporaneous and independent publications from the laboratories of George Sprague (Printen and Sprague 1994) and Michael H. Wigler (Marcus et al. 1994) further corroborated all of these conclusions, as well as extended these observations by showing, for example, that Ste20 did not detectably interact with Ste5 and by mapping the domain of Ste11 responsible for its association with Ste5. In subsequent work, it was shown that Ste20 is responsible for phosphorylation and activation of Ste11 (Wu et al. 1995; Drogen et al. 2000). Despite some initially misleading results from deletion analysis of the apparent Cdc42-binding site in Ste20 that suggested association of Cdc42 with Ste20 was not required for its function in mating (Peter et al. 1996; Leberer et al. 1997b), it, first, was subsequently shown that such alterations also destroyed the function of an autoinhibitory domain, leading to significant activation, and, then, clearly demonstrated that binding of the GTP-bound form of Cdc42 is essential for Ste20 function in mating (Moskow et al. 2000; Lamson et al. 2002).

Taken together, these results suggested that Ste5 serves a unique and previously undescribed role as a molecular matchmaker. It was explicitly proposed by all three groups that

Ste5 acts as a scaffold to facilitate interactions among members of the MAPK kinase cascade. It was further suggested that, in this role as a facilitator, Ste5 makes both signal propagation and, perhaps, signal attenuation more efficient. It was also suggested that Ste5 may also help minimize cross-talk with other MAPK cascades and thus ensure integrity and fidelity in the pheromone response pathway. All of these functions of Ste5 have been amply borne out in subsequent studies of native Ste5 (Yashar et al. 1995; Inouye et al. 1997b; Sette et al. 2000; Bardwell et al. 2001). Moreover, the now well-recognized scaffold function of Ste5 (Elion 2001) set a precedent that made it much easier for other investigators studying diverse organisms to recognize the roles of such linker, adapter, and anchoring proteins in other signaling pathways (Pawson and Scott 1997; Yasuda et al. 1999; Catling et al. 2001; Jordan et al. 2003). In addition, studies using reengineered versions of Ste5, in which its normal binding sites have been replaced with other modules that permit interaction with correspondingly reengineered versions of the MAPK cascade constituents, reinforce the conclusion that the most critical role of scaffold proteins is simply to increase the effective local concentration of the pathway components (Harris et al. 2001; Park et al. 2003). The success of this human tinkering suggests a mechanism by which novel signaling pathways might have evolved and then been subjected to natural selection. Specifically, by recombining simple modular binding interactions in different combinations, cells could construct scaffold proteins that achieve both molecular diversity and high specificity without the need for change in the enzymatic machinery involved in signaling per se (Ferrell and Cimprich 2003).

Despite the power of the two-hybrid approach in revealing the scaffold function of Ste5, all three groups missed the fact that Ste5 associates with Gβγ, even though the potential interaction between Ste4 and Ste5 was explicitly tested (Choi et al. 1994; Printen and Sprague 1994). However, compelling evidence for direct binding of Gβγ to Ste5 was obtained by the same and other methods (Whiteway et al. 1995; Inouye et al. 1997a; Feng et al. 1998; Pryciak and Huntress 1998), thereby providing an explanation for how Ste5 delivers Ste11 and the other MAPK constituents to the plasma membrane, so that they encounter Ste20. Likewise, how Ste20 is also recruited to Gβγ (Leeuw et al. 1998), how GTP-bound Cdc42 is generated in response to presence of Gβγ (Butty et al. 1998; Nern and Arkowitz 1999; Toenjes et al. 1999), and how Cdc42 gets converted to its active state (Zheng et al. 1994) were all addressed in other studies.

SUMMARY

As should be clear from the above recounting, the acquisition of our current picture of how the yeast pheromone response pathway operates has taken the concerted effort of numerous investigators applying the tools of genetics, molecular biology, biochemistry, and cell biology over the course of nearly half a century. Just as many of the constituent proteins of this pathway, and their interrelationships and regulation, were first identified as described above, other molecules that participate in the yeast pheromone response pathway continue to be discovered and still remain to be discovered. These novel components will, most likely, carry out homologous or analogous roles in metazoans and thus will continue to inform and illuminate our understanding of signal transduction mechanisms in all cells.

ACKNOWLEDGMENTS

I would like to take this opportunity to thank the many colleagues, too numerous to name, who have made it such a pleasant, exciting, and stimulating endeavor to labor in this field of research by generously sharing information and research materials, whenever asked. I also thank the many members of my own laboratory (past and present) whose work contributed to our current level of understanding of the yeast mating pheromone response pathway, especially Elena Ciejek, George Fehrenbacher, David Julius, Monica Flessel, Johanna Reneke, Rachel Sterne-Marr, Mimi Hasson, Sofie Salama, Jean Cook, Lindsay Garrenton, Dan Ballon, Y'Vonne Jones-Brown, Hans Liao, Tony Brake, Buff Blair, Gary Stetler, Scott van Arsdell, Rich Freedman, Bob Fuller, Bill Courchesne, Ken Blumer, Karl Kuchler, Lis Barfod, Doreen Ma, Henrik Dohlman, Josh Trueheart, Markus Künzler, Claudio Sette, Carla Inouye, Namrita Dhillon, Lee Bardwell, Judy Zhu-Shimoni, Dagmar Truckses, Alma Saviñon-Tejeda, Riyo Kunisawa, and Françoise Roelants. Our research on pheromone signaling as a model signal transduction network has been supported by National Institutes of Health research grant GM21841.

STUDY QUESTIONS

1. Define the genetic relationship known as "epistasis" between two different mutations and describe how it can be applied, in practice, for the purpose of ordering the function of two gene products that operate in the same pathway or process.

2. The pheromone response pathway involves both a small Ras-related GTPase, Cdc42, as well as a large GTPase, Gpa1 (α subunit of a heterotrimeric G-protein). Describe the known functions of each of these two GTPases. What gene products are responsible for promoting conversion of each of these proteins to their active (GTP-bound) state and what gene products are responsible for promoting conversion of each of these proteins to their inactive (GDP-bound) state?

3. Why should yeast cells arrest in G_1 before entering mating?

4. The pheromone response pathway activates a multitiered cascade of protein kinases. What traditional methods have been used to infer that a protein is the direct physiological target of a given protein kinase in the cell? What more recently developed methods have been devised to provide more unequivocal demonstrations that a protein is the direct substrate of a given protein kinase in the cell? You might consult the following articles: Zhu et al. (2000), Bishop et al. (2001), and Ficarro et al. (2002).

5. Describe in what ways the scaffold protein Ste5 promotes the specificity, efficiency, and fidelity of signal propagation in the pheromone response pathway by acting as an adapter, a platform, a linker, an insulator, and a membrane anchor.

6. Quite a number of the proteins that participate in the pheromone response pathway are posttranslationally modified with lipophilic substituents that give these proteins an affinity for binding to membranes. What are these modifications and on which gene products are they located?

7. Because mating requires the expression of new genes, one ultimate consequence of the pheromone response pathway must be induction of the transcription of those genes. Given that the transcription factor Ste12 seems to reside in the nucleus constitutively, propose mechanisms by which activation of the MAPK (Fus3) leads to the expression of pheromone-induced genes.

REFERENCES

Anderegg R.J., Betz R., Carr S.A., and Crabb J.W. 1988. Structure of *Saccharomyces cerevisiae* mating hormone **a**-factor. Identification of S-farnesyl cysteine as a structural component. *J. Biol. Chem.* **263:** 18236–18240.

Bardwell A.J., Flatauer L.J., Matsukuma K., Thorner J., and Bardwell L. 2001. A conserved docking site in MEKs mediates high-affinity binding to MAP kinases and cooperates with a scaffold protein to enhance signal transmission. *J. Biol. Chem.* **276:** 10374–10386.

Bardwell L. and Thorner J. 1996. A conserved motif at the amino termini of MEKs might mediate high-affinity interaction with the cognate MAPKs. *Trends Biochem. Sci.* **21:** 373–374.

Betz R., MacKay V., and Düntze W. 1977. **a**-Factor from *Saccharomyces cerevisiae:* Partial characterization of a mating hormone produced by cells of mating type **a**. *J. Bacteriol.* **132:** 462–472.

Birnbaumer L. 1987. Which G protein subunits are the active mediators of signal transduction? *Trends Pharmacol. Sci.* **8:** 209–211.

Bishop A.C., Buzko O., and Shokat K.M. 2001. Magic bullets for protein kinases. *Trends Cell Biol.* **11:** 167–172.

Blinder D., Bouvier S., and Jenness D.D. 1989. Constitutive mutants in the yeast pheromone response: Ordered function of the gene products. *Cell* **56:** 479–486.

Blumer K.J. and Johnson G.L. 1994. Diversity in function and regulation of MAP kinase pathways. *Trends Biochem. Sci.* **19:** 236–240.

Blumer K.J., Reneke J.E., and Thorner J. 1988. The *STE2* gene product is the ligand-binding component of the alpha-factor receptor of *Saccharomyces cerevisiae. J. Biol. Chem.* **263:** 10836–10842.

Boguski M.S. 1994. Bioinformatics. *Curr. Opin. Genet. Dev.* **4:** 383–388.

Botstein D., Falco S.C., Stewart S.E., Brennan M., Scherer S., Stinchcomb D.T., Struhl K., and Davis R.W. 1979. Sterile host yeasts (SHY): A eukaryotic system of biological containment for recombinant DNA experiments. *Gene* **8:** 17–24.

Boulton T.G., Yancopoulos G.D., Gregory J.S., Slaughter C., Moomaw C., Hsu J., and Cobb M.H. 1990. An insulin-stimulated protein kinase similar to yeast kinases involved in cell cycle control. *Science* **249:** 64–67.

Brake A.J., Liao H.H., Thorner J., and Nasmyth K. 1981. Analysis of the role of the *STE5* gene product in the yeast mating response using the cloned *STE5* gene. In Abstracts from the *Molecular Biology of Yeast* Meeting, p. 5. Cold Spring Harbor Laboratory, Cold Spring Harbor, New York.

Bücking-Throm E., Düntze W., Hartwell L.H., and Manney T.R. 1973. Reversible arrest of haploid yeast cells in the initiation of DNA synthesis by a diffusible sex factor. *Exp. Cell Res.* **76:** 99–110.

Burkholder A.C. and Hartwell L.H. 1985. The yeast alpha-factor receptor: Structural properties deduced from the sequence of the *STE2* gene. *Nucleic Acids Res.* **13:** 8463–8475.

Butty A.C., Pryciak P.M., Huang L.S., Herskowitz I., and Peter M. 1998. The role of Far1p in linking the heterotrimeric G protein to polarity establishment proteins during yeast mating. *Science* **282:** 1511–1516.

Cairns B.R., Ramer S.W., and Kornberg R.D. 1992. Order of action of components in the yeast pheromone response pathway revealed with a dominant allele of the STE11 kinase and the multiple phosphorylation of the STE7 kinase. *Genes Dev.* **6:** 1305–1318.

Catling A.D., Eblen S.T., Schaeffer H.J., and Weber M.J. 2001. Scaffold protein regulation of mitogen-activated protein kinase cascade. *Methods Enzymol.* **332:** 368–387.

Chan R.K. and Otte C.A. 1982. Physiological characterization of *Saccharomyces cerevisiae* mutants supersensitive to G_1 arrest by **a** factor and α factor pheromones. *Mol. Cell. Biol.* **2:** 21–29.

Chien C.T., Bartel P.L., Sternglanz R., and Fields S. 1991. The two-hybrid system: A method to identify and clone genes for proteins that interact with a protein of interest. *Proc. Natl. Acad. Sci.* **88:** 9578–9582.

Choi K-Y., Satterberg B., Lyons D.M., and Elion E.A. 1994. Ste5 tethers multiple protein kinases in the MAP kinase cascade required for mating in *S. cerevisiae*. *Cell* **78:** 499–512.

Ciejek E. and Thorner J. 1979. Recovery of *S. cerevisiae* **a** cells from G1 arrest by alpha factor pheromone requires endopeptidase action. *Cell* **18:** 623–635.

Ciejek E., Thorner J., and Geier M. 1977. Solid phase peptide synthesis of alpha-factor, a yeast mating pheromone. *Biochem. Biophys. Res. Commun.* **78:** 952–961.

Clapham D.E. and Neer E.J. 1997. G protein beta gamma subunits. *Annu. Rev. Pharmacol. Toxicol.* **37:** 167–203.

Courchesne W.E., Kunisawa R., and Thorner J. 1989. A putative protein kinase overcomes pheromone-induced arrest of cell cycling in *S. cerevisiae*. *Cell* **58:** 1107–1119.

Crews C.M. and Erikson R.L. 1992. Purification of a murine protein-tyrosine/threonine kinase that phosphorylates and activates the Erk-1 gene product: Relationship to the fission yeast *byr1* gene product. *Proc. Natl. Acad. Sci.* **89:** 8205–8209.

Dietzel C. and Kurjan J. 1987. The yeast *SCG1* gene: A Gα-like protein implicated in the **a**- and α-factor response pathway. *Cell* **50:** 1001–1010.

Dixon R.A., Kobilka B.K., Strader D.J., Benovic J.L., Dohlman H.G., Frielle T., Bolanowski M.A., Bennett C.D., Rands E., Diehl R.E., Mumford R.A., Slater E.E., Sigal I.S., Caron M.G., Lefkowitz R.J., and Strader C.D. 1986. Cloning of the gene and cDNA for mammalian beta-adrenergic receptor and homology with rhodopsin. *Nature* **321:** 75–79.

Dohlman H.G. and Thorner J.W. 1997. RGS proteins and signaling by heterotrimeric G proteins. *J. Biol. Chem.* **272:** 3871–3874.

———. 2001. Regulation of G protein-initiated signal transduction in yeast: Paradigms and principles. *Annu. Rev. Biochem.* **70:** 703–754.

Dohlman H.G., Thorner J., Caron M.G., and Lefkowitz R.J. 1991. Model systems for the study of seven-transmembrane-segment receptors. *Annu. Rev. Biochem.* **60:** 653–688.

Dolan J.W., Kirkman C., and Fields S. 1989. The yeast *STE12* protein binds to the DNA sequence mediating pheromone induction. *Proc. Natl. Acad. Sci.* **86:** 5703–5707.

Drogen F., O'Rourke S.M., Stucke V.M., Jaquenoud M., Neiman A.M., and Peter M. 2000. Phosphorylation of the MEKK Ste11p by the PAK-like kinase Ste20p is required for MAP kinase signaling *in vivo*. *Curr. Biol.* **10:** 630–639.

Düntze W., MacKay V., and Manney T.R. 1970. *Saccharomyces cerevisiae*: A diffusible sex factor. *Science* **168:** 1472–1473.

Elion E.A. 2000. Pheromone response, mating and cell biology. *Curr. Opin. Microbiol.* **3:** 573–581.

———. 2001. The Ste5p scaffold. *J. Cell Sci.* **114:** 3967–3978.

Elion E.A., Brill J.A., and Fink G.R. 1991. FUS3 represses CLN1 and CLN2 and in concert with KSS1 promotes signal transduction. *Proc. Natl. Acad. Sci.* **88:** 9392–9396.

Elion E.A., Grisafi P.L., and Fink G.R. 1990. *FUS3* encodes a cdc2+/CDC28-related kinase required for the transition from mitosis into conjugation. *Cell* **60:** 649–664.

Errede B., Gartner A., Zhou A., Nasmyth K., and Ammerer G. 1993. MAP kinase-related FUS3 from *S. cerevisiae* is activated by STE7 *in vitro*. *Nature* **362:** 261–264.

Feng Y., Song L.Y., Kincaid E., Mahanty S.K., and Elion E.A. 1998. Functional binding between Gβ and the LIM domain of Ste5 is required to activate the MEKK Ste11. *Curr. Biol.* **8:** 267–278.

Ferrell J.E.J. and Cimprich K.A. 2003. Enforced proximity in the function of a famous scaffold. *Mol. Cell. Biol.* **11:** 289–291.

Ficarro S.B., McCleland M.L., Stukenberg P.T., Burke D.J., Ross M.M., Shabanowitz J., Hunt D.F., and White F.M. 2002. Phosphoproteome analysis by mass spectrometry and its application to *Saccharomyces cerevisiae*. *Nat. Biotechnol.* **20:** 301–305.

Fields S. and Herskowitz I. 1985. The yeast *STE12* product is required for expression of two sets of cell-type specific genes. *Cell* **42:** 923–930.

Fields S. and Song O. 1989. A novel genetic system to detect protein-protein interactions. *Nature* **340:** 245–246.

Gartner A., Nasmyth K., and Ammerer G. 1992. Signal transduction in *Saccharomyces cerevisiae* requires tyrosine and threonine phosphorylation of FUS3 and KSS1. *Genes Dev.* **6:** 1280–1292.

Gilman A.G. 1987. G proteins: Transducers of receptor-generated signals. *Annu. Rev. Biochem.* **56:** 615–649.

Goffeau A., Barrell B.G., Bussey H., Davis R.W., Dujon B., Feldmann H., Galibert F., Hoheisel J.D., Jacq C.,

Johnston M., Louis E.J., Mewes H.W., Murakami Y., Philippsen P., Tettelin H., and Oliver S.G. 1996. Life with 6000 genes. *Science* **274:** 546–567.

Gulli M.P. and Peter M. 2001. Temporal and spatial regulation of Rho-type guanine-nucleotide exchange factors: The yeast perspective. *Genes Dev.* **15:** 365–379.

Hagen D.C., McCaffrey G., and Sprague G.F., Jr. 1986. Evidence the yeast *STE3* gene encodes a receptor for the peptide pheromone **a**-factor: Gene sequence and implications for the structure of the presumed receptor. *Proc. Natl. Acad. Sci.* **83:** 1418–1422.

Harris K., Lamson R.E., Nelson B., Hughes T.R., Marton M.J., Roberts C.J., Boone C., and Pryciak P.M. 2001. Role of scaffolds in MAP kinase pathway specificity revealed by custom design of pathway-dedicated signaling proteins. *Curr. Biol.* **11:** 1815–1824.

Hartwell L.H. 1980. Mutants of *Saccharomyces cerevisiae* unresponsive to cell division control by polypeptide mating hormone. *J. Cell Biol.* **85:** 811–822.

Hasson M.S., Blinder D., Thorner J., and Jenness D.D. 1994. Mutational activation of the *STE5* gene product bypasses the requirement for G protein beta and gamma subunits in the yeast pheromone response pathway. *Mol. Cell. Biol.* **14:** 1054–1065.

Haystead C.M., Wu J., Gregory P., Sturgill T.W., and Haystead T.A. 1993. Functional expression of a MAP kinase kinase in COS cells and recognition by an anti-STE7/byr1 antibody. *FEBS Lett.* **317:** 12–16.

Herskowitz I. and Oshima Y. 1981. Control of cell type in *Saccharomyces cerevisiae:* Mating type and mating-type interconversion. In *The molecular biology of the yeast* Saccharomyces: *Life cycle and inheritance* (ed. J.N. Strathern et al.), pp. 181–209. Cold Spring Harbor Laboratory Press, Cold Spring Harbor, New York.

Hicks J.B. and Herskowitz I. 1976. Evidence for a new diffusible element of mating pheromones in yeast. *Nature* **260:** 246–248.

Hindley J. and Phear G.A. 1984. Sequence of the cell division gene CDC2 from *Schizosaccharomyces pombe;* patterns of splicing and homology to protein kinases. *Gene* **31:** 129–134.

Hinnen A., Hicks J.B., and Fink G.R. 1978. Transformation of yeast. *Proc. Natl. Acad. Sci.* **75:** 1929–1933.

Inouye C., Dhillon N., and Thorner J. 1997a. Ste5 RING-H2 domain: Role in Ste4-promoted oligomerization for yeast pheromone signaling. *Science* **278:** 103–106.

Inouye C., Dhillon N., Durfee T., Zambryski P.C., and Thorner J. 1997b. Mutational analysis of *STE5* in the yeast *Saccharomyces cerevisiae:* Application of a differential interaction trap assay for examining protein-protein interactions. *Genetics* **147:** 479–492.

Jones S., Vignais M.L., and Broach J.R. 1991. The CDC25 protein of *Saccharomyces cerevisiae* promotes exchange of guanine nucleotides bound to ras. *Mol. Cell. Biol.* **11:** 2641–2646.

Jordan M.S., Singer A.L., and Koretzky G.A. 2003. Adaptors as central mediators of signal transduction in immune cells. *Nat. Immunol.* **4:** 110–116.

Kosako H., Nishida E., and Gotoh Y. 1993. cDNA cloning of MAP kinase kinase reveals kinase cascade pathways in yeasts to vertebrates. *EMBO J.* **12:** 787–794.

Lamson R.E., Winters M.J., and Pryciak P.M. 2002. Cdc42 regulation of kinase activity and signaling by the yeast p21-activated kinase Ste20. *Mol. Cell. Biol.* **22:** 2939–2951.

Lange-Carter C.A., Pleiman C.M., Gardner A.M., Blumer K.J., and Johnson G.L. 1993. A divergence in the MAP kinase regulatory network defined by MEK kinase and Raf. *Science* **260:** 315–319.

Leberer E., Thomas D.Y., and Whiteway M. 1997a. Pheromone signalling and polarized morphogenesis in yeast. *Curr. Opin. Genet. Dev.* **7:** 59–66.

Leberer E., Dignard D., Harcus S., Thomas D.Y., and Whiteway M. 1992. The protein kinase homologue Ste20p is required to link the yeast pheromone response G-protein beta gamma subunits to downstream signalling components. *EMBO J.* **11:** 4815–4824.

Leberer E., Dignard D., Harcus D., Hougan L., Whiteway M., and Thomas D.Y. 1993. Cloning of *Saccharomyces cerevisiae STE5* as a suppressor of a Ste20 protein kinase mutant: Structural and functional similarity of Ste5 to Far1. *Mol. Gen. Genet.* **241:** 241–254.

Leberer E., Wu C., Leeuw T., Fourest-Lieuvin A., Segall J.E., and Thomas D.Y. 1997b. Functional characterization of the Cdc42p binding domain of yeast Ste20p protein kinase. *EMBO J.* **16:** 83–97.

Leeuw T., Wu C., Schrag J.D., Whiteway M., Thomas D.Y., and Leberer E. 1998. Interaction of a G-protein beta-subunit with a conserved sequence in Ste20/PAK family protein kinases. *Nature* **391:** 191–195.

Levi J.D. 1956. Mating reaction in yeast. *Nature* **177:** 753–754.

Lim L., Manser E., Leung T., and Hall C. 1996. Regulation of phosphorylation pathways by p21 GTPases. The p21 Ras-related Rho subfamily and its role in phosphorylation signalling pathways. *Eur. J. Biochem.* **242:** 171–185.

Logothetis D.E., Kurachi Y., Galper J., Neer E.J., and Clapham D.E. 1987. The beta gamma subunits of GTP-binding proteins activate the muscarinic K+ channel in heart. *Nature* **325:** 321–326.

Lorincz A.T. and Reed S.I. 1984. Primary structure homology between the product of yeast cell division control gene *CDC28* and vertebrate oncogenes. *Nature* **307:** 183–185.

Ma D., Cook J.G., and Thorner J. 1995. Phosphorylation and localization of Kss1, a MAP kinase of the *Saccharomyces cerevisiae* pheromone response pathway. *Mol. Biol. Cell* **6:** 889–909.

MacKay V.L. 1983. Cloning of yeast *STE* genes in 2 micron vectors. *Methods Enzymol.* **101:** 325–343.

MacKay V. and Manney T.R. 1974a. Mutations affecting sexual conjugation and related processes in *Saccharomyces cerevisiae*. I. Isolation and phenotypic characterization of nonmating mutants. *Genetics* **76:** 255–271.

———. 1974b. Mutations affecting sexual conjugation and related processes in *Saccharomyces cerevisiae*. II. Genetic analysis of nonmating mutants. *Genetics* **76:** 273–288.

Marcus S., Polverino A., Barr M., and Wigler M. 1994. Complexes between STE5 and components of the pheromone-responsive mitogen-activated protein kinase module. *Proc. Natl. Acad. Sci.* **91:** 7762–7776.

May D.C., Ross E.M., Gilman A.G., and Smigel M.D. 1985. Reconstitution of catecholamine-stimulated adenylate cyclase activity using three purified proteins. *J. Biol. Chem.* **260:** 15829–15833.

Miyajima I., Nakafuku M., Nakayama N., Brenner C., Miyajima A., Kaibuchi K., Arai K., Kaziro Y., and Matsumoto K. 1987. *GPA1*, a haploid-specific essential gene, encodes a yeast homolog of mammalian G protein which may be involved in mating factor signal transduction. *Cell* **50:** 1011–1019.

Moore S.A. 1983. Comparison of dose-response curves for alpha factor-induced cell division arrest, agglutination, and projection formation of yeast cells. Implication for the mechanism of alpha factor action. *J. Biol. Chem.* **258:** 13849–13856.

Morrow J.F., Cohen S.N., Chang A.C., Boyer H.W., Goodman H.M., and Helling R.B. 1974. Replication and transcription of eukaryotic DNA in *Escherichia coli*. *Proc. Natl. Acad. Sci.* **71:** 1743–1747.

Mortimer R.K. and Hawthorne D.C. 1966. Yeast genetics. *Annu. Rev. Microbiol.* **20:** 151–168.

Moskow J.J., Gladfelter A.S., Lamson R.E., Pryciak P.M., and Lew D.J. 2000. Role of Cdc42p in pheromone-stimulated signal transduction in *Saccharomyces cerevisiae*. *Mol. Cell. Biol.* **20:** 7559–7571.

Nakafuku M., Itoh H., Nakamura S., and Kaziro Y. 1987. Occurrence in *Saccharomyces cerevisiae* of a gene homologous to the cDNA coding for the alpha subunit of mammalian G proteins. *Proc. Natl. Acad. Sci.* **84:** 2140–2144.

Nakayama N., Miyajima A., and Arai K. 1985. Nucleotide sequences of *STE2* and *STE3*, cell type-specific sterile genes from *Saccharomyces cerevisiae*. *EMBO J.* **4:** 2643–2648.

Nasmyth K.A. 1982. Molecular genetics of yeast mating type. *Annu. Rev. Genet.* **16:** 439–500.

Nathans J. and Hogness D.S. 1983. Isolation, sequence analysis, and intron-exon arrangement of the gene encoding bovine rhodopsin. *Cell* **34:** 807–814.

Neiman A.M. and Herskowitz I. 1994. Reconstitution of a yeast protein kinase cascade in vitro: Activation of the yeast MEK homologue STE7 by STE11. *Proc. Natl. Acad. Sci.* **91:** 3398–3404.

Neiman A.M., Stevenson B.J., Xu H.P., Sprague G.F., Jr., Herskowitz I., Wigler M., and Marcus S. 1993. Functional homology of protein kinases required for sexual differentiation in *Schizosaccharomyces pombe* and *Saccharomyces cerevisiae* suggests a conserved signal transduction module in eukaryotic organisms. *Mol. Biol. Cell* **4:** 107–120.

Nern A. and Arkowitz R.A. 1999. A Cdc24p-Far1p-Gβγ protein complex required for yeast orientation during mating. *J. Cell Biol.* **144:** 1187–1202.

Park S.H., Zarrinpar A., and Lim W.A. 2003. Rewiring MAP kinase pathways using alternative scaffold assembly mechanisms. *Science* **299:** 1061–1064.

Pawson T. and Scott J.D. 1997. Signaling through scaffold, anchoring, and adaptor proteins. *Science* **278:** 2075–2080.

Pearson G., Robinson F., Beers Gibson T., Xu B.E., Karandikar M., Berman K., and Cobb M.H. 2001. Mitogen-activated protein (MAP) kinase pathways: Regulation and physiological functions. *Endocr. Rev.* **22:** 153–183.

Perlman R., Yablonski D., Simchen G., and Levitzki A. 1993. Cloning of the *STE5* gene of *Saccharomyces cere-*

visiae as a suppressor of the mating defect of *cdc25* temperature-sensitive mutants. *Proc. Natl. Acad. Sci.* **90:** 5474–5478.

Peter M., Neiman A.M., Park H.O., van Lohuizen M., and Herskowitz I. 1996. Functional analysis of the interaction between the small GTP binding protein Cdc42 and the Ste20 protein kinase in yeast. *EMBO J.* **15:** 7046–7059.

Petes T.D., Broach J.R., Wensink P.C., Hereford L.M., Fink G.R., and Botstein D. 1978. Isolation and analysis of recombinant DNA molecules containing yeast DNA. *Gene* **4:** 37–49.

Pierce K.L., Premont R.T., and Lefkowitz R.J. 2002. Seven-transmembrane receptors. *Nat. Rev. Mol. Cell Biol.* **3:** 639–650.

Powers A., O'Neill K., and Wigler M. 1989. Dominant yeast and mammalian RAS mutants that interfere with the *CDC25*-dependent activation of wild-type RAS in *Saccharomyces cerevisiae*. *Mol. Cell. Biol.* **9:** 390–395.

Printen J.A. and Sprague G.F., Jr. 1994. Protein-protein interactions in the yeast pheromone response pathway: Ste5p interacts with all members of the MAP kinase cascade. *Genetics* **138:** 609–619.

Pryciak P.M. and Huntress F.A. 1998. Membrane recruitment of the kinase cascade scaffold protein Ste5 by the Gβγ complex underlies activation of the yeast pheromone response pathway. *Genes Dev.* **12:** 2684–2697.

Ramer S.W. and Davis R.W. 1993. A dominant truncation allele identifies a gene, STE20, that encodes a putative protein kinase necessary for mating in *Saccharomyces cerevisiae*. *Proc. Natl. Acad. Sci.* **90:** 452–456.

Reid B.J. and Hartwell L.H. 1977. Regulation of mating in the cell cycle of *Saccharomyces cerevisiae*. *J. Cell Biol.* **75:** 355–365.

Rhodes N., Connell L., and Errede B. 1990. STE11 is a protein kinase required for cell-type-specific transcription and signal transduction in yeast. *Genes Dev.* **4:** 1862–1874.

Roberts C.J., Nelson B., Marton M.J., Stoughton R., Meyer M.R., Bennett H.A., He Y.D., Dai H., Walker W.L., Hughes T.R., Tyers M., Boone C., and Friend S.H. 2000. Signaling and circuitry of multiple MAPK pathways revealed by a matrix of global gene expression profiles. *Science* **287:** 873–880.

Seger R., Seger D., Lozeman F.J., Ahn N.G., Graves L.M., Campbell J.S., Ericsson L., Harrylock M., Jensen A.M., and Krebs E.G. 1992. Human T-cell mitogen-activated protein kinase kinases are related to yeast signal transduction kinases. *J. Biol. Chem.* **267:** 25628–25631.

Sette C., Inouye C.J., Stroschein S.L., Iaquinta P., and Thorner J. 2000. Mutational analysis suggests that activation of the yeast pheromone response mitogen-activated protein kinase pathway involves conformational changes in the Ste5 scaffold protein. *Mol. Biol. Cell* **11:** 4033–4049.

Siekhaus D.E. and Drubin D.G. 2003. Spontaneous receptor-independent heterotrimeric G-protein signalling in an RGS mutant. *Nat. Cell Biol.* **5:** 231–235.

Sprague G.F., Jr. and Thorner J.W. 1992. Pheromone response and signal transduction during the mating process of *Saccharomyces cerevisiae*. In *The molecular and cellular biology of the yeast* Saccharomyces. 2. *Gene expression* (ed. E.W. Jones et al.), pp. 657–744. Cold Spring Harbor Laboratory Press, Cold Spring Harbor, New York.

Sprague G.F., Jr., Blair L.C., and Thorner J. 1983. Cell interactions and regulation of cell type in the yeast *Saccharomyces cerevisiae*. *Annu. Rev. Microbiol.* **37:** 623–660.

Stevenson B.J., Rhodes N., Errede B., and Sprague G.F., Jr. 1992. Constitutive mutants of the protein kinase STE11 activate the yeast pheromone response pathway in the absence of the G protein. *Genes Dev.* **6:** 1293–1304.

Stryer L., Hurley J.B., and Fung B.K. 1983. Transducin and the cyclic GMP phosphodiesterase of retinal rod outer segments. *Methods Enzymol.* **96:** 617–627.

Tanoue T. and Nishida E. 2002. Docking interactions in the mitogen-activated protein kinase cascades. *Pharmacol. Ther.* **93:** 193–202.

Teague M.A., Chaleff D.T., and Errede B. 1986. Nucleotide sequence of the yeast regulatory gene *STE7* predicts a protein homologous to protein kinases. *Proc. Natl. Acad. Sci.* **83:** 7371–7375.

Thorner J. 1981. Pheromonal regulation of development in *Saccharomyces cerevisiae*. In *The molecular biology of the yeast* Saccharomyces: *Life cycle and inheritance* (ed. J.N. Strathern et al.), pp. 143–180. Cold Spring Harbor Laboratory Press, Cold Spring Harbor, New York.

Toenjes K.A., Sawyer M.M., and Johnson D.I. 1999. The guanine-nucleotide-exchange factor Cdc24p is targeted to the nucleus and polarized growth sites. *Curr. Biol.* **9:** 1183–1186.

van Drogen F. and Peter M. 2001. MAP kinase dynamics in yeast. *Biol. Cell* **93:** 63–70.

Whiteway M.S., Wu C., Leeuw T., Clark K., Fourest-Lieuvin A., Thomas D.Y., and Leberer E. 1995. Association

of the yeast pheromone response G protein beta gamma subunits with the MAP kinase scaffold Ste5p. *Science* **269:** 1572–1575.
Whiteway M., Hougan L., Dignard D., Bell L., Saari G., Grant F., O'Hara P., MacKay V.L., and Thomas D.Y. 1988. Function of the *STE4* and *STE18* genes in mating pheromone signal transduction in *Saccharomyces cerevisiae*. *Cold Spring Harbor Symp. Quant. Biol.* **53:** 585–590.
Whiteway M., Hougan L., Dignard D., Thomas D.Y., Bell L., Saari G.C., Grant F.J., O'Hara P., and MacKay V.L. 1989. The *STE4* and *STE18* genes of yeast encode potential β and γ subunits of the mating factor receptor-coupled G protein. *Cell* **56:** 467–477.
Wu C., Whiteway M., Thomas D.Y., and Leberer E. 1995. Molecular characterization of Ste20p, a potential mitogen-activated protein or extracellular signal-regulated kinase kinase (MEK) kinase kinase from *Saccharomyces cerevisiae*. *J. Biol. Chem.* **270:** 15984–15992.
Yashar B., Irie K., Printen J.A., Stevenson B.J., Sprague G.F., Jr., Matsumoto K., and Errede B. 1995. Yeast MEK-dependent signal transduction: Response thresholds and parameters affecting fidelity. *Mol. Cell. Biol.* **15:** 6545–6553.
Yasuda J., Whitmarsh A.J., Cavanagh J., Sharma M., and Davis R.J. 1999. The JIP group of mitogen-activated protein kinase scaffold proteins. *Mol. Cell. Biol.* **19:** 7245–7254.
Zheng Y., Cerione R., and Bender A. 1994. Control of the yeast bud-site assembly GTPase Cdc42. Catalysis of guanine nucleotide exchange by Cdc24 and stimulation of GTPase activity by Bem3. *J. Biol. Chem.* **269:** 2369–2372.
Zhou Z., Gartner A., Cade R., Ammerer G., and Errede B. 1993. Pheromone-induced signal transduction in *Saccharomyces cerevisiae* requires the sequential function of three protein kinases. *Mol. Cell. Biol.* **13:** 2069–2080.
Zhu H., Klemic J.F., Chang S., Bertone P., Casamayor A., Klemic K.G., Smith D., Gerstein M., Reed M.A., and Snyder M. 2000. Analysis of yeast protein kinases using protein chips. *Nat. Genet.* **26:** 283–289.

11

Cytoskeleton and Morphogenesis

John R. Pringle
Department of Biology and Program in Molecular Biology and Biotechnology
The University of North Carolina
Chapel Hill, North Carolina 27599-3280

The early days, 212
   Byers B. and Goetsch L. 1975. Behavior of spindles and spindle plaques in the cell cycle and conjugation of *Saccharomyces cerevisiae*. *J. Bacteriol.* **124:** 511–523.

The spindle-pole body, 214
   Rout M.P. and Kilmartin J.V. 1990. Components of the yeast spindle and spindle pole body. *J. Cell Biol.* **111:** 1913–1927.

The cytoplasmic microtubules, 216
   Jacobs C.W., Adams A.E.M., Szaniszlo P.J., and Pringle J.R. 1988. Functions of microtubules in the *Saccharomyces cerevisiae* cell cycle. *J. Cell Biol.* **107:** 1409–1426.

The actin cytoskeleton, 218
   Kilmartin J.V. and Adams A.E.M. 1984. Structural rearrangements of tubulin and actin during the cell cycle of the yeast *Saccharomyces*. *J. Cell Biol.* **98:** 922–933.

Bud-site selection and polarization signaling, 222
   Sloat B.F., Adams A., and Pringle J.R. 1981. Roles of the CDC24 gene product in cellular morphogenesis during the *Saccharomyces cerevisiae* cell cycle. *J. Cell Biol.* **89:** 395–405.

   Chant J. and Herskowitz I. 1991. Genetic control of bud site selection in yeast by a set of gene products that constitute a morphogenetic pathway. *Cell* **65:** 1203–1212.

   Lew D.J. and Reed S.I. 1993. Morphogenesis in the yeast cell cycle: Regulation by Cdc28 and cyclins. *J. Cell Biol.* **120:** 1305–1320.

Note: The landmark papers listed above are those discussed in this chapter. Each landmark paper is preceded by the name of the section (with starting page number) where the paper is first discussed in detail.

THE EARLY DAYS

In a magisterial 1969 review, Matile, Moor, and Robinow summarized what was then known of yeast cell biology (Matile et al. 1969). Remarkably, of the various topics covered in the present chapter, Matile and his colleagues had very little to say. They noted that the nucleus contains microtubules in a structure that has at least some resemblance to the mitotic spindles of other eukaryotic cells, and they presented some respectable electron micrographs (EMs), but only a rather vague description, of the spindle-pole body (SPB). They described bud scars and birth scars and presented some of Streiblová's and Beran's early fluorescence images of these structures, but they did not describe specific budding patterns and overlooked the early papers (Winge 1935; Barton 1950; Freifelder 1960) that had done so. Two important papers showing the polarization of cell-surface growth to the bud tip (Johnson and Gibson 1966; Johnson 1968) were also overlooked. The cytoplasmic microtubules (cMTs), the actin cytoskeleton, the septins, and the signaling systems controlling cytoskeletal organization had not yet been discovered. Thus, the modern fields of cytoskeletal function and cellular morphogenesis in yeast have developed almost entirely since 1970.

In fact, most of this development has occurred since 1980. However, several major developments during the 1970s should also be highlighted. Of particular importance were several meticulous EM studies (Moens and Rapport 1971; **Byers and Goetsch** 1974, **1975**, 1976a; Peterson and Ris 1976; Byers 1981a, b). These studies provided a largely accurate description of the structure of the SPB and its changes during the life cycle, considerable detail about the organization and behavior of the spindle microtubules (sMTs), the first descriptions of the cMTs and their behavior during the life cycle, and the first views of the septin-associated "neck filaments." The demonstration that the yeast spindle is fundamentally similar to other spindles helped to convince the many remaining skeptics that yeast is a bona fide eukaryote whose study should yield cell biological insights of general significance. This work also showed that informative thin-section EM studies could be done on yeast despite the presence of a cell wall and the high concentration of ribosomes in the cytoplasm.

Other important papers during the 1970s provided details of the pattern of bud growth that allowed later elucidation of its cell-cycle control (Tkacz and Lampen 1972; Farkas et al. 1974), showed that the *cdc24* mutant provided an entrée to studies of the genetic control of cell polarization (Hartwell et al. 1974; Sloat and Pringle 1978), and described the axial and bipolar budding patterns and their control by mating type (Hicks et al. 1977). Finally, it should be stressed that the isolation of many mutants and the development of molecular genetic techniques during the 1970s laid essential groundwork for the rapid progress that began in 1980.

Here, I have attempted to summarize the major intellectual developments and highlight some key papers that have contributed to our present understanding of the various facets of cytoskeletal function and morphogenesis in vegetative (budding) cells. Because of space constraints, I have not attempted to discuss the equally interesting topics of filamentous growth, mating, and sporulation. The focus is on developments during the 1980s and 1990s, and references to some more recent papers are intended only to indicate the directions of current research; they should not be mistaken for systematic reviews of the areas in question or an attempt to identify the most important of recent papers. Readers should also be aware that the progress in these fields has involved many more papers than the editors have allowed me to cite and, indeed, many more "landmark papers" than I have been allowed to designate as such. Thus, I must apologize to those who may feel that the importance of their own work has been overlooked or insufficiently recognized. Finally, readers should note that most of the papers designated as landmarks are cited in more than one section of this chapter, so that their full significance may not be apparent from any single citation.

THE MITOTIC SPINDLE

The EM studies of the 1970s had described correctly the formation of the yeast spindle coincident with the separation of the SPBs, the persistence of a short spindle for a considerable period before anaphase onset, the presence of both pole-to-chromosome and pole-to-pole microtubules, and the presence of a single microtubule per chromatid. The first immunofluorescence studies of the spindle (Kilmartin and Adams 1984) were largely confirmatory of earlier conclusions but did also provide an easier way to monitor spindle behavior under various conditions. The use of tubulin mutants and an efficient microtubule-depolymerizing drug then showed that the sMTs, but not the cMTs, are necessary for spindle elongation and that SPB separation, but not SPB duplication, is microtubule dependent (Huffaker et al. 1988; Jacobs et al. 1988; Sullivan and Huffaker 1992). Although there had been some previous time-lapse observations of nuclear division, the modern era in time-lapse microscopy of yeast began when Bloom, Salmon, Yeh, and their co-workers applied digitally enhanced differential-interference-contrast (DIC) microscopy to reveal new aspects of spindle dynamics through the cell cycle (Yeh et al. 1995). Real-time fluorescence microscopy soon revealed further aspects of spindle and cMT dynamics while also helping to usher in the era of green fluorescent protein (GFP) tagging in yeast (Carminati and Stearns 1997; Shaw et al. 1997).

Meanwhile, a new generation of EM studies also made important contributions. By computer-aided reconstruction of spindles from serial cross-section images, Winey, McIntosh, and their co-workers both confirmed many of the earlier conclusions and reached important new ones (Winey et al. 1995). For example, they showed that both anaphase A (chromosome-to-pole) and anaphase B (pole-separation) movements contribute to chromosome segregation. They also found an antiparallel overlap of nonkinetochore microtubules, suggesting that sMT sliding, as well as sMT elongation, contributes to anaphase B. Left unresolved was whether yeast mitosis involves a proper metaphase. When the existence of metaphase was later demonstrated using combinations of fluorescent markers (Pearson et al. 2001; Winey and O'Toole 2001), it became clear that the puzzling preanaphase separation between the presumed kinetochore microtubules (Winey et al. 1995) reflected the separation between sister centromeres that occurs during metaphase.

These studies established that the structure and behavior of the yeast spindle are mostly similar to those of other spindles. Meanwhile, genetic analyses were giving yeast a leading role in studies of mitotic mechanisms. Mutations affecting mitosis in yeast (Meluh and Rose 1990) and the fungus *Aspergillus* (Enos and Morris 1990) proved to affect proteins related to the microtubule motor protein kinesin, which had been identified by virtue of its role in vesicle transport in squid axons. These results established that kinesins are involved in mitosis and thus launched the modern study of mitotic motors. Additional genetic analyses (Hoyt et al. 1990, 1992) and sequence-based searches (Roof et al. 1992) soon led to the identification of additional yeast kinesins, which led in turn to the recognition of two general principles of spindle function. First, kinesins of opposite polarity (i.e., that translocate toward the plus or minus ends of microtubules) function in dynamic tension during spindle assembly (Saunders and Hoyt 1992; Hoyt et al. 1993). Second, kinesins function redundantly to produce spindle elongation and chromosome movements (Hoyt et al. 1992; Roof et al. 1992). Meanwhile, analysis of mutants lacking dynein revealed that this other major microtubule motor protein is important for the function of the cMTs but not of the spindle, at least in yeast (Eshel et al. 1993; Li et al. 1993; Yeh et al. 1995). Given the complex interactions among multiple motor proteins during mitosis, it is not surprising that genetic studies in yeast continue to have a major role in the efforts to elucidate these interactions (Cottingham et al. 1999; Sharp et al. 2000; McIntosh et al. 2002).

Microtubule-associated proteins (MAPs) that regulate the dynamics of microtubule assembly and disassembly also have essential roles in spindle function and have complex interactions with the motor proteins (McIntosh et al. 2002). Genetic studies in yeast have also been central to the identification and analysis of such MAPs. For example, the first member of the "CLIP-170 family" of proteins to be described was actually the yeast Bik1 protein (Trueheart et al. 1987), whose role in microtubule stability was also recognized early (Berlin et al. 1990). Studies of Bik1 remain central to efforts to understand the functions of this protein family (Brunner and Nurse 2000; Lin et al. 2001). Similarly, the role of EB1-family proteins in regulating microtubule dynamics (Rogers et al. 2002) was first recognized in studies of yeast Bim1 (Schwartz et al. 1997; Tirnauer et al. 1999) and fission yeast Mal3 (Beinhauer et al. 1997). Finally, despite many years of in vitro studies of *Xenopus* XMAP215 and related proteins, it was genetic analysis of the yeast homolog Stu2 (Wang and Huffaker 1997) that revealed the role of these proteins as microtubule-destabilizing factors in vivo (Kosco et al. 2001; van Breugel et al. 2003).

THE SPINDLE-POLE BODY

The EM studies of the early 1970s had revealed the basic structure of the SPB and its changes during the cell cycle, mating, and sporulation. Among other factors, these studies had suggested that the SPB "satellite" is an intermediate in a "conservative" process of SPB duplication and that the modified SPB of sporulating cells is involved in organizing the prospore membrane, two hypotheses that have held up well (Winey and O'Toole 2001; Nickas and Neiman 2002). Studies of the SPB then made little progress for 15 years, except for the characterization of two mutants, *cdc31* (Byers 1981b) and *kar1* (Conde and Fink 1976; Rose and Fink 1987), with defects in SPB duplication. This long dry spell was broken dramatically by the paper of **Rout and Kilmartin (1990)**, which capped 14 years of often-frustrating efforts by Kilmartin and his co-workers. Rout and Kilmartin partially purified SPBs, generated monoclonal antibodies that recognized specific proteins in the preparation, and used these antibodies in immunolocalization experiments on both whole cells and purified SPBs. These studies showed that two of the proteins were indeed components of the SPB, whereas a third was a component of the kinetochore (which, along with the associated sMTs, copurified with the SPBs). Remarkably, Rout and Kilmartin also predicted accurately the functions of the SPB components from their substructural localizations. Meanwhile, genetic studies in *Aspergillus* had identified γ-tubulin (Weil et al. 1986; Oakley et al. 1990), which is responsible for the microtubule-nucleating activities of SPBs and other microtubule-organizing centers. Together, these studies set the stage for rapid progress in understanding of the SPB.

Some of the important papers that followed further elucidated SPB composition and structure. First, Kilmartin et al. (1993) used gene truncations and high-quality EM to show that Spc110 (Rout and Kilmartin 1990) functions as a spacer between the central and inner SPB plaques. Second, Schiebel and co-workers were led by suppressor analyses to the discovery that the yeast γ-tubulin, Tub4, is found at both faces of the SPB in a tight complex with Spc98 (Rout and Kilmartin 1990) and Spc97 (Geissler et al. 1996; Knop and Schiebel 1997); this in turn led to the realization that animal γ-tubulins form similar complexes (Murphy et al. 1998). Third, improved EM and image-processing techniques provided a clearer picture of the several layers of the SPB, their organization around a crystalline core of Spc42, and their relationships to the nuclear envelope (Bullitt et al. 1997; O'Toole et al. 1999). Finally, Wigge et al. (1998) improved the purification of SPBs sufficiently to identify multiple new components of the SPB by mass spectrometry; this study provided one of the first illustrations of the ability of modern mass spectrometry to identify the proteins in a complex mixture when a genome sequence is available. The combination of biochemical, genetic, and morphological approaches has by now yielded a long list of SPB components and many details about their organization into subcomplexes and localization within the overall structure. Although more detailed structural information from X-ray crystallography does not seem imminent, additional genetic analyses (see, e.g., Nguyen et al. 1998) and fluorescence-resonance-energy-transfer (FRET) methods (Hailey et al. 2002) should allow further clarification of protein organization within the SPB.

Other important papers further elucidated the reproduction and function of the SPB. First, localization of a Kar1–LacZ fusion protein was used to show the asymmetric segregation of the old and newly formed SPBs (Vallen et al. 1992). Recent evidence suggests that the polarity of this segregation is the opposite of that originally thought, so that it is generally the old SPB that enters the bud (Pereira et al. 2001). Nonetheless, knowledge of the asymmetry was key to recognizing the central role of the SPB (and thus also the animal centrosome) in controlling the exit from mitosis and entry into cytokinesis (Pereira and Schiebel 2001). Second, Geiser et al. (1993) used the convergence of two distinct genetic approaches (suppressor and two-hybrid analyses), with appropriate biochemical and cytological follow-up, to identify the interaction between Spc110 and calmodulin. This study revealed that calmodulin was present in the SPB and also much about its (and Spc110's) function there; it thus explained, at least in part, the previously known importance of calmodulin for mitosis. Third, genetic, biochemical, and immunoelectron microscopy analyses showed that Kar1 and Cdc31 are components of the "half-bridge" portion of the unduplicated SPB, thus confirming the previously hypothesized importance of this structure in SPB duplication (Byers 1981b; Rose and Fink 1987; Spang et al. 1993; Biggins and Rose 1994; Vallen et al. 1994). Subsequent studies have further clarified the roles of the half-bridge and associated satellite in SPB duplication (Adams and Kilmartin 1999). Finally, the analysis of *mps* (*m*onopolar *s*pindle) mutants has revealed additional aspects of SPB duplication, such as the role of Mps1-dependent phosphorylation of Spc42 in the assembly of this core SPB component during duplication (Winey et al. 1991; Castillo et al. 2002).

THE CYTOPLASMIC MICROTUBULES

Hartwell and co-workers realized in the early 1970s that the migration of the nucleus to the mother-bud neck and its extension into the bud are distinct steps in the cell cycle that required explanation (Culotti and Hartwell 1971; Hartwell et al. 1974). However, recognition of the central role of the cMTs in these processes did not come until the mid 1980s. In the interval, thinking about the cMTs was dominated by an alternative hypothesis about their function. On the basis of multiple temporal and spatial correlations and the evidence from other organisms that cMTs could be responsible for vesicle transport, Byers and Goetsch had hypothesized, very plausibly, that the SPB and cMTs control the selection of the bud site and the polarized delivery of secretory vesicles into the bud (Byers and Goetsch 1974, 1975; Byers 1981a,b). The first studies using fluorescence microscopy to examine cytoskeletal organization showed that the distribution of actin also correlates strongly with the regions of cell-surface growth (Adams and Pringle 1984; Kilmartin and Adams 1984). However, ironically, by allowing more complete visualization of the cMTs than had been possible by EM, these studies also strengthened the correlations on which the hypothesis of a cMT role in bud growth was based!

Thus, the confusion continued until it was shown (by immunofluorescence) that the drug nocodazole produces a highly effective dissolution of yeast microtubules (Pringle et al. 1986; **Jacobs et al. 1988**). In the drug-treated cells, "...the selection of nonrandom budding sites, the formation of chitin rings and rings of 10-nm [septin] filaments at those sites, bud emergence, differential bud enlargement, and apical bud growth appeared to proceed normally" (Jacobs et al. 1988); thus, it seemed clear that the cMTs and SPB could not be responsible for these processes. In contrast, "...nuclei did not migrate to the mother-bud necks...[and]...SPBs... were often not oriented toward the budding sites..." in the drug-treated cells, suggesting that the cMTs are responsible for nuclear migration and spindle orientation and providing a satisfying alternative interpretation for the correlations that had suggested a role in bud growth. Independent studies using cold-sensitive tubulin mutants led to many of the same conclusions (Huffaker et al. 1988). It was subsequently shown that cortical proteins normally assemble at the new budding site before the SPB and cMTs orient toward that site (Snyder et al. 1991), consistent with the hypothesis that the cMTs interact with a previously established site of polarization.

This story illustrates both the power (the actin side of the story, as discussed further below) and the danger of inferring causal relationships from temporal and spatial correlations. It also illustrates the pros and cons of inhibitor experiments relative to those with mutants. Because nocodazole so effectively abolished the microtubules, the experiments with the drug were simple to perform and very clear-cut; thus, Jacobs et al. (1988) were able to compile a detailed list of processes that *do not* involve the cMTs. In contrast, because nocodazole affected both cMTs and sMTs, and because of the danger of an undetected and relevant side effect, conclusions about what the cMTs might actually *do* needed to be tentative. In contrast, Huffaker et al. (1988) were able to identify particular *TUB2* alleles that differentially affected the cMTs or sMTs and thus discriminate between their functions (see also Sullivan and Huffaker 1992); they could also be certain that the mutations directly affected only one protein in the cell.

With the central role of the cMTs identified, the stage was set for exciting progress in understanding the detailed mechanisms of nuclear migration and spindle orientation.

Multiple, complexly interwoven issues have been identified and partially elucidated during the past decade, as follows.

1. *Interactions between the cMT and actin cytoskeletons.* Palmer et al. (1992) first demonstrated that nuclear migration and orientation involve actin as well as the cMTs. Subsequent studies have shown that actin is involved specifically in the early (pre-anaphase) stage of the normal process (Theesfeld et al. 1999) and that the actin role has at least two aspects. First, the dynamic, SPB-distal plus ends of the cMTs can be captured by cortical protein complexes whose own asymmetric localization depends on actin (see below). Second, the plus ends can attach to actin cables by a linkage involving Bim1, Kar9, and the myosin Myo2, which can then deliver them to cortical-attachment complexes at the bud site and the tip of the bud (Yin et al. 2000; Hwang et al. 2003).

2. *Roles of microtubule motor proteins.* Mutational analysis showed that dynein (Eshel et al. 1993; Li et al. 1993) and its accessory complex dynactin (Clark and Meyer 1994; Muhua et al. 1994) are important (but not essential) for nuclear migration and orientation and function primarily late in the cell cycle (Yeh et al. 1995, 2000; DeZwaan et al. 1997). Lateral sliding interactions of the cMTs with the cell cortex (Carminati and Stearns 1997; Adames and Cooper 2000) apparently allow the minus-end-directed motor activity to pull one SPB into the bud. The roles of kinesins were demonstrated only somewhat later (Cottingham and Hoyt 1997; DeZwaan et al. 1997; Miller et al. 1998; Yeh et al. 2000; Maekawa et al. 2003), although kinesin involvement could have been anticipated from the role of Kar3 in nuclear migration during mating (Meluh and Rose 1990). The kinesins function primarily early in the cell cycle, and kinesin mutations have genetic interactions that suggest both overlapping and antagonistic roles for the several proteins. The mutant phenotypes suggest that the kinesins do not function by pulling on the cMTs but rather (i) by controlling the dynamics of the plus ends, at least in part by delivering proteins (like Bik1 and Bim1) that control microtubule stability and (ii) by delivering to the plus ends proteins (like Bim1 and Kar9) that allow association with actin cables and/or cortical-attachment sites (see above and below).

3. *Attachment of the cMTs to cortical factors.* The cMTs can apparently find cortical attachment sites either by a search-and-capture mechanism based on their dynamic instability (see below) or by being led there (together with components of the attachment sites) along actin cables (see above). In either case, attachment of the cMT plus ends at the bud tip involves a complex of proteins (including Bni1, Bud6, and Spa2) that is also involved in the assembly of polarized actin cables (Lee et al. 1999; Miller et al. 1999; Segal et al. 2000a, 2002) and the dedicated protein Kar9, which localizes to the bud tip in an actin-dependent manner and binds to the cMTs through Bim1 (Miller and Rose 1998; Miller et al. 1999; Korinek et al. 2000; Lee et al. 2000). After attachment, depolymerization of the plus ends can pull the SPB in the appropriate direction (Adames and Cooper 2000; Yeh et al. 2000). In contrast, the lateral interactions of cMTs with the cortex that are important for dynein function (see above) involve distinct cortical factors including the dynein-interacting Num1 (Kormanec et al. 1991; Farkasovsky and Küntzel 2001).

4. *The roles of cMT dynamic instability.* The dynamic instability of microtubule plus ends (Kirschner and Mitchison 1986) appears to be important in two ways for nuclear

migration and orientation in yeast: It allows a search for cortical attachment sites (Carminati and Stearns 1997; Shaw et al. 1997), and it allows attached cMTs to exert force on the SPB by depolymerizing (see above). As for the sMTs (see above), the stability of the cMTs appears to be regulated by a variety of factors including the proteins Bik1 and Bim1 (Berlin et al. 1990; Schwartz et al. 1997; Miller et al. 1998; Tirnauer et al. 1999; Tirnauer and Bierer 2000).

5. *Establishment of spindle/SPB asymmetry*. Successful segregation of just one of the two daughter nuclei into the bud requires that the two SPBs and their attached cMTs be differentiated in some way. Although the mechanisms of such differentiation are incompletely understood, they appear to involve phosphorylation of SPB/cMT-associated proteins, perhaps including Kar9, by the cell-cycle-controlling kinase Cdc28 in conjunction with particular B-type cyclins (Segal et al. 2000b; Maekawa et al. 2003).

Both the general mechanisms and many of the specific proteins discussed above have been widely conserved; thus, the work on yeast has also been deeply informative about similar processes in other cell types (Gundersen 2002). A major lesson appears to be that processes as critical as spindle orientation may typically be controlled by multiple mechanisms with partially redundant functions, thus providing some fail-safe protection for the cell (cf. the discussion of spindle function, above).

THE ACTIN CYTOSKELETON

By 1979, it was known that actin was not just a muscle protein. However, it was not clear whether it would be present, or important, in a nonmotile cell like yeast. Thus, the identification of an actin protein (Koteliansky et al. 1979; Water et al. 1980; Greer and Schekman 1982), the independent and nearly simultaneous cloning of the *ACT1* gene (Gallwitz and Seidel 1980; Ng and Abelson 1980), and the demonstration that *ACT1* is essential (Shortle et al. 1982) were major developments. The last of these papers had added impact because it also described a new and simple gene-disruption technique and thus was pivotal in ushering in the era of reverse genetics.

Nonetheless, it remained unclear *why* actin was essential in yeast. However, when immunofluorescence and fluorescent small-molecule probes were used to visualize the intracellular distribution of actin, it was found to be present in cytoplasmic cables and cortical patches, both of which were strikingly associated with regions of localized cell surface growth, both during the normal cell cycle and in various morphogenetic mutants (Adams and Pringle 1984; **Kilmartin and Adams 1984**). This immediately suggested that actin was involved in the polarized delivery of secretory vesicles during both bud growth and cytokinesis. As in the case of Shortle et al. (1982), the significance of these 1984 papers was not limited to their contributions to understanding of actin and microtubule function, because they also made a major methodological contribution. It had not previously been clear that fluorescence methods would be practicable and effective in revealing internal structures in small, round, wall-enclosed cells like yeast. By showing how much could be seen (and how easily, in comparison to the EM), the 1984 papers opened the door for fluorescence studies of other intracellular systems in yeast.

For most of the next decade, studies of the yeast actin cytoskeleton focused on mutations in actin itself and on the identification of actin-interacting proteins. First, Novick and Botstein (1985) showed that conditional *act1* mutants indeed had phenotypes consistent with the roles inferred from spatial and temporal correlations in 1984. The *act1* mutations then provided a basis for some of the many genetic and biochemical approaches that were used to identify and analyze the complex set of actin-interacting proteins (Pruyne and Bretscher 2000b). The discovery by Adams et al. (1989, 1991) that the *act1* suppressor *SAC6* encodes a protein (fimbrin) that interacts physically with actin was particularly important, because its linkage of genetics to biochemistry helped to validate the genetic approach to study of the cytoskeleton. Systematic site-directed mutagenesis then yielded new *ACT1* alleles with phenotypes suggesting several previously unrecognized roles for the yeast actin cytoskeleton (Wertman et al. 1992; Drubin et al. 1993).

Studies in this field then took off in multiple directions that both elucidated actin function in yeast and gave yeast a central role in studies of the actin cytoskeleton more generally. These directions have included:

1. *Characterization of multiple myosins with distinct functions.* MYO1, encoding a type II myosin, was identified and disrupted early (Watts et al. 1987; Rodriguez and Paterson 1990), but its role remained unclear until it was found along with actin in the cytokinetic contractile ring (Bi et al. 1998; Lippincott and Li 1998). *MYO2*, encoding a type V myosin, was identified by a mutant that produces large, unbudded cells (Johnston et al. 1991), suggesting a role in the polarized transport of secretory vesicles into the bud. This hypothesis has since been confirmed (Govindan et al. 1995; Karpova et al. 2000; Schott et al. 2002) and extended by showing that Myo2 is also responsible for the polarized transport of other organelles (Hill et al. 1996; Hoepfner et al. 2001). A second type V myosin, Myo4, was found to be involved specifically in the polarized delivery of certain mRNAs into the bud (Jansen et al. 1996; Long et al. 1997; Takizawa et al. 1997, 2000). Early evidence that Myo3 (Goodson and Spudich 1995) and Myo5 (Geli and Riezman 1996; Goodson et al. 1996) are involved in endocytosis now appears to reflect the role of type I myosins in Arp2/3-stimulated actin polymerization and hence in formation of the actin patches that function in endocytosis (see below).

2. *Analysis of the conserved cMT-actin-Myo2 interactions* during nuclear migration and spindle orientation (see discussion of cMT function, above).

3. *Clarification of the roles of actin cables and patches.* The correlations between cable and patch behavior were further characterized (Mulholland et al. 1994), and it was shown (in two of the first studies of GFP-tagged proteins in living yeast) that the patches are highly dynamic (Doyle and Botstein 1996; Waddle et al. 1996). However, despite early evidence that actin is involved in endocytosis (Kübler and Riezman 1993; Geli and Riezman 1996), the relationship between cable and patch function remained obscure until the elegant studies of Pruyne et al. (1998), who used conditional tropomyosin mutants to demonstrate that the cables are directly responsible for the polarization of secretion and cell surface growth. This conclusion was further supported by studies of a mutant lacking patch polarization (Karpova et al. 2000). Pruyne et al. further suggested that the patches might be sites of membrane retrieval through endocytosis, a hypothesis that appears to be correct (Engqvist-Goldstein and Drubin 2003).

4. *Identification of a drug that depolymerizes yeast actin.* Latrunculin-A (Ayscough et al. 1997) has allowed many insights into actin assembly dynamics and their modulation by various proteins; it has also provided a critical reagent (often better than the best available mutants) for investigating other aspects of actin function.

5. *Characterization of Arp2/3 complex-nucleated actin polymerization.* Although much of the progress in this field has come from other systems (see Higgs and Pollard 2001), yeast has made important contributions. For example, Arp2 and Arp3 proteins were first identified in yeasts (Lees-Miller et al. 1992; Schwob and Martin 1992). Yeast has also contributed (i) genetic dissection of the roles of individual Arp2/3-complex subunits and of the WASP-family protein Las17/Bee1, an Arp2/3-complex activator like its mammalian homologs (Moreau et al. 1996; Li 1997; Madania et al. 1999; Winter et al. 1999a,b); (ii) the discovery of a myosin I role in Arp2/3-complex function (Evangelista et al. 2000; Lechler et al. 2000); and (iii) the recognition of an apparent Arp2/3 role in the movement of mitochondria into the bud along actin-cable tracks (Drubin et al. 1993; V.R. Simon et al. 1995; Boldogh et al. 2001). Finally, the localization of Arp2/3 and Las17 to actin patches and the effects of Arp2/3-complex mutations on the formation of actin patches but not cables (Moreau et al. 1996; Li 1997; Winter et al. 1997, 1999a) contributed to the eventual discrimination between the roles of Arp2/3 and the formins in nucleating actin assembly in the patches and cables, respectively (see below).

6. *Identification of the formins as Arp2/3-independent nucleators of actin polymerization.* BNI1 was identified in several independent genetic screens that pointed to its interaction with Rho GTPases and role in cell polarization (Fares 1995; Jansen et al. 1996; Kohno et al. 1996; Zahner et al. 1996), and its sequence helped to define the formin family of proteins (Frazier and Field 1997; Wasserman 1998). A second yeast formin, Bnr1, was soon also identified and shown to overlap in function with Bni1 (Imamura et al. 1997; Kamei et al. 1998; Vallen et al. 2000). Meanwhile, analyses of protein localization, protein interactions, and mutant phenotypes led Evangelista et al. (1997) to conclude that "Bni1p may occur as part of a complex that directs the assembly of actin filaments in response to Cdc42p signaling during polarized morphogenesis." Studies in both yeast and other systems then confirmed and extended these conclusions, and it now seems clear (i) that a domain of Bni1, together with the actin-binding protein profilin, can directly nucleate actin-filament formation and (ii) that Bni1 and Bnr1 are responsible for the assembly of actin cables, whereas the Arp2/3 complex is responsible for the assembly of actin patches (Evangelista et al. 2002; Pruyne et al. 2002; Sagot et al. 2002a,b). The nucleation of distinct sets of actin filaments by Arp2/3 and formins is probably ubiquitous (see, e.g., Severson et al. 2002). In yeast, at least, the formins appear to function in a complex containing also Bud6, Spa2, and Pea2 (Evangelista et al. 1997; Fujiwara et al. 1998; Sheu et al. 1998; Ozaki-Kuroda et al. 2001).

THE SEPTINS

Four of the earliest cell-cycle mutants (*cdc3*, *cdc10*, *cdc11*, and *cdc12*) continued to form buds and traverse the nuclear cycle but were defective in cytokinesis (Hartwell 1971). EM studies showed that these mutants lacked an apparent ring of filaments that was found in wild-type cells adjacent to the plasma membrane in the mother-bud neck (Byers and Goetsch 1976a,b; Byers 1981a). There was then little progress for nearly a decade, except for the observations that both the actin cytoskeleton and bud growth are hyperpolarized in these mutants (Adams and Pringle 1984) and that *CDC10* expression increases dramatically during sporulation (Kaback and Feldberg 1985). In the late 1980s, cloning and sequencing of the four genes revealed that their products comprise a family, named the septins for their role in septation, while immunofluorescence studies showed that the four proteins localize, in both wild-type and mutant cells, as if they are components of the "neck filaments" (Haarer and Pringle 1987; Kim et al. 1991; Longtine et al. 1996). Three more yeast septins were found later; two of these are involved only in spore formation and one is present at the neck, but nonessential, in vegetative cells (De Virgilio et al. 1996; Fares et al. 1996; Gladfelter et al. 2001). Meanwhile, it also became clear that septins are found widely, and probably ubiquitously, in other fungi and in animals (Longtine et al. 1996; Kinoshita 2003).

There has been some progress in determining how the septins function at the molecular level. The first clue was the observation that the axial budding pattern of haploid cells is lost in some septin mutants (Flescher et al. 1993), which was later explained by the discovery that axial budding depends on a spatial marker that is recruited to the neck in a septin-dependent manner (Chant and Pringle 1995; Chant et al. 1995; see also below). The hypothesis that the septins provide a scaffold for the assembly of other proteins at the neck was then supported by the observations that chitin synthase localization to the bud site depends on the septins (DeMarini et al. 1997) and that several mutations in other genes that alter septin organization also alter the organization of the associated proteins (Longtine et al. 1998, 2000). Although it remains unclear exactly how the recruited proteins interact with the septin ring and how proteins can associate asymmetrically with the mother or bud side of the ring, the scaffold model seems to account for most aspects of septin function in yeast (Gladfelter et al. 2001). There is now also good evidence that the septins provide a barrier that separates the membranes of mother cell and bud (Barral et al. 2000; Takizawa et al. 2000), but it remains unclear whether they perform this role directly or by recruiting another protein(s) that forms the actual barrier. Another open question is whether either a scaffold or a barrier model can explain various aspects of septin function in animal cells (see, e.g., Kinoshita et al. 2002; Surka et al. 2002).

Studies of the septins now involve many topics including cell-cycle control (Barral et al. 1999; Shulewitz et al. 1999; Longtine et al. 2000; McMillan et al. 2002), other signaling functions (Robinson et al. 1999), nuclear movement and orientation (Segal et al. 2000a), secretion (Beites et al. 1999; Dent et al. 2002), neuronal function (Kinoshita et al. 1998; Xue et al. 2000), and apoptosis (Larisch et al. 2000). However, it must be stressed that important earlier questions also remain unanswered. Most prominently, it remains unclear what the septins really do in cytokinesis. In yeast, the septins are required for formation of the actomyosin contractile ring (Bi et al. 1998; Lippincott and Li 1998), but this does not appear to be true in most other organisms. Moreover, the septins must have at least one other role, because the actomyosin ring itself is not essential for cytokinesis in yeast (Rodriguez and Paterson 1990; Bi et al. 1998; Schmidt et al. 2002). To make matters worse, although the septins also localize to the division site in most or all other fungal and animal cells, they are not essential for cytokinesis in some cases (Longtine et al. 1996; Adam et al. 2000; Nguyen et al. 2000). Thus, it is now necessary to identify the other septin role(s) and to determine whether this role is redundant with that of a non-septin system in some cell types (Hales et al. 1999; Vallen et al. 2000; Lippincott et al. 2001; Roh et al. 2002). Other outstanding questions include how the septins are initially recruited to the presumptive bud site (see below), the roles of GTP binding and hydrolysis by the septins (Field et al. 1996; Mitchison and Field 2002), the organization of the septins and other proteins in the higher-order structures seen as the "neck filaments" in the EM, and the role (if any) of assembly into these higher-order structures (Frazier et al. 1998; Longtine et al. 1998, 2000).

BUD-SITE SELECTION AND POLARIZATION SIGNALING

One of the early cell-cycle mutants, *cdc24-1*, continued the nuclear cycle although it was unable to bud. As many other mutants could form buds without progressing through the nuclear cycle, this suggested that the events of the yeast cell cycle were organized into two independent pathways (Hartwell et al. 1974). To explore this model further, *cdc24* mutants were studied in more detail, and additional mutants with budding defects were sought. When *cdc24* cells were stained with the fluorescent dye Calcofluor, which had been shown to stain the chitin-rich bud scars of normal cells (Hayashibe and Katohda 1973), it was found that chitin was deposited randomly over the entire cell surface (Sloat and Pringle 1978; **Sloat et al. 1981**). This suggested a defect in the polarization of cell-surface growth, a conclusion that was soon confirmed using other methods (Field and Schekman 1980; Sloat et al. 1981). Subsequently, staining of actin revealed that *cdc24* cells were also defective in polarizing the actin cytoskeleton (Adams and Pringle 1984).

The search for additional *cdc24*-like mutants was discouragingly fruitless for several years but eventually yielded one new mutant, *cdc42-1* (Pringle and Hartwell 1981; Adams and Pringle 1984; Pringle et al. 1986; Adams et al. 1990), that proved to be the key to deeper understanding. The sequence of Cdc42 revealed that it is a highly conserved (~80% identical to its animal homologs) member of the Rho family of small GTPases (Johnson and Pringle 1990; Johnson 1999). This immediately established that the spatial control of cytoskeletal organization and morphogenesis in yeast involves signaling as well as cytoskeletal proteins. It also suggested that the recently discovered Rho proteins might be generally involved in controlling cytoskeletal organization, a suggestion that has since been amply confirmed (Hall 1992, 1998; Johnson 1999; Nelson 2003).

The genetic screen that found *cdc42-1* also yielded *cdc24-4*, a mutant that budded in random locations at permissive temperature, regardless of its mating type (Sloat et al. 1981). This showed that bud-site (and hence polarization-axis) selection was under genetic control and also that Cdc24 itself was involved in generating both the bipolar budding pattern of **a**/α cells and the axial budding pattern of **a** or α cells. A screen for dosage suppressors of the temperature-sensitive lethality of *cdc24-4* yielded both *CDC42* (suggesting that Cdc24 and Cdc42 might interact closely) and a new gene, *RSR1*, which encodes a small GTPase in the Ras family (Bender and Pringle 1989). Deletion of *RSR1* did not affect growth or bud formation but randomized budding pattern in **a**, α, and **a**/α cells.

The hypothesis that Rsr1 and perhaps other proteins might be dedicated to bud-site selection then received strong support from a direct screen for mutants with normal growth but altered budding patterns (**Chant and Herskowitz 1991**; Chant et al. 1991); this screen identified four additional genes (*BUD2–BUD5*). Concern that these gene products might be functionally redundant for an essential role(s) in cell polarization was further reduced by the finding that a dominant-negative mutation of *RSR1* also did not affect growth or bud formation (Ruggieri et al. 1992). In addition, it was shown both biochemically and by elegant genetic analyses that Bud5 and Bud2 are, respectively, the positive (guanine-nucleotide-exchange factor [GEF]) and negative (GTPase-activating protein [GAP]) regulatory factors for Rsr1 (Bender 1993; Park et al. 1993; Zheng et al. 1995). The identification of proteins dedicated to bud-site selection showed that there must exist a functional hierarchy in which these proteins transmit positional information to the proteins, like Cdc24 and Cdc42, that are actually responsible for polarity establishment. Moreover, because the *bud3* and *bud4* mutations disrupted only axial budding, whereas the *rsr1*, *bud2*, and *bud5* mutations disrupted both axial and bipolar budding, it appeared that the bud-site-selection proteins themselves must comprise a hierarchy of at least two steps. However, the precise form of this hierarchy remained obscure until the eventual identification of proteins dedicated to bipolar budding (see below).

Earlier descriptions of the budding patterns (Freifelder 1960; Hicks et al. 1977) led to the suggestion that bud-site selection might involve a marker that remained at the division site after cytokinesis (Chant and Herskowitz 1991; Snyder et al. 1991; Flescher et al. 1993). Detailed descriptive analyses by Chant and Pringle (1995) then led to the explicit prediction that the axial pattern depends on a transient landmark that marks the immediately preceding division site on both mother and daughter cells. Bud3, Bud4, and the newly identified Axl2/Bud10 were soon shown to behave as predicted for this landmark (Chant et al. 1995; Halme et al. 1996; Roemer et al. 1996; Sanders and Herskowitz 1996). In contrast, the bipolar pattern was predicted to depend on a distinct set of persistent landmarks that marked both cell poles (Chant and Pringle 1995). This prediction led to the isolation of mutants apparently defective in components of these landmarks (Zahner et al. 1996), and the hierarchic model for the spatial control of polarity establishment during budding assumed its modern form (Fig. 1) (Pringle et al. 1995; Pruyne and Bretscher 2000a).

Meanwhile, the important issues of the temporal control of morphogenesis and its coordination with the cell cycle were also addressed. **Lew and Reed (1993)** showed that each of the major morphogenetic transitions (polarization toward the bud site, switch from apical to isotropic bud growth, and repolarization to the neck) is triggered by a distinct state of the cell-cycle-controlling kinase Cdc28. Despite considerable further effort and some progress (see Longtine et al. 2000; Gulli and Peter 2001), the precise pathways by which the Cdc28-cyclin complexes trigger the morphogenetic transitions remain unknown. However, in another important advance, Lew and Reed (1995) showed that coordination between morphogenesis and the nuclear cycle is further ensured by a checkpoint that delays the nuclear cycle when bud formation is blocked. There has since been progress in identifying both the aspect(s) of morphogenesis that the checkpoint might monitor and how it controls the nuclear cycle (by controlling the phosphorylation, and hence the activity, of Cdc28), but many questions remain in both areas (see Theesfeld et al. 2003).

Other studies during the past decade have furthered our understanding of yeast cell polarization in a variety of ways.

1. *Bud-site-selection landmarks.* *AXL1* was identified as a fourth gene in which mutations do not affect **a**/α cells but cause **a** or α cells to bud bipolarly (Fujita et al. 1994). Unlike the other known bud-site-selection genes, *AXL1* is expressed only in **a** and α cells and

6. *Cdc42 effectors*. Ste20 was identified during genetic analysis of the pheromone-response pathway (Leberer et al. 1992), and a connection to Cdc42 was suggested by biochemical studies showing that a related mammalian kinase (a "p21-activated kinase," or PAK) is activated by specific binding to GTP-bound Cdc42 (Manser et al. 1994). Both Ste20 and the related Cla4 were soon shown also to behave like Cdc42 effectors (Cvrcková et al. 1995; M.-N. Simon et al. 1995; Zhao et al. 1995). Current evidence suggests that Cdc42-dependent activation of Ste20 stimulates several MAP-kinase pathways (see above), that Cla4 is among several factors participating in the Cdc42-dependent formation of the septin ring (Cvrcková et al. 1995; Longtine et al. 2000; Weiss et al. 2000; Caviston et al. 2003), and that Cla4 has some additional role(s) late in the cell cycle (Benton et al. 1997; Tjandra et al. 1998; Mitchell and Sprague 2001; Höfken and Schiebel 2002). However, despite considerable effort, it remains unclear whether the PAKs are involved in organizing the actin cytoskeleton. Although evidence suggests that PAK-mediated phosphorylation of the type I myosins is important for actin-patch assembly (Wu et al. 1997; Lechler et al. 2001), *cla4 ste20* double mutants (Cvrcková et al. 1995; Holly and Blumer 1999; Weiss et al. 2000), and even *cdc24* and *cdc42* mutants (Adams and Pringle 1984; Adams et al. 1990), form abundant actin patches. Similarly, although some evidence suggests that the PAKs are important for actin polarization (Holly and Blumer 1999; Sheu et al. 2000), other evidence suggests that loss of PAK function causes *hyper*-polarization of the actin cytoskeleton and hence of cell-surface growth (Cvrcková et al. 1995; Weiss et al. 2000). These apparent contradictions may reflect, at least in part, indirect effects resulting from the multiplicity of Cdc42 roles and the involvement of the PAKs in the feedback loops mentioned above.

   The PAKs were found to bind to Cdc42-GTP by means of a domain termed "CRIB" (Burbelo et al. 1995). The yeast genome encodes only two other proteins with CRIB domains, the related Gic1 and Gic2 proteins, and there is good evidence that the Gic proteins are indeed Cdc42 effectors that function in cell polarization (Brown et al. 1997; Chen et al. 1997; Bi et al. 2000; Jaquenoud and Peter 2000). However, it remains unclear precisely what the Gic proteins do, whether they function together with or in parallel to the formins, and how their function relates to that of the proteins Msb3 and Msb4 (Bi et al. 2000). In addition, although it now seems clear that the formins nucleate the assembly of actin cables (see above), it remains unclear whether the formins themselves are directly activated by Cdc42 (Evangelista et al. 1997; Ozaki-Kuroda et al. 2001), by another Rho protein (Kohno et al. 1996; Kamei et al. 1998), by some other mechanism, or perhaps by different mechanisms at different points in the life cycle. Although there is also evidence that the Cdc42-GAPs have effector as well as regulatory function (see above), it also seems likely that additional Cdc42 effectors await discovery.

7. *Other Rho proteins*. Cdc42 is one of six Rho GTPases in yeast (Madaule et al. 1987; Matsui and Toh-e 1992; Roumanie et al. 2001), all of which have been implicated in cell polarization. *RHO1* is essential but partially redundant in function with the nonessential *RHO2* (Ozaki et al. 1996; Schmidt et al. 1997). Rho1-deficient cells typically arrest with small buds and polarized actin (Yamochi et al. 1994; Helliwell et al. 1998), the localization of Rho1 to the bud site depends on actin (Ayscough et al. 1999), and the localized activation of Bni1 at the bud site does not normally depend on Rho1

(Ozaki-Kuroda et al. 2001; Dong et al. 2003). Thus, Rho1 and Rho2 do not appear to function directly in actin polarization toward the bud site. Instead, they appear to function through the protein kinase Pkc1 and associated MAP-kinase cascade to control actin organization as one part of a program for maintaining cell integrity during environmental stress and perhaps also during normal bud growth (Qadota et al. 1994; Nonaka et al. 1995; Helliwell et al. 1998; Delley and Hall 1999; Dong et al. 2003). Rho1 also contributes to cell integrity by functioning as an activator of the cell-wall-synthetic enzyme 1,3-β-glucan synthase (Drgonová et al. 1996; Qadota et al. 1996), and it may also have a direct role in the polarization of secretory-pathway components (Guo et al. 2001). Rho3 and Rho4 appear to have overlapping roles both in organization of the actin cytoskeleton (Matsui and Toh-e 1992; Imai et al. 1996; Dong et al. 2003) and in polarized exocytosis (Adamo et al. 1999); their functions may also overlap with those of Cdc42 (Matsui and Toh-e 1992; Adamo et al. 2001). Because cells deficient for Rho3/Rho4 appear to initiate budding before arresting development (Matsui and Toh-e 1992; Imai et al. 1996), it seems possible that Cdc42 and Rho3/Rho4 function sequentially in stimulating the formin-dependent formation of actin cables.

CONCLUDING REMARKS

In the early 1970s, it seemed likely that the budding mode of reproduction in yeast would involve many highly idiosyncratic morphogenetic mechanisms. It has thus been somewhat surprising, highly instructive, and not a little gratifying to see the degree to which fundamental mechanisms of cellular morphogenesis have been conserved during eukaryotic evolution. This conservation has allowed yeast to have a leading role in deciphering many aspects of morphogenetic signaling and cytoskeletal function, a role that seems certain to continue for many years to come.

ACKNOWLEDGMENTS

Given the complexities of the multiple fields covered in this chapter, I needed considerable help in sorting out the histories and identifying the most important papers. I am very grateful to Erfei Bi, Kerry Bloom, Charlie Boone, David Botstein, Tony Bretscher, Breck Byers, John Cooper, Trisha Davis, David Drubin, the late Ira Herskowitz, Andy Hoyt, John Kilmartin, Danny Lew, Dick McIntosh, Tom Pollard, Mike Snyder, and Tim Stearns for providing very helpful comments. These correspondents had a variety of distinct perspectives, but, reassuringly, there was also considerable overlap in their comments. Nonetheless, any blame for errors of fact or misplaced emphasis lies only on me. I also thank the many coworkers and colleagues who have made my own work in this field so enjoyable over the years.

STUDY QUESTIONS

1. As in most other areas of biology, progress in understanding the cytoskeleton and morphogenesis has been driven in part by the technical advances that have allowed new kinds of observations and experiments. From the many examples described in this chapter, pick five such advances and explain why each was so important to further progress in the cytoskeleton/morphogenesis field.

2. Temporal and spatial correlations can suggest causal relationships, but cannot prove them, and can indeed be highly misleading. From the examples described in this chapter, outline briefly two cases in which a causal relationship suggested by correlative data was subsequently confirmed by functional tests and one case in which subsequent functional tests showed that the inference from correlative data was incorrect.

3. According to the current model, the spatial control of polarization in S. cerevisiae involves a hierarchy of proteins and types of functions (Fig. 1). What exactly does this mean; i.e., what is the logical basis for the contention that these proteins/functions function in a hierarchic manner? What is the key experimental evidence that supports both the existence of a hierarchy and the position of particular proteins/functions within it?

4. There is good evidence that the temporal control of morphogenesis and its coordination with the cell cycle are governed, at least in part, by different activation states of the Cdc28/cyclin complexes. However, the precise effectors through which this control operates are not well understood. Given what is known of the hierarchy of proteins/functions that provides spatial control of morphogenesis (Fig. 1), suggest plausible targets (direct or indirect) for the Cdc28/cyclin complexes that control (i) the initial polarization of the cell toward the presumptive bud site late in the G_1 phase; (ii) the apical-isotropic growth transition at the G_2/M transition; and (iii) the repolarization of the cell toward the neck at the end of anaphase. Explain your answers.

REFERENCES

Adam J.C., Pringle J.R., and Peifer M. 2000. Evidence for functional differentiation among *Drosophila* septins in cytokinesis and cellularization. *Mol. Biol. Cell* **11:** 3123–3135.

Adames N.R. and Cooper J.A. 2000. Microtubule interactions with the cell cortex causing nuclear movements in *Saccharomyces cerevisiae*. *J. Cell Biol.* **149:** 863–874.

Adames N., Blundell K., Ashby M.N., and Boone C. 1995. Role of yeast insulin-degrading enzyme homologs in propheromone processing and bud site selection. *Science* **270:** 464–467.

Adamo J.E., Rossi G., and Brennwald P. 1999. The Rho GTPase Rho3 has a direct role in exocytosis that is distinct from its role in actin polarity. *Mol. Biol. Cell* **10:** 4121–4133.

Adamo J.E., Moskow J.J., Gladfelter A.S., Viterbo D., Lew D.J., and Brennwald P.J. 2001. Yeast Cdc42 functions at a late step in exocytosis, specifically during polarized growth of the emerging bud. *J. Cell Biol.* **155:** 581–592.

Adams A.E.M. and Pringle J.R. 1984. Relationship of actin and tubulin distribution to bud growth in wild-type and morphogenetic-mutant *Saccharomyces cerevisiae*. *J. Cell Biol.* **98:** 934–945.

Adams A.E.M., Botstein D., and Drubin D.G. 1989. A yeast actin-binding protein is encoded by *SAC6*, a gene found by suppression of an actin mutation. *Science* **243:** 231–233.

———. 1991. Requirement of yeast fimbrin for actin organization and morphogenesis *in vivo*. *Nature* **354:** 404–408.

Adams A.E.M., Johnson D.I., Longnecker R.M., Sloat B.F., and Pringle J.R. 1990. *CDC42* and *CDC43*, two additional genes involved in budding and the establishment of cell polarity in the yeast *Saccharomyces cerevisiae*. *J. Cell Biol.* **111:** 131–142.

Adams I.R. and Kilmartin J.V. 1999. Localization of core spindle pole body (SPB) components during SPB duplication in *Saccharomyces cerevisiae*. *J. Cell Biol.* **145:** 809–823.

Amberg D.C., Zahner J.E., Mulholland J.W., Pringle J.R., and Botstein D. 1997. Aip3p/Bud6p, a yeast actin-interacting protein that is involved in morphogenesis and the selection of bipolar budding sites. *Mol. Biol. Cell* **8:** 729–753.

Ayscough K.R., Eby J.J., Lila T., Dewar H., Kozminski K.G., and Drubin D.G. 1999. Sla1p is a functionally modular component of the yeast cortical actin cytoskeleton required for correct localization of both Rho1p-GTPase and Sla2p, a protein with talin homology. *Mol. Biol. Cell* **10:** 1061–1075.

Ayscough K.R., Stryker J., Pokala N., Sanders M., Crews P., and Drubin D.G. 1997. High rates of actin filament turnover in budding yeast and roles for actin in establishment and maintenance of cell polarity revealed using the actin inhibitor latrunculin-A. *J. Cell Biol.* **137:** 399–416.

Barral Y., Mermall V., Mooseker M.S., and Snyder M. 2000. Compartmentalization of the cell cortex by septins is required for maintenance of cell polarity in yeast. *Mol. Cell* **5:** 841–851.

Barral Y., Parra M., Bidlingmaier S., and Snyder M. 1999. Nim1-related kinases coordinate cell cycle progression with the organization of the peripheral cytoskeleton in yeast. *Genes Dev.* **13:** 176–187.

Barton A.A. 1950. Some aspects of cell division in *Saccharomyces cerevisiae*. *J. Gen. Microbiol.* **4:** 84–86.

Beinhauer J.D., Hagan I.M., Hegemann J.H., and Fleig U. 1997. Mal3, the fission yeast homologue of the human APC-interacting protein EB-1, is required for microtubule integrity and the maintenance of cell form. *J. Cell Biol.* **139:** 717–728.

Beites C.L., Xie H., Bowser R., and Trimble W.S. 1999. The septin CDCrel-1 binds syntaxin and inhibits exocytosis. *Nat. Neurosci.* **2:** 434–439.

Bender A. 1993. Genetic evidence for the roles of the bud-site-selection genes *BUD5* and *BUD2* in control of the Rsr1p (Bud1p) GTPase in yeast. *Proc. Natl. Acad. Sci.* **90:** 9926–9929.

Bender A. and Pringle J.R. 1989. Multicopy suppression of the *cdc24* budding defect in yeast by *CDC42* and three newly identified genes including the *ras*-related gene *RSR1*. *Proc. Natl. Acad. Sci.* **86:** 9976–9980.

———. 1991. Use of a screen for synthetic lethal and multicopy suppresser mutants to identify two new genes involved in morphogenesis in *Saccharomyces cerevisiae*. *Mol. Cell. Biol.* **11:** 1295–1305.

Benton B.K., Tinkelenberg A., Gonzalez I., and Cross F.R. 1997. Cla4p, a *Saccharomyces cerevisiae* Cdc42p-activated kinase involved in cytokinesis, is activated at mitosis. *Mol. Cell. Biol.* **17:** 5067–5076.

Berlin V., Styles C.A., and Fink G.R. 1990. BIK1, a protein required for microtubule function during mating and mitosis in *Saccharomyces cerevisiae*, colocalizes with tubulin. *J. Cell Biol.* **111:** 2573–2586.

Bi E., Chiavetta J.B., Chen H., Chen G.-C., Chan C.S.M., and Pringle J.R. 2000. Identification of novel, evolutionarily conserved Cdc42p-interacting proteins and of redundant pathways linking Cdc24p and Cdc42p to actin polarization in yeast. *Mol. Biol. Cell* **11:** 773–793.

Bi E., Maddox P., Lew D.J., Salmon E.D., McMillan J.N., Yeh E., and Pringle J.R. 1998. Involvement of an actomyosin contractile ring in *Saccharomyces cerevisiae* cytokinesis. *J. Cell Biol.* **142:** 1301–1312.

Biggins S. and Rose M.D. 1994. Direct interaction between yeast spindle pole body components: Kar1p is required for Cdc31p localization to the spindle pole body. *J. Cell Biol.* **125:** 843–852.

Boldogh I.R., Yang H.-C., Nowakowski W.D., Karmon S.L., Hays L.G., Yates J.R., III, and Pon L.A. 2001. Arp2/3 complex and actin dynamics are required for actin-based mitochondrial motility in yeast. *Proc. Natl. Acad. Sci.* **98:** 3162–3167.

Brown J.L., Jaquenoud M., Gulli M.-P., Chant J., and Peter M. 1997. Novel Cdc42-binding proteins Gic1 and Gic2 control cell polarity in yeast. *Genes Dev.* **11:** 2972–2982.

Brunner D. and Nurse P. 2000. CLIP170-like Tip1p spatially organizes microtubular dynamics in fission yeast. *Cell* **102:** 695–704.

Bullitt E., Rout M.P., Kilmartin J.V., and Akey C.W. 1997. The yeast spindle pole body is assembled around a central crystal of Spc42p. *Cell* **89:** 1077–1086.

Burack W.R. and Shaw A.S. 2000. Signal transduction: Hanging on a scaffold. *Curr. Opin. Cell Biol.* **12:** 211–216.

Burbelo P.D., Drechsel D., and Hall A. 1995. A conserved binding motif defines numerous candidate target proteins for both Cdc42 and Rac GTPases. *J. Biol. Chem.* **270:** 29071–29074.

Butty A.-C., Pryciak P.M., Huang L.S., Herskowitz I., and Peter M. 1998. The role of Far1p in linking the heterotrimeric G protein to polarity establishment proteins during yeast mating. *Science* **282:** 1511–1516.

Butty A.-C., Perrinjaquet N., Petit A., Jaquenoud M., Segall J.E., Hofmann K., Zwahlen C., and Peter M. 2002. A positive feedback loop stabilizes the guanine-nucleotide exchange factor Cdc24 at sites of polarization. *EMBO J.* **21:** 1565–1576.

Byers B. 1981a. Cytology of the yeast life cycle. In *The molecular biology of the yeast* Saccharomyces: *Life cycle and inheritance* (ed. J.N. Strathern et al.), pp. 59–96. Cold Spring Harbor Laboratory Press, Cold Spring Harbor, New York.

———. 1981b. Multiple roles of the spindle pole bodies in the life cycle of *Saccharomyces cerevisiae*. *Alfred Benzon Symp.* **16:** 119–133.

Byers B. and Goetsch L. 1974. Duplication of spindle plaques and integration of the yeast cell cycle. *Cold Spring Harbor Symp. Quant. Biol.* **38:** 123–131.

———. 1975. Behavior of spindles and spindle plaques in the cell cycle and conjugation of *Saccharomyces cerevisiae*. *J. Bacteriol.* **124:** 511–523.

———. 1976a. A highly ordered ring of membrane-associated filaments in budding yeast. *J. Cell Biol.* **69:** 717–721.

———. 1976b. Loss of the filamentous ring in cytokinesis-defective mutants of budding yeast. *J. Cell Biol.* **70:** 35a. (Abstr.)

Carminati J.L. and Stearns T. 1997. Microtubules orient the mitotic spindle in yeast through dynein-dependent interactions with the cell cortex. *J. Cell Biol.* **138:** 629–641.

Castillo A.R., Meehl J.B., Morgan G., Schutz-Geschwender A., and Winey M. 2002. The yeast protein kinase Mps1p is required for assembly of the integral spindle pole body component Spc42p. *J. Cell Biol.* **156:** 453–465.

Caviston J.P., Longtine M., Pringle J.R., and Bi E. 2003. The role of Cdc42p GTPase-activating proteins in assembly of the septin ring in yeast. *Mol. Biol. Cell* **14:** 4051–4066.

Chant J. and Herskowitz I. 1991. Genetic control of bud site selection in yeast by a set of gene products that constitute a morphogenetic pathway. *Cell* **65:** 1203–1212.

Chant J. and Pringle J.R. 1995. Patterns of bud-site selection in the yeast *Saccharomyces cerevisiae*. *J. Cell Biol.* **129:** 751–765.

Chant J., Corrado K., Pringle J.R., and Herskowitz I. 1991. Yeast *BUD5*, encoding a putative GDP-GTP exchange factor, is necessary for bud site selection and interacts with bud formation gene *BEM1*. *Cell* **65:** 1213–1224.

Chant J., Mischke M., Mitchell E., Herskowitz I., and Pringle J.R. 1995. Role of Bud3p in producing the axial budding pattern of yeast. *J. Cell Biol.* **129:** 767–778.

Chen G.-C., Kim Y.-J., and Chan C.S.M. 1997. The Cdc42 GTPase-associated proteins Gic1 and Gic2 are required for polarized cell growth in *Saccharomyces cerevisiae*. *Genes Dev.* **11:** 2958–2971.

Chen G.-C., Zheng L., and Chan C.S.M. 1996. The LIM domain-containing Dbm1 GTPase-activating protein is required for normal cellular morphogenesis in *Saccharomyces cerevisiae*. *Mol. Cell. Biol.* **16:** 1376–1390.

Chen T., Hiroko T., Chaudhuri A., Inose F., Lord M., Tanaka S., Chant J., and Fujita A. 2000. Multigenerational cortical inheritance of the Rax2 protein in orienting polarity and division in yeast. *Science* **290:** 1975–1978.

Clark S.W. and Meyer D.I. 1994. *ACT3*: A putative centractin homologue in *S. cerevisiae* is required for proper orientation of the mitotic spindle. *J. Cell Biol.* **127:** 129–138.

Conde J. and Fink G.R. 1976. A mutant of *Saccharomyces cerevisiae* defective for nuclear fusion. *Proc. Natl. Acad. Sci.* **73:** 3651–3655.

Cottingham F.R. and Hoyt M.A. 1997. Mitotic spindle positioning in *Saccharomyces cerevisiae* is accomplished by antagonistically acting microtubule motor proteins. *J. Cell Biol.* **138:** 1041–1053.

Cottingham F.R., Gheber L., Miller D.L., and Hoyt M.A. 1999. Novel roles for *Saccharomyces cerevisiae* mitotic spindle motors. *J. Cell Biol.* **147:** 335–349.

Culotti J. and Hartwell L.H. 1971. Genetic control of the cell division cycle in yeast. III. Seven genes controlling nuclear division. *Exp. Cell Res.* **67:** 389–401.

Cvrcková F., De Virgilio C., Manser E., Pringle J.R., and Nasmyth K. 1995. Ste20-like protein kinases are required for normal localization of cell growth and for cytokinesis in budding yeast. *Genes Dev.* **9:** 1817–1830.

Delley P.-A. and Hall M.N. 1999. Cell wall stress depolarizes cell growth via hyperactivation of RHO1. *J. Cell Biol.* **147:** 163–174.

DeMarini D.J., Adams A.E.M., Fares H., De Virgilio C., Valle G., Chuang J.S., and Pringle J.R. 1997. A septin-based hierarchy of proteins required for localized deposition of chitin in the *Saccharomyces cerevisiae* cell wall. *J. Cell Biol.* **139:** 75–93.

Dent J., Kato K., Peng X.-R., Martinez C., Cattaneo M., Poujol C., Nurden P., Nurden A., Trimble W.S., and Ware J. 2002. A prototypic platelet septin and its participation in secretion. *Proc. Natl. Acad. Sci.* **99:** 3064–3069.

De Virgilio C., DeMarini D.J., and Pringle J.R. 1996. SPR28, a sixth member of the septin gene family in

Saccharomyces cerevisiae that is expressed specifically in sporulating cells. *Microbiology* **142:** 2897–2905.

DeZwaan T.M., Ellingson E., Pellman D., and Roof D.M. 1997. Kinesin-related *KIP3* of *Saccharomyces cerevisiae* is required for a distinct step in nuclear migration. *J. Cell Biol.* **138:** 1023–1040.

Dong Y., Pruyne D., and Bretscher A. 2003. Formin-dependent actin assembly is regulated by distinct modes of Rho signaling in yeast. *J. Cell Biol.* **161:** 1081–1092.

Doyle T. and Botstein D. 1996. Movement of yeast cortical actin cytoskeleton visualized *in vivo*. *Proc. Natl. Acad. Sci.* **93:** 3886–3891.

Drgonová J., Drgon T., Tanaka K., Kollár R., Chen G.-C., Ford R.A., Chan C.S.M., Takai Y., and Cabib E. 1996. Rho1p, a yeast protein at the interface between cell polarization and morphogenesis. *Science* **272:** 277–279.

Drubin D.G., Jones H.D., and Wertman K.F. 1993. Actin structure and function: Roles in mitochondrial organization and morphogenesis in budding yeast and identification of the phalloidin-binding site. *Mol. Biol. Cell* **4:** 1277–1294.

Eitzen G., Thorngren N., and Wickner W. 2001. Rho1p and Cdc42p act after Ypt7p to regulate vacuole docking. *EMBO J.* **20:** 5650–5656.

Enos A.P. and Morris N.R. 1990. Mutation of a gene that encodes a kinesin-like protein blocks nuclear division in *A. nidulans*. *Cell* **60:** 1019–1027.

Engqvist-Goldstein Å.E.Y. and Drubin D. 2003. Actin assembly and endocytosis: From yeast to mammals. *Annu. Rev. Cell Dev. Biol.* **19:** 287–332.

Erickson J.W. and Cerione R.A. 2001. Multiple roles for Cdc42 in cell regulation. *Curr. Opin. Cell Biol.* **13:** 153–157.

Eshel D., Urrestarazu L.A., Vissers S., Jauniaux J.-C., van Vliet-Reedijk J.C., Planta R.J., and Gibbons I.R. 1993. Cytoplasmic dynein is required for normal nuclear segregation in yeast. *Proc. Natl. Acad. Sci.* **90:** 11172–11176.

Evangelista M., Pruyne D., Amberg D.C., Boone C., and Bretscher A. 2002. Formins direct Arp2/3-independent actin filament assembly to polarize cell growth in yeast. *Nat. Cell Biol.* **4:** 260–269.

Evangelista M., Blundell K., Longtine M.S., Chow C.J., Adames N., Pringle J.R., Peter M., and Boone C. 1997. Bni1p, a yeast formin linking Cdc42p and the actin cytoskeleton during polarized morphogenesis. *Science* **276:** 118–122.

Evangelista M., Klebl B.M., Tong A.H.Y., Webb B.A., Leeuw T., Leberer E., Whiteway M., Thomas D.Y., and Boone C. 2000. A role for myosin-I in actin assembly through interactions with Vrp1p, Bee1p, and the Arp2/3 complex. *J. Cell Biol.* **148:** 353–362.

Fares H. 1995. "Functional and comparative studies of septins." Ph.D. thesis, University of North Carolina, Chapel Hill.

Fares H., Goetsch L., and Pringle J.R. 1996. Identification of a developmentally regulated septin and involvement of the septins in spore formation in *Saccharomyces cerevisiae*. *J. Cell Biol.* **132:** 399–411.

Farkas V., Kovarík J., Kosinová A., and Bauer S. 1974. Autoradiographic study of mannan incorporation into the growing cell walls of *Saccharomyces cerevisiae*. *J. Bacteriol.* **117:** 265–269.

Farkasovsky M. and Küntzel H. 2001. Cortical Num1p interacts with the dynein intermediate chain Pac11p and cytoplasmic microtubules in budding yeast. *J. Cell Biol.* **152:** 251–262.

Field C. and Schekman R. 1980. Localized secretion of acid phosphatase reflects the pattern of cell surface growth in *Saccharomyces cerevisiae*. *J. Cell Biol.* **86:** 123–128.

Field C.M., Al-Awar O., Rosenblatt J., Wong M.L., Alberts B., and Mitchison T.J. 1996. A purified *Drosophila* septin complex forms filaments and exhibits GTPase activity. *J. Cell Biol.* **133:** 605–616.

Finegold A.A., Johnson D.I., Farnsworth C.C., Gelb M.H., Judd S.R., Glomset J.A., and Tamanoi F. 1991. Protein geranylgeranyltransferase of *Saccharomyces cerevisiae* is specific for Cys-Xaa-Xaa-Leu motif proteins and requires the *CDC43* gene product but not the *DPR1* gene product. *Proc. Natl. Acad. Sci.* **88:** 4448–4452.

Flescher E.G., Madden K., and Snyder M. 1993. Components required for cytokinesis are important for bud site selection in yeast. *J. Cell Biol.* **122:** 373–386.

Frazier J.A. and Field C.M. 1997. Actin cytoskeleton: Are FH proteins local organizers? *Curr. Biol.* **7:** R414–R417.

Frazier J.A., Wong M.L., Longtine M.S., Pringle J.R., Mann M., Mitchison T.J., and Field C. 1998. Polymerization of purified yeast septins: Evidence that organized filament arrays may not be required for septin function. *J. Cell Biol.* **143:** 737–749.

Freifelder D. 1960. Bud position in *Saccharomyces cerevisiae*. *J. Bacteriol.* **80:** 567–568.

Fujita A., Oka C., Arikawa Y., Katagai T., Tonouchi A., Kuhara S., and Misumi Y. 1994. A yeast gene necessary for bud-site selection encodes a protein similar to insulin-degrading enzymes. *Nature* **372:** 567–570.

Fujiwara T., Tanaka K., Mino A., Kikyo M., Takahashi K., Shimizu K., and Takai Y. 1998. Rho1p-Bni1p-Spa2p interactions: Implication in localization of Bni1p at the bud site and regulation of the actin cytoskeleton in *Saccharomyces cerevisiae*. *Mol. Biol. Cell* **9:** 1221–1233.

Gallwitz D. and Seidel R. 1980. Molecular cloning of the actin gene from yeast *Saccharomyces cerevisiae*. *Nucleic Acids Res.* **8:** 1043–1059.

Geiser J.R., Sundberg H.A., Chang B.H., Muller E.G.D., and Davis T.N. 1993. The essential mitotic target of calmodulin is the 110-kilodalton component of the spindle pole body in *Saccharomyces cerevisiae*. *Mol. Cell. Biol.* **13:** 7913–7924.

Geissler S., Pereira G., Spang A., Knop M., Souès S., Kilmartin J., and Schiebel E. 1996. The spindle pole body component Spc98p interacts with the γ-tubulin-like Tub4p of *Saccharomyces cerevisiae* at the sites of microtubule attachment. *EMBO J.* **15:** 3899–3911.

Geli M.I. and Riezman H. 1996. Role of type I myosins in receptor-mediated endocytosis in yeast. *Science* **272:** 533–535.

Gladfelter A.S., Pringle J.R., and Lew D.J. 2001. The septin cortex at the yeast mother-bud neck. *Curr. Opin. Microbiol.* **4:** 681–689.

Goodson H.V. and Spudich J.A. 1995. Identification and molecular characterization of a yeast myosin I. *Cell Motil. Cytoskel.* **30:** 73–84.

Goodson H.V., Anderson B.L., Warrick H.M., Pon L.A., and Spudich J.A. 1996. Synthetic lethality screen identifies a novel yeast myosin I gene (*MYO5*): Myosin I proteins are required for polarization of the actin cytoskeleton. *J. Cell Biol.* **133:** 1277–1291.

Govindan B., Bowser R., and Novick P. 1995. The role of Myo2, a yeast class V myosin, in vesicular transport. *J. Cell Biol.* **128:** 1055–1068.

Greer C. and Schekman R. 1982. Actin from *Saccharomyces cerevisiae*. *Mol. Cell. Biol.* **2:** 1270–1278.

Gulli M.-P. and Peter M. 2001. Temporal and spatial regulation of Rho-type guanine-nucleotide exchange factors: The yeast perspective. *Genes Dev.* **15:** 365–379.

Gundersen G.G. 2002. Evolutionary conservation of microtubule-capture mechanisms. *Nat. Rev. Mol. Cell Biol.* **3:** 296–304.

Guo W., Tamanoi F., and Novick P. 2001. Spatial regulation of the exocyst complex by Rho1 GTPase. *Nat. Cell Biol.* **3:** 353–360.

Haarer B.K. and Pringle J.R. 1987. Immunofluorescence localization of the *Saccharomyces cerevisiae CDC12* gene product to the vicinity of the 10-nm filaments in the mother-bud neck. *Mol. Cell. Biol.* **7:** 3678–3687.

Hailey D.W., Davis T.N., and Muller E.G. 2002. Fluorescence resonance energy transfer using color variants of green fluorescent protein. *Methods Enzymol.* **351:** 34–49.

Hales K.G., Bi E., Wu J.-Q., Adam J.C., Yu I.-C., and Pringle J.R. 1999. Cytokinesis: An emerging unified theory for eukaryotes? *Curr. Opin. Cell Biol.* **11:** 717–725.

Hall A. 1992. Ras-related GTPases and the cytoskeleton. *Mol. Biol. Cell* **3:** 475–479.

———. 1998. Rho GTPases and the actin cytoskeleton. *Science* **279:** 509–514.

Halme A., Michelitch M., Mitchell E.L., and Chant J. 1996. Bud10p directs axial cell polarization in budding yeast and resembles a transmembrane receptor. *Curr. Biol.* **6:** 570–579.

Harkins H.A., Pagé N., Schenkman L.R., DeVirgilio C., Shaw S., Bussey H., and Pringle J.R. 2001. Bud8p and Bud9p, proteins that may mark the sites for bipolar budding in yeast. *Mol. Biol. Cell* **12:** 2497–2518.

Hart M.J., Eva A., Evans T., Aaronson S.A., and Cerione R.A. 1991. Catalysis of guanine nucleotide exchange on the CDC42Hs protein by the *dbl* oncogene product. *Nature* **354:** 311–314.

Hartwell L.H. 1971. Genetic control of the cell division cycle in yeast. IV. Genes controlling bud emergence and cytokinesis. *Exp. Cell Res.* **69:** 265–276.

Hartwell L.H., Culotti J., Pringle J.R., and Reid B.J. 1974. Genetic control of the cell division cycle in yeast. *Science* **183:** 46–51.

Hayashibe M. and Katohda S. 1973. Initiation of budding and chitin ring. *J. Gen. Appl. Microbiol.* **19:** 23–39.

Helliwell S.B., Schmidt A., Ohya Y., and Hall M.N. 1998. The Rho1 effector Pkc1, but not Bni1, mediates signalling from Tor2 to the actin cytoskeleton. *Curr. Biol.* **8:** 1211–1214.

Hicks J.B., Strathern J.N., and Herskowitz I. 1977. Interconversion of yeast mating types. III. Action of the homothallism (*HO*) gene in cells homozygous for the mating type locus. *Genetics* **85:** 395–405.

Higgs H.N. and Pollard T.D. 2001. Regulation of actin filament network formation through ARP2/3 complex: Activation by a diverse array of proteins. *Annu. Rev. Biochem.* **70:** 649–676.

Hill K.L., Catlett N.L., and Weisman L.S. 1996. Actin and myosin function in directed vacuole movement during cell division in *Saccharomyces cerevisiae. J. Cell Biol.* **135:** 1535–1549.

Hoepfner D., van den Berg M., Philippsen P., Tabak H.F., and Hettema E.H. 2001. A role for Vps1p, actin, and the Myo2p motor in peroxisome abundance and inheritance in *Saccharomyces cerevisiae. J. Cell Biol.* **155:** 979–990.

Höfken T. and Schiebel E. 2002. A role for cell polarity proteins in mitotic exit. *EMBO J.* **21:** 4851–4862.

Holly S.P. and Blumer K.J. 1999. PAK-family kinases regulate cell and actin polarization throughout the cell cycle of *Saccharomyces cerevisiae. J. Cell Biol.* **147:** 845–856.

Hoyt M.A., Stearns T., and Botstein D. 1990. Chromosome instability mutants of *Saccharomyces cerevisiae* that are defective in microtubule-mediated processes. *Mol. Cell. Biol.* **10:** 223–234.

Hoyt M.A., He L., Loo K.K., and Saunders W.S. 1992. Two *Saccharomyces cerevisiae* kinesin-related gene products required for mitotic spindle assembly. *J. Cell Biol.* **118:** 109–120.

Hoyt M.A., He L., Totis L., and Saunders W.S. 1993. Loss of function of *Saccharomyces cerevisiae* kinesin-related *CIN8* and *KIP1* is suppressed by *KAR3* motor domain mutations. *Genetics* **135:** 35–44.

Huffaker T.C., Thomas J.H., and Botstein D. 1988. Diverse effects of β-tubulin mutations on microtubule formation and function. *J. Cell Biol.* **106:** 1997–2010.

Hwang E., Kusch J., Barral Y., and Huffaker T.C. 2003. Spindle orientation in *Saccharomyces cerevisiae* depends on the transport of microtubule ends along polarized actin cables. *J. Cell Biol.* **161:** 483–488.

Imai J., Toh-e A., and Matsui Y. 1996. Genetic analysis of the *Saccharomyces cerevisiae RHO3* gene, encoding a Rho-type small GTPase, provides evidence for a role in bud formation. *Genetics* **142:** 359–369.

Imamura H., Tanaka K., Hihara T., Umikawa M., Kamei T., Takahashi K., Sasaki T., and Takai Y. 1997. Bni1p and Bnr1p: Downstream targets of the Rho family small G-proteins which interact with profilin and regulate actin cytoskeleton in *Saccharomyces cerevisiae. EMBO J.* **16:** 2745–2755.

Jacobs C.W., Adams A.E.M., Szaniszlo P.J., and Pringle J.R. 1988. Functions of microtubules in the *Saccharomyces cerevisiae* cell cycle. *J. Cell Biol.* **107:** 1409–1426.

Jansen R.-P., Dowzer C., Michaelis C., Galova M., and Nasmyth K. 1996. Mother cell-specific *HO* expression in budding yeast depends on the unconventional myosin Myo4p and other cytoplasmic proteins. *Cell* **84:** 687–697.

Jaquenoud M. and Peter M. 2000. Gic2p may link activated Cdc42p to components involved in actin polarization, including Bni1p and Bud6p (Aip3p). *Mol. Cell. Biol.* **20:** 6244–6258.

Johnson B.F. 1968. Lysis of yeast cell walls induced by 2-deoxyglucose at their sites of glucan synthesis. *J. Bacteriol.* **95:** 1169–1172.

Johnson B.F. and Gibson E.J. 1966. Autoradiographic analysis of regional cell wall growth of yeasts. III. *Saccharomyces cerevisiae. Exp. Cell Res.* **41:** 580–591.

Johnson D.I. 1999. Cdc42: An essential Rho-type GTPase controlling eukaryotic cell polarity. *Microbiol. Mol. Biol. Rev.* **63:** 54–105.

Johnson D.I. and Pringle J.R. 1990. Molecular characterization of *CDC42*, a *Saccharomyces cerevisiae* gene involved in the development of cell polarity. *J. Cell Biol.* **111:** 143–152.

Johnston G.C., Prendergast J.A., and Singer R.A. 1991. The *Saccharomyces cerevisiae MYO2* gene encodes an essential myosin for vectorial transport of vesicles. *J. Cell Biol.* **113:** 539–551.

Kaback D.B. and Feldberg L.R. 1985. *Saccharomyces cerevisiae* exhibits a sporulation-specific temporal pattern of transcript accumulation. *Mol. Cell. Biol.* **5:** 751–761.

Kamei T., Tanaka K., Hihara T., Umikawa M., Imamura H., Kikyo M., Ozaki K., and Takai Y. 1998. Interaction of Bnr1p with a novel Src homology 3 domain-containing Hof1p. Implication in cytokinesis in *Saccharomyces cerevisiae. J. Biol. Chem.* **273:** 28341–28345.

Kang P.J., Sanson A., Lee B., and Park H.-O. 2001. A GDP/GTP exchange factor involved in linking a spatial landmark to cell polarity. *Science* **292:** 1376–1378.

Karpova T.S., Reck-Peterson S.L., Elkind N.B., Mooseker M.S., Novick P.J., and Cooper J.A. 2000. Role of actin and Myo2p in polarized secretion and growth of *Saccharomyces cerevisiae. Mol. Biol. Cell* **11:** 1727–1737.

Kilmartin J.V. and Adams A.E.M. 1984. Structural rearrangements of tubulin and actin during the cell cycle of the yeast *Saccharomyces. J. Cell Biol.* **98:** 922–933.

Kilmartin J.V., Dyos S.L., Kershaw D., and Finch J.T. 1993. A spacer protein in the *Saccharomyces cerevisiae* spindle pole body whose transcript is cell cycle-regulated. *J. Cell Biol.* **123:** 1175–1184.

Kim H.B., Haarer B.K., and Pringle J.R. 1991. Cellular morphogenesis in the *Saccharomyces cerevisiae* cell cycle: Localization of the CDC3 gene product and the timing of events at the budding site. *J. Cell Biol.* **112:** 535–544.

Kinoshita M. 2003. The septins. *Genome Biol.* **4:** 236.1–236.9.

Kinoshita M., Field C.M., Coughlin M.L., Straight A.F., and Mitchison T.J. 2002. Self- and actin-templated assembly of mammalian septins. *Dev. Cell* **3:** 791–802.

Kinoshita A., Kinoshita M., Akiyama H., Tomimoto H., Akiguchi I., Kumar S., Noda M., and Kimura J. 1998. Identification of septins in neurofibrillary tangles in Alzheimer's disease. *Am. J. Pathol.* **153:** 1551–1560.

Kirschner M. and Mitchison T.J. 1986. Beyond self-assembly: From microtubules to morphogenesis. *Cell* **45:** 329–342.

Knop M. and Schiebel E. 1997. Spc98p and Spc97p of the yeast γ-tubulin complex mediate binding to the spindle pole body via their interaction with Spc110p. *EMBO J.* **16:** 6985–6995.

Kohno H., Tanaka K., Mino A., Umikawa M., Imamura H., Fujiwara T., Fujita Y., Hotta K., Qadota H., Watanabe T., Ohya Y., and Takai Y. 1996. Bni1 implicated in cytoskeletal control is a putative target of Rho1p small GTP binding protein in *Saccharomyces cerevisiae*. *EMBO J.* **15:** 6060–6068.

Korinek W.S., Copeland M.J., Chaudhuri A., and Chant J. 2000. Molecular linkage underlying microtubule orientation toward cortical sites in yeast. *Science* **287:** 2257–2259.

Kormanec J., Schaaff-Gerstenschläger I., Zimmermann F.K., Perecko D., and Küntzel H. 1991. Nuclear migration in *Saccharomyces cerevisiae* is controlled by the highly repetitive 313 kDa NUM1 protein. *Mol. Gen. Genet.* **230:** 277–287.

Kosco K.A., Pearson C.G., Maddox P.S., Wang P.J., Adams I.R., Salmon E.D., Bloom K., and Huffaker T.C. 2001. Control of microtubule dynamics by Stu2p is essential for spindle orientation and metaphase chromosome alignment in yeast. *Mol. Biol. Cell* **12:** 2870–2880.

Koteliansky V.E., Glukhova M.A., Bejanian M.V., Surguchov A.P., and Smirnov V.N. 1979. Isolation and characterization of actin-like protein from yeast *Saccharomyces cerevisiae*. *FEBS Lett.* **102:** 55–58.

Kübler E. and Riezman H. 1993. Actin and fimbrin are required for the internalization step of endocytosis in yeast. *EMBO J.* **12:** 2855–2862.

Lamson R.E., Winters M.J., and Pryciak P.M. 2002. Cdc42 regulation of kinase activity and signaling by the yeast p21-activated kinase Ste20. *Mol. Cell. Biol.* **22:** 2939–2951.

Larisch S., Yi Y., Lotan R., Kerner H., Eimerl S., Tony Parks W., Gottfried Y., Birkey Reffey S., de Caestecker M.P., Danielpour D., Book-Melamed N., Timberg R., Duckett C.S., Lechleider R.J., Steller H., Orly J., Kim S.J., and Roberts A.B. 2000. A novel mitochondrial septin-like protein, ARTS, mediates apoptosis dependent on its P-loop motif. *Nat. Cell Biol.* **2:** 915–921.

Leberer E., Dignard D., Harcus D., Thomas D.Y., and Whiteway M. 1992. The protein kinase homologue Ste20p is required to link the yeast pheromone response G-protein βγ subunits to downstream signalling components. *EMBO J.* **11:** 4815–4824.

Lechler T., Shevchenko A., Shevchenko A., and Li R. 2000. Direct involvement of yeast type I myosins in Cdc42-dependent actin polymerization. *J. Cell Biol.* **148:** 363–373.

Lechler T., Jonsdottir G.A., Klee S.K., Pellman D., and Li R. 2001. A two-tiered mechanism by which Cdc42 controls the localization and activation of an Arp2/3-activating motor complex in yeast. *J. Cell Biol.* **155:** 261–270.

Lee L., Klee S.K., Evangelista M., Boone C., and Pellman D. 1999. Control of mitotic spindle position by the *Saccharomyces cerevisiae* formin Bni1p. *J. Cell Biol.* **144:** 947–961.

Lee L., Tirnauer J.S., Li J., Schuyler S.C., Liu J.Y., and Pellman D. 2000. Positioning of the mitotic spindle by a cortical-microtubule capture mechanism. *Science* **287:** 2260–2262.

Lees-Miller J.P., Henry G., and Helfman D.M. 1992. Identification of *act2*, an essential gene in the fission yeast *Schizosaccharomyces pombe* that encodes a protein related to actin. *Proc. Natl. Acad. Sci.* **89:** 80–83.

Lew D.J. and Reed S.I. 1993. Morphogenesis in the yeast cell cycle: Regulation by Cdc28 and cyclins. *J. Cell Biol.* **120:** 1305–1320.

———. 1995. A cell cycle checkpoint monitors cell morphogenesis in budding yeast. *J. Cell Biol.* **129:** 739–749.

Li R. 1997. Bee1, a yeast protein with homology to Wiscott-Aldrich syndrome protein, is critical for the assembly of cortical actin cytoskeleton. *J. Cell Biol.* **136:** 649–658.

Li Y.-Y., Yeh E., Hays T., and Bloom K. 1993. Disruption of mitotic spindle orientation in a yeast dynein mutant.

Proc. Natl. Acad. Sci. **90:** 10096–10100.

Lin H., de Carvalho P., Kho D., Tai C.-Y., Pierre P., Fink G.R., and Pellman D. 2001. Polyploids require Bik1 for kinetochore-microtubule attachment. *J. Cell Biol.* **155:** 1173–1184.

Lippincott J. and Li R. 1998. Sequential assembly of myosin II, an IQGAP-like protein, and filamentous actin to a ring structure involved in budding yeast cytokinesis. *J. Cell Biol.* **140:** 355–366.

Lippincott J., Shannon K.B., Shou W., Deshaies R.J., and Li R. 2001. The Tem1 small GTPase controls actomyosin and septin dynamics during cytokinesis. *J. Cell Sci.* **114:** 1379–1386.

Long R.M., Singer R.H., Meng X., Gonzalez I., Nasmyth K., and Jansen R.-P. 1997. Mating type switching in yeast controlled by asymmetric localization of ASH1 mRNA. *Science* **277:** 383–387.

Longtine M.S., Fares H., and Pringle J.R. 1998. Role of the yeast Gin4p protein kinase in septin assembly and the relationship between septin assembly and septin function. *J. Cell Biol.* **143:** 719–736.

Longtine M.S., Theesfeld C.L., McMillan J.N., Weaver E., Pringle J.R., and Lew D.J. 2000. Septin-dependent assembly of a cell cycle-regulatory module in *Saccharomyces cerevisiae*. *Mol. Cell. Biol.* **20:** 4049–4061.

Longtine M.S., DeMarini D.J., Valencik M.L., Al-Awar O.S., Fares H., DeVirgilio C., and Pringle J.R. 1996. The septins: Roles in cytokinesis and other processes. *Curr. Opin. Cell Biol.* **8:** 106–119.

Lord M., Yang M.C., Mischke M., and Chant J. 2000. Cell cycle programs of gene expression control morphogenetic protein localization. *J. Cell Biol.* **151:** 1501–1511.

Lord M., Inose F., Hiroko T., Hata T., Fujita A., and Chant J. 2002. Subcellular localization of Axl1, the cell type-specific regulator of polarity. *Curr. Biol.* **12:** 1347–1352.

Lyons D.M., Mahanty S.K., Choi K.-Y., Manandhar M., and Elion E.A. 1996. The SH3-domain protein Bem1 coordinates mitogen-activated protein kinase cascade activation with cell cycle control in *Saccharomyces cerevisiae*. *Mol. Cell. Biol.* **16:** 4095–4106.

Mack D., Nishimura K., Dennehey B.K., Arbogast T., Parkinson J., Toh-e A., Pringle J.R., Bender A., and Matsui Y. 1996. Identification of the bud emergence gene *BEM4* and its interactions with Rho-type GTPases in *Saccharomyces cerevisiae*. *Mol. Cell. Biol.* **16:** 4387–4395.

Madania A., Dumoulin P., Grava S., Kitamoto H., Schärer-Brodbeck C., Soulard A., Moreau V., and Winsor B. 1999. The *Saccharomyces cerevisiae* homologue of human Wiskott-Aldrich syndrome protein Las17p interacts with the Arp2/3 complex. *Mol. Biol. Cell* **10:** 3521–3538.

Madaule P., Axel R., and Myers A.M. 1987. Characterization of two members of the *rho* gene family from the yeast *Saccharomyces cerevisiae*. *Proc. Natl. Acad. Sci.* **84:** 779–783.

Maekawa H., Usui T., Knop M., and Schiebel E. 2003. Yeast Cdk1 translocates to the plus end of cytoplasmic microtubules to regulate bud cortex interactions. *EMBO J.* **22:** 438–449.

Manser E., Leung T., Salihuddin H., Zhao Z., and Lim L. 1994. A brain serine/threonine protein kinase activated by Cdc42 and Rac1. *Nature* **367:** 40–46.

Marquitz A.R., Harrison J.C., Bose I., Zyla T.R., McMillan J.N., and Lew D.J. 2002. The Rho-GAP Bem2p plays a GAP-independent role in the morphogenesis checkpoint. *EMBO J.* **21:** 4012–4025.

Marston A.L., Chen T., Yang M.C., Belhumeur P., and Chant J. 2001. A localized GTPase exchange factor, Bud5, determines the orientation of division axes in yeast. *Curr. Biol.* **11:** 803–807.

Matile P., Moor H., and Robinow C.F. 1969. Yeast cytology. In *The yeasts: Biology of yeasts* (ed. A.H. Rose and J.S. Harrison), vol. 1, pp. 219–302. Academic Press, London.

Matsui Y. and Toh-e A. 1992. Yeast *RHO3* and *RHO4* ras superfamily genes are necessary for bud growth, and their defect is suppressed by a high dose of bud formation genes *CDC42* and *BEM1*. *Mol. Cell. Biol.* **12:** 5690–5699.

Matsui Y., Matsui R., Akada R., and Toh-e A. 1996. Yeast *src* homology region 3 domain-binding proteins involved in bud formation. *J. Cell Biol.* **133:** 865–878.

McIntosh J.R., Grishchuk E.L., and West R.R. 2002. Chromosome-microtubule interactions during mitosis. *Annu. Rev. Cell Dev. Biol.* **18:** 193–219.

McMillan J.N., Theesfeld C.L., Harrison J.C., Bardes E.S.G., and Lew D.J. 2002. Determinants of Swe1p degradation in *Saccharomyces cerevisiae*. *Mol. Biol. Cell* **13:** 3560–3575.

Meluh P.B. and Rose M.D. 1990. *KAR3*, a kinesin-related gene required for yeast nuclear fusion. *Cell* **60:** 1029–1041.

Miller R.K. and Rose M.D. 1998. Kar9p is a novel cortical protein required for cytoplasmic microtubule orientation in yeast. *J. Cell Biol.* **140:** 377–390.

Miller R.K., Matheos D., and Rose M.D. 1999. The cortical localization of the microtubule orientation protein, Kar9p, is dependent upon actin and proteins required for polarization. *J. Cell Biol.* **144:** 963–975.

Miller R.K., Heller K.K., Frisèn L., Wallack D.L., Loayza D., Gammie A.E., and Rose M.D. 1998. The kinesin-related proteins, Kip2p and Kip3p, function differently in nuclear migration in yeast. *Mol. Biol. Cell* **9:** 2051–2068.

Mitchell D.A. and Sprague G.F., Jr. 2001. The phosphotyrosyl phosphatase activator, Ncs1p (Rrd1p), functions with Cla4p to regulate the G_2/M transition in *Saccharomyces cerevisiae*. *Mol. Cell. Biol.* **21:** 488–500.

Mitchison T.J. and Field C.M. 2002. Cytoskeleton: What does GTP do for septins? *Curr. Biol.* **12:** R788–R790.

Moens P.B. and Rapport E. 1971. Spindles, spindle plaques, and meiosis in the yeast *Saccharomyces cerevisiae* (Hansen). *J. Cell Biol.* **50:** 344–361.

Moreau V., Madania A., Martin R.P., and Winsor B. 1996. The *Saccharomyces cerevisiae* actin-related protein Arp2 is involved in the actin cytoskeleton. *J. Cell Biol.* **134:** 117–132.

Mösch H.-U., Roberts R.L., and Fink G.R. 1996. Ras2 signals via the Cdc42/Ste20/mitogen-activated protein kinase module to induce filamentous growth in *Saccharomyces cerevisiae*. *Proc. Natl. Acad. Sci.* **93:** 5352–5356.

Moskow J.J., Gladfelter A.S., Lamson R.E., Pryciak P.M., and Lew D.J. 2000. Role of Cdc42p in pheromone-stimulated signal transduction in *Saccharomyces cerevisiae*. *Mol. Cell. Biol.* **20:** 7559–7571.

Muhua L., Karpova T.S., and Cooper J.A. 1994. A yeast actin-related protein homologous to that in vertebrate dynactin complex is important for spindle orientation and nuclear migration. *Cell* **78:** 669–679.

Mulholland J., Preuss D., Moon A., Wong A., Drubin D., and Botstein D. 1994. Ultrastructure of the yeast actin cytoskeleton and its association with the plasma membrane. *J. Cell Biol.* **125:** 381–391.

Murphy S.M., Urbani L., and Stearns T. 1998. The mammalian γ-tubulin complex contains homologues of the yeast spindle pole body components Spc97p and Spc98p. *J. Cell Biol.* **141:** 663–674.

Nelson W.J. 2003. Adaptation of core mechanisms to generate cell polarity. *Nature* **422:** 766–774.

Nern A. and Arkowitz R.A. 2000. Nucleocytoplasmic shuttling of the Cdc42p exchange factor Cdc24p. *J. Cell Biol.* **148:** 1115–1122.

Ng R. and Abelson J. 1980. Isolation and sequence of the gene for actin in *Saccharomyces cerevisiae*. *Proc. Natl. Acad. Sci.* **77:** 3912–3916.

Nguyen T., Vinh D.B.N., Crawford D.K., and Davis T.N. 1998. A genetic analysis of interactions with Spc110p reveals distinct functions of Spc97p and Spc98p, components of the yeast γ-tubulin complex. *Mol. Biol. Cell* **9:** 2201–2216.

Nguyen T.Q., Sawa H., Okano H., and White J.G. 2000. The *C. elegans* septin genes, *unc-59* and *unc-61*, are required for normal postembryonic cytokineses and morphogenesis but have no essential function in embryogenesis. *J. Cell Sci.* **113:** 3825–3837.

Ni L. and Snyder M. 2001. A genomic study of the bipolar bud site selection pattern in *Saccharomyces cerevisiae*. *Mol. Biol. Cell* **12:** 2147–2170.

Nickas M.E. and Neiman A.M. 2002. Ady3p links spindle pole body function to spore wall synthesis in *Saccharomyces cerevisiae*. *Genetics* **160:** 1439–1450.

Nonaka H., Tanaka K., Hirano H., Fujiwara T., Kohno H., Umikawa M., Mino A., and Takai Y. 1995. A downstream target of *RHO1* small GTP-binding protein is *PKC1*, a homolog of protein kinase C, which leads to activation of the MAP kinase cascade in *Saccharomyces cerevisiae*. *EMBO J.* **14:** 5931–5938.

Novick P. and Botstein D. 1985. Phenotypic analysis of temperature-sensitive yeast actin mutants. *Cell* **40:** 405–416.

Oakley B.R., Oakley C.E., Yoon Y., and Jung M.K. 1990. γ-tubulin is a component of the spindle pole body that is essential for microtubule function in *Aspergillus nidulans*. *Cell* **61:** 1289–1301.

Ohya Y., Qadota H., Anraku Y., Pringle J.R., and Botstein D. 1993. Suppression of yeast geranylgeranyl transferase I defect by alternative prenylation of two target GTPases, Rho1p and Cdc42p. *Mol. Biol. Cell* **4:** 1017–1025.

O'Toole E.T., Winey M., and McIntosh J.R. 1999. High-voltage electron tomography of spindle pole bodies and early mitotic spindles in the yeast *Saccharomyces cerevisiae*. *Mol. Biol. Cell* **10:** 2017–2031.

Ozaki K., Tanaka K., Imamura H., Hihara T., Kameyama T., Nonaka H., Hirano H., Matsuura Y., and Takai Y. 1996. Rom1p and Rom2p are GDP/GTP exchange proteins (GEPs) for the Rho1p small GTP binding protein in *Saccharomyces cerevisiae*. *EMBO J.* **15:** 2196–2207.

Ozaki-Kuroda K., Yamamoto Y., Nohara H., Kinoshita M., Fujiwara T., Irie K., and Takai Y. 2001. Dynamic localization and function of Bni1p at the sites of directed growth in *Saccharomyces cerevisiae*. *Mol. Cell. Biol.* **21:** 827–839.

Palmer R.E., Sullivan D.S., Huffaker T., and Koshland D. 1992. Role of astral microtubules and actin in spindle orientation and migration in the budding yeast, *Saccharomyces cerevisiae*. *J. Cell Biol.* **119:** 583–593.

Park H.-O., Chant J., and Herskowitz I. 1993. *BUD2* encodes a GTPase-activating protein for Bud1/Rsr1 necessary for proper bud-site selection in yeast. *Nature* **365:** 269–274.

Park H.-O., Kang P.J., and Rachfal A.W. 2002. Localization of the Rsr1/Bud1 GTPase involved in selection of a proper growth site in yeast. *J. Biol. Chem.* **277:** 26721–26724.

Park H.-O., Sanson A., and Herskowitz I. 1999. Localization of Bud2p, a GTPase-activating protein necessary for programming cell polarity in yeast to the presumptive bud site. *Genes Dev.* **13:** 1912–1917.

Pearson C.G., Maddox P.S., Salmon E.D., and Bloom K. 2001. Budding yeast chromosome structure and dynamics during mitosis. *J. Cell Biol.* **152:** 1255–1266.

Pereira G. and Schiebel E. 2001. The role of the yeast spindle pole body and the mammalian centrosome in regulating late mitotic events. *Curr. Opin. Cell Biol.* **13:** 762–769.

Pereira G., Tanaka T.U., Nasmyth K., and Schiebel E. 2001. Modes of spindle pole body inheritance and segregation of the Bfa1p-Bub2p checkpoint protein complex. *EMBO J.* **20:** 6359–6370.

Peterson J.B. and Ris H. 1976. Electron-microscopic study of the spindle and chromosome movement in the yeast *Saccharomyces cerevisiae*. *J. Cell Sci.* **22:** 219–242.

Peterson J., Zheng Y., Bender L., Myers A., Cerione R., and Bender A. 1994. Interactions between the bud emergence proteins Bem1p and Bem2p and Rho-type GTPases in yeast. *J. Cell Biol.* **127:** 1395–1406.

Pringle J.R. and Hartwell L.H. 1981. The *Saccharomyces cerevisiae* cell cycle. In *The molecular biology of the yeast Saccharomyces: Life cycle and inheritance* (ed. J.N. Strathern et al.), pp. 97–142. Cold Spring Harbor Laboratory Press, Cold Spring Harbor, New York.

Pringle J.R., Bi E., Harkins H.A., Zahner J.E., De Virgilio C., Chant J., Corrado K., and Fares H. 1995. Establishment of cell polarity in yeast. *Cold Spring Harbor Symp. Quant. Biol.* **60:** 729–744.

Pringle J.R., Lillie S.H., Adams A.E.M., Jacobs C.W., Haarer B.K., Coleman K.G., Robinson J.S., Bloom L., and Preston R.A. 1986. Cellular morphogenesis in the yeast cell cycle. In *Yeast cell biology* (ed. J. Hicks), pp. 47–80. A.R. Liss, New York.

Pruyne D. and Bretscher A. 2000a. Polarization of cell growth in yeast. I. Establishment and maintenance of polarity states. *J. Cell Sci.* **113:** 365–375.

———. 2000b. Polarization of cell growth in yeast. II. The role of the cortical actin cytoskeleton. *J. Cell Sci.* **113:** 571–585.

Pruyne D.W., Schott D.H., and Bretscher A. 1998. Tropomyosin-containing actin cables direct the Myo2p-dependent polarized delivery of secretory vesicles in budding yeast. *J. Cell Biol.* **143:** 1931–1945.

Pruyne D., Evangelista M., Yang C., Bi E., Zigmond S., Bretscher A., and Boone C. 2002. Role of formins in actin assembly: Nucleation and barbed-end association. *Science* **297:** 612–615.

Qadota H., Anraku Y., Botstein D., and Ohya Y. 1994. Conditional lethality of a yeast strain expressing human *RHOA* in place of *RHO1*. *Proc. Natl. Acad. Sci.* **91:** 9317–9321.

Qadota H., Python C.P., Inoue S.B., Arisawa M., Anraku Y., Zheng Y., Watanabe T., Levin D.E., and Ohya Y. 1996. Identification of yeast Rho1p GTPase as a regulatory subunit of 1,3-β-glucan synthase. *Science* **272:** 279–281.

Raitt D.C., Posas F., and Saito H. 2000. Yeast Cdc42 GTPase and Ste20 PAK-like kinase regulate Sho1-dependent activation of the Hog1 MAPK pathway. *EMBO J.* **19:** 4623–4631.

Reiser V., Salah S.M., and Ammerer G. 2000. Polarized localization of yeast Pbs2 depends on osmostress, the membrane protein Sho1 and Cdc42. *Nat. Cell Biol.* **2:** 620–627.

Richman T.J., Sawyer M.M., and Johnson D.I. 2002. *Saccharomyces cerevisiae* Cdc42p localizes to cellular membranes and clusters at sites of polarized growth. *Eukaryot. Cell* **1:** 458–468.

Robinson L.C., Bradley C., Bryan J.D., Jerome A., Kweon Y., and Panek H.R. 1999. The Yck2 yeast casein kinase 1 isoform shows cell cycle-specific localization to sites of polarized growth and is required for proper septin organization. *Mol. Biol. Cell* **10:** 1077–1092.

Rodriguez J.R. and Paterson B.M. 1990. Yeast myosin heavy chain mutant: Maintenance of the cell type specific budding pattern and the normal deposition of chitin and cell wall components requires an intact myosin heavy chain gene. *Cell Motil. Cytoskel.* **17:** 301–308.

Roemer T., Madden K., Chang J., and Snyder M. 1996. Selection of axial growth sites in yeast requires Axl2p, a novel plasma membrane glycoprotein. *Genes Dev.* **10:** 777–793.

Rogers S.L., Rogers G.C., Sharp D.J., and Vale R.D. 2002. *Drosophila* EB1 is important for proper assembly, dynamics, and positioning of the mitotic spindle. *J. Cell Biol.* **158:** 873–884.

Roh D.-H., Bowers B., Schmidt M., and Cabib E. 2002. The septation apparatus, an autonomous system in budding yeast. *Mol. Biol. Cell* **13:** 2747–2759.

Roof D.M., Meluh P.B., and Rose M.D. 1992. Kinesin-related proteins required for assembly of the mitotic spindle. *J. Cell Biol.* **118:** 95–108.

Rose M.D. and Fink G.R. 1987. *KAR1*, a gene required for function of both intranuclear and extranuclear microtubules in yeast. *Cell* **48:** 1047–1060.

Roumanie O., Weinachter C., Larrieu I., Crouzet M., and Doignon F. 2001. Functional characterization of the Bag7, Lrg1 and Rgd2 RhoGAP proteins from *Saccharomyces cerevisiae*. *FEBS Lett.* **506:** 149–156.

Rout M.P. and Kilmartin J.V. 1990. Components of the yeast spindle and spindle pole body. *J. Cell Biol.* **111:** 1913–1927.

Ruggieri R., Bender A., Matsui Y., Powers S., Takai Y., Pringle J.R., and Matsumoto K. 1992. *RSR1*, a *ras*-like gene homologous to K*rev*-1 (*smg21A/rap1A*): Role in the development of cell polarity and interactions with the Ras pathway in *Saccharomyces cerevisiae*. *Mol. Cell. Biol.* **12:** 758–766.

Sagot I., Klee S.K., and Pellman D. 2002a. Yeast formins regulate cell polarity by controlling the assembly of actin cables. *Nat. Cell Biol.* **4:** 42–50.

Sagot I., Rodal A.A., Moseley J., Goode B.L., and Pellman D. 2002b. An actin nucleation mechanism mediated by Bni1 and profilin. *Nat. Cell Biol.* **4:** 626–631.

Sanders S.L. and Herskowitz I. 1996. The Bud4 protein of yeast, required for axial budding, is localized to the mother/bud neck in a cell cycle-dependent manner. *J. Cell Biol.* **134:** 413–427.

Saunders W.S. and Hoyt M.A. 1992. Kinesin-related proteins required for structural integrity of the mitotic spindle. *Cell* **70:** 451–458.

Schenkman L.R., Caruso C., Pagé N., and Pringle J.R. 2002. The role of cell cycle-regulated expression in the localization of spatial landmark proteins in yeast. *J. Cell Biol.* **156:** 829–841.

Schmidt A., Bickle M., Beck T., and Hall M.N. 1997. The yeast phosphatidylinositol kinase homolog TOR2 activates RHO1 and RHO2 via the exchange factor ROM2. *Cell* **88:** 531–542.

Schmidt M., Bowers B., Varma A., Roh D.-H., and Cabib E. 2002. In budding yeast, contraction of the actomyosin ring and formation of the primary septum at cytokinesis depend on each other. *J. Cell Sci.* **115:** 293–302.

Schott D.H., Collins R.N., and Bretscher A. 2002. Secretory vesicle transport velocity in living cells depends on the myosin-V lever arm length. *J. Cell Biol.* **156:** 35–39.

Schwartz K., Richards K., and Botstein D. 1997. *BMI1* encodes a microtubule-binding protein in yeast. *Mol. Biol. Cell* **8:** 2677–2691.

Schwob E. and Martin R.P. 1992. New yeast actin-like gene required late in the cell cycle. *Nature* **355:** 179–182.

Segal M., Bloom K., and Reed S.I. 2000a. Bud6 directs sequential microtubule interactions with the bud tip and bud neck during spindle morphogenesis in *Saccharomyces cerevisiae*. *Mol. Biol. Cell* **11:** 3689–3702.

———. 2002. Kar9p-independent microtubule capture at Bud6p cortical sites primes spindle polarity before bud emergence in *Saccharomyces cerevisiae*. *Mol. Biol. Cell* **13:** 4141–4155.

Segal M., Clarke D.J., Maddox P., Salmon E.D., Bloom K., and Reed S.I. 2000b. Coordinated spindle assembly and orientation requires Clb5p-dependent kinase in budding yeast. *J. Cell Biol.* **148:** 441–451.

Severson A.F., Baillie D.L., and Bowerman B. 2002. A formin homology protein and a profilin are required for cytokinesis and Arp2/3-independent assembly of cortical microfilaments in *C. elegans*. *Curr. Biol.* **12:** 2066–2075.

Sharp D.J., Rogers G.C., and Scholey J.M. 2000. Microtubule motors in mitosis. *Nature* **407:** 41–47.

Shaw S.L., Yeh E., Maddox P., Salmon E.D., and Bloom K. 1997. Astral microtubule dynamics in yeast: A microtubule-based searching mechanism for spindle orientation and nuclear migration into the bud. *J. Cell Biol.* **139:** 985–994.

Sheu Y.-J., Barral Y., and Snyder M. 2000. Polarized growth controls cell shape and bipolar bud site selection in *Saccharomyces cerevisiae*. *Mol. Cell. Biol.* **20:** 5235–5247.

Sheu Y.-J., Santos B., Fortin N., Costigan C., and Snyder M. 1998. Spa2p interacts with cell polarity proteins and signaling components involved in yeast cell morphogenesis. *Mol. Cell. Biol.* **18:** 4053–4069.

Shortle D., Haber J.E., and Botstein D. 1982. Lethal disruption of the yeast actin gene by integrative DNA transformation. *Science* **217:** 371–373.

Shulewitz M.J., Inouye C.J., and Thorner J. 1999. Hsl7 localizes to a septin ring and serves as an adapter in a regulatory pathway that relieves tyrosine phosphorylation of Cdc28 protein kinase in *Saccharomyces cerevisiae*. *Mol. Cell. Biol.* **19:** 7123–7137.

Simon M.-N., De Virgilio C., Souza B., Pringle J.R., Abo A., and Reed S.I. 1995. Role for the Rho-family GTPase Cdc42 in yeast mating-pheromone signal pathway. *Nature* **376:** 702–705.

Simon V.R., Swayne T.C., and Pon L.A. 1995. Actin-dependent mitochondrial motility in mitotic yeast and cell-free systems: Identification of a motor activity on the mitochondrial surface. *J. Cell Biol.* **130:** 345–354.

Sloat B.F. and Pringle J.R. 1978. A mutant of yeast defective in cellular morphogenesis. *Science* **200:** 1171–1173.

Sloat B.F., Adams A., and Pringle J.R. 1981. Roles of the *CDC24* gene product in cellular morphogenesis during the *Saccharomyces cerevisiae* cell cycle. *J. Cell Biol.* **89:** 395–405.

Smith G.R., Givan S.A., Cullen P., and Sprague G.F., Jr. 2002. GTPase-activating proteins for Cdc42. *Eukaryot. Cell* **1:** 469–480.

Snyder M., Gehrung S., and Page B.D. 1991. Studies concerning the temporal and genetic control of cell polarity in *Saccharomyces cerevisiae*. *J. Cell Biol.* **114:** 515–532.

Spang A., Courtney I., Fackler U., Matzner M., and Schiebel E. 1993. The calcium-binding protein cell division cycle 31 of *Saccharomyces cerevisiae* is a component of the half bridge of the spindle pole body. *J. Cell Biol.* **123:** 405–416.

Stevenson B.J., Ferguson B., De Virgilio C., Bi E., Pringle J.R., Ammerer G., and Sprague G.F., Jr. 1995. Mutation of *RGA1*, which encodes a putative GTPase-activating protein for the polarity-establishment protein Cdc42p, activates the pheromone-response pathway in the yeast *Saccharomyces cerevisiae*. *Genes Dev.* **9:** 2949–2963.

Sullivan D.S. and Huffaker T.C. 1992. Astral microtubules are not required for anaphase B in *Saccharomyces cerevisiae*. *J. Cell Biol.* **119:** 379–388.

Surka M.C., Tsang C.W., and Trimble W.S. 2002. The mammalian septin MSF localizes with microtubules and is required for completion of cytokinesis. *Mol. Biol. Cell* **13:** 3532–3545.

Takizawa P.A., DeRisi J.L., Wilhelm J.E., and Vale R.D. 2000. Plasma membrane compartmentalization in yeast by messenger RNA transport and a septin diffusion barrier. *Science* **290:** 341–344.

Takizawa P.A., Sil A., Swedlow J.R., Herskowitz I., and Vale R.D. 1997. Actin-dependent localization of an RNA encoding a cell-fate determinant in yeast. *Nature* **389:** 90–93.

Theesfeld C.L., Irazoqui J.E., Bloom K., and Lew D.J. 1999. The role of actin in spindle orientation changes during the *Saccharomyces cerevisiae* cell cycle. *J. Cell Biol.* **146:** 1019–1031.

Theesfeld C.L., Zyla T.R., Bardes E.G.S., and Lew D.J. 2003. A monitor for bud emergence in the yeast morphogenesis checkpoint. *Mol. Biol. Cell* **14:** 3280–3291.

Tirnauer J.S. and Bierer B.E. 2000. EB1 proteins regulate microtubule dynamics, cell polarity, and chromosome stability. *J. Cell Biol.* **149:** 761–766.

Tirnauer J.S., O'Toole E., Berrueta L., Bierer B.E., and Pellman D. 1999. Yeast Bim1p promotes the G1-specific dynamics of microtubules. *J. Cell Biol.* **145:** 993–1007.

Tjandra H., Compton J., and Kellogg D. 1998. Control of mitotic events by the Cdc42 GTPase, the Clb2 cyclin and a member of the PAK kinase family. *Curr. Biol.* **8:** 991–1000.

Tkacz J.S. and Lampen J.O. 1972. Wall replication in *Saccharomyces* species: Use of fluorescein-conjugated concanavalin A to reveal the site of mannan insertion. *J. Gen. Microbiol.* **72:** 243–247.

Toenjes K.A., Sawyer M.M., and Johnson D.I. 1999. The guanine-nucleotide-exchange factor Cdc24p is targeted to the nucleus and polarized growth sites. *Curr. Biol.* **9:** 1183–1186.

Trueheart J., Boeke J.D., and Fink G.R. 1987. Two genes required for cell fusion during yeast conjugation: Evidence for a pheromone-induced surface protein. *Mol. Cell. Biol.* **7:** 2316–2328.

Vallen E.A., Caviston J., and Bi E. 2000. Roles of Hof1p, Bni1p, Bnr1p, and Myo1p in cytokinesis in *Saccharomyces cerevisiae*. *Mol. Biol. Cell* **11:** 593–611.

Vallen E.A., Ho W., Winey M., and Rose M.D. 1994. Genetic interactions between *CDC31* and *KAR1*, two genes required for duplication of the microtubule organizing center in *Saccharomyces cerevisiae*. *Genetics* **137:** 407–422.

Vallen E.A., Scherson T.Y., Roberts T., van Zee K., and Rose M.D. 1992. Asymmetric mitotic segregation of the yeast spindle pole body. *Cell* **69:** 505–515.

Valtz N. and Herskowitz I. 1996. Pea2 protein of yeast is localized to sites of polarized growth and is required for efficient mating and bipolar budding. *J. Cell Biol.* **135:** 725–739.

van Breugel M., Drechsel D., and Hyman A. 2003. Stu2p, the budding yeast member of the conserved

Dis1/XMAP215 family of microtubule-associated proteins is a plus end-binding microtubule destabilizer. *J. Cell Biol.* **161:** 359–369.
Waddle J.A., Karpova T.S., Waterston R.H., and Cooper J.A. 1996. Movement of cortical actin patches in yeast. *J. Cell Biol.* **132:** 861–870.
Wang P.J. and Huffaker T.C. 1997. Stu2p: A microtubule-binding protein that is an essential component of the yeast spindle pole body. *J. Cell Biol.* **139:** 1271–1280.
Wasserman S. 1998. FH proteins as cytoskeletal organizers. *Trends Cell Biol.* **8:** 111–115.
Water R.D., Pringle J.R., and Kleinsmith L.J. 1980. Identification of an actin-like protein and of its messenger ribonucleic acid in *Saccharomyces cerevisiae*. *J. Bacteriol.* **144:** 1143–1151.
Watts F.Z., Shiels G., and Orr E. 1987. The yeast *MYO1* gene encoding a myosin-like protein required for cell division. *EMBO J.* **6:** 3499–3505.
Weil C.F., Oakley C.E., and Oakley B.R. 1986. Isolation of *mip* (microtubule-interacting protein) mutations of *Aspergillus nidulans*. *Mol. Cell. Biol.* **6:** 2963–2968.
Weiss E.L., Bishop A.C., Shokat K.M., and Drubin D.G. 2000. Chemical genetic analysis of the budding-yeast p21-activated kinase Cla4p. *Nat. Cell Biol.* **2:** 677–685.
Wertman K.F., Drubin D.G., and Botstein D. 1992. Systematic mutational analysis of the yeast *ACT1* gene. *Genetics* **132:** 337–350.
Wigge P.A., Jensen O.N., Holmes S., Souès S., Mann M., and Kilmartin J.V. 1998. Analysis of the *Saccharomyces cerevisiae* spindle pole by matrix-assisted laser desorption/ionization (MALDI) mass spectrometry. *J. Cell Biol.* **141:** 967–977.
Winey M. and O'Toole E.T. 2001. The spindle cycle in budding yeast. *Nat. Cell Biol.* **3:** E23–E27.
Winey M., Goetsch L., Baum P., and Byers B. 1991. *MPS1* and *MPS2:* Novel yeast genes defining distinct steps of spindle pole body duplication. *J. Cell Biol.* **114:** 745–754.
Winey M., Mamay C.L., O'Toole E.T., Mastronarde D.N., Giddings T.H., Jr., McDonald K.L., and McIntosh J.R. 1995. Three-dimensional ultrastructural analysis of the *Saccharomyces cerevisiae* mitotic spindle. *J. Cell Biol.* **129:** 1601–1615.
Winge Ø. 1935. On haplophase and diplophase in some *Saccharomycetes*. *C.R. Trav. Lab. Carlsberg Ser. Physiol.* **21:** 77–111.
Winter D.C., Choe E.Y., and Li R. 1999a. Genetic dissection of the budding yeast Arp2/3 complex: A comparison of the *in vivo* and structural roles of individual subunits. *Proc. Natl. Acad. Sci.* **96:** 7288–7293.
Winter D., Lechler T., and Li R. 1999b. Activation of the yeast Arp2/3 complex by Bee1p, a WASP-family protein. *Curr. Biol.* **9:** 501–504.
Winter D., Podtelejnikov A.V., Mann M., and Li R. 1997. The complex containing actin-related proteins Arp2 and Arp3 is required for the motility and integrity of yeast actin patches. *Curr. Biol.* **7:** 519–529.
Wu C., Lytvyn V., Thomas D.Y., and Leberer E. 1997. The phosphorylation site for Ste20p-like protein kinases is essential for the function of myosin-I in yeast. *J. Biol. Chem.* **272:** 30623–30626.
Xue J., Wang X., Malladi C.S., Kinoshita M., Milburn P.J., Lengyel I., Rostas J.A.P., and Robinson P.J. 2000. Phosphorylation of a new brain-specific septin, G-septin, by cGMP-dependent protein kinase. *J. Biol. Chem.* **275:** 10047–10056.
Yamochi W., Tanaka K., Nonaka H., Maeda A., Musha T., and Takai Y. 1994. Growth site localization of Rho1 small GTP-binding protein and its involvement in bud formation in *Saccharomyces cerevisiae*. *J. Cell Biol.* **125:** 1077–1093.
Yang S., Ayscough K.R., and Drubin D.G. 1997. A role for the actin cytoskeleton of *Saccharomyces cerevisiae* in bipolar bud-site selection. *J. Cell Biol.* **136:** 111–123.
Yeh E., Skibbens R.V., Cheng J.W., Salmon E.D., and Bloom K. 1995. Spindle dynamics and cell cycle regulation of dynein in the budding yeast, *Saccharomyces cerevisiae*. *J. Cell Biol.* **130:** 687–700.
Yeh E., Yang C., Chin E., Maddox P., Salmon E.D., Lew D.J., and Bloom K. 2000. Dynamic positioning of mitotic spindles in yeast: Role of microtubule motors and cortical determinants. *Mol. Biol. Cell* **11:** 3949–3961.
Yin H., Pruyne D., Huffaker T.C., and Bretscher A. 2000. Myosin V orientates the mitotic spindle in yeast. *Nature* **406:** 1013–1015.
Zahner J.E., Harkins H.A., and Pringle J.R. 1996. Genetic analysis of the bipolar pattern of bud site selection in the yeast *Saccharomyces cerevisiae*. *Mol. Cell. Biol.* **16:** 1857–1870.

Zhang X., Bi E., Novick P., Du L., Kozminski K.G., Lipschutz J.H., and Guo W. 2001. Cdc42 interacts with the exocyst and regulates polarized secretion. *J. Biol. Chem.* **276:** 46745–46750.

Zhao Z.-S., Leung T., Manser E., and Lim L. 1995. Pheromone signalling in *Saccharomyces cerevisiae* requires the small GTP-binding protein Cdc42p and its activator *CDC24*. *Mol. Cell. Biol.* **15:** 5246–5257.

Zheng Y., Bender A., and Cerione R.A. 1995. Interactions among proteins involved in bud-site selection and bud-site assembly in *Saccharomyces cerevisiae*. *J. Biol. Chem.* **270:** 626–630.

Zheng Y., Cerione R., and Bender A. 1994. Control of the yeast bud-site assembly GTPase Cdc42. Catalysis of guanine nucleotide exchange by Cdc24 and stimulation of GTPase activity by Bem3. *J. Biol. Chem.* **269:** 2369–2372.

Ziman M., O'Brien J.M., Ouellette L.A., Church W.R., and Johnson D.I. 1991. Mutational analysis of *CDC42Sc*, a *Saccharomyces cerevisiae* gene that encodes a putative GTP-binding protein involved in the control of cell polarity. *Mol. Cell. Biol.* **11:** 3537–3544.

Ziman M., Preuss D., Mulholland J., O'Brien J.M., Botstein D., and Johnson D.I. 1993. Subcellular localization of Cdc42p, a *Saccharomyces cerevisiae* GTP-binding protein involved in the control of cell polarity. *Mol. Biol. Cell* **4:** 1307–1316.

12

Membrane Traffic

Randy Schekman
Department of Molecular and Cell Biology
University of California
Berkeley, California 94720-3202

Revealing the nuts and bolts of the secretory pathway, 245

Novick P., Field C., and Schekman R. 1980. Identification of 23 complementation groups required for post-translational events in the yeast secretory pathway. *Cell* **21**: 205–215.

The mechanics of vesicular traffic, 245

Salminen A. and Novick P.J. 1987. A *ras*-like protein is required for a post-Golgi event in yeast secretion. *Cell* **49**: 527–538.

Schu P.V., Takegawa K., Fry M.J., Stack J.H., Waterfield M.D., and Emr S.D. 1993. Phosphatidylinositol 3-kinase encoded by yeast *VPS34* gene essential for protein sorting. *Science* **260**: 88–91.

Maintaining organelle composition: How the cell avoids secreting the secretory pathway, 248

Lewis M.J., Sweet D.J., and Pelham H.R.B. 1990. The *ERD2* gene determines the specificity of the luminal ER protein retention system. *Cell* **61**: 1359–1363.

Note: The landmark papers listed above are those discussed in this chapter. Each landmark paper is preceded by the name of the section (with starting page number) where the paper is first discussed in detail.

EUKARYOTIC CELLS ELABORATE AN EXTENSIVE NETWORK of membranes, many of which interrelate by a flow of vesicles constituting the secretory pathway. This pathway is intimately involved in the full range of physiologic responses of cells, as well as in the regulation, growth, and division of all eukaryotic cells. The study of membrane traffic in *Saccharomyces cerevisiae* has developed into a major theme of membrane cell biology. However, for many years, the subject was viewed with skepticism by the larger cell biology community that dealt with traditional approaches to the inspection and fractionation of intracellular membranes. Yeast cells were difficult to preserve for ultrastructural analysis, cell sections revealed nondescript internal architecture, and yeast cells were considered too small to allow serious morphological inspection, certainly at the level of fluorescence microscopy. Cell fractionation was considered difficult because of the refractory cell wall that could be digested only with a crude lytic enzyme harvested from snail gut. Yeast lysates are loaded with proteolytic enzymes that would damage membrane proteins and organelles. And even when proteolytic activity could be controlled through the use of mutants or inhibitors, the inventory of useful marker proteins and antibodies was quite limited, at least in comparison to the more traditional cell specimens such as the pancreatic acinar cell and hepatocytes.

Nonetheless, a few hearty souls embarked on the analysis of membrane function and assembly in yeast. By the mid 1970s, Gottfried Schatz and Walter Neupert had begun their pioneering work on mitochondrial assembly (Criddle and Schatz 1969; Neupert et al. 1969), Elizabeth Jones had initiated genetic studies on the control of yeast proteases (Jones 1977), Clinton Ballou had developed chemical and genetic methods to map the synthesis and structure of yeast glycoproteins (Ballou et al. 1973), and Enrico Cabib had developed a biochemical analysis of yeast cell wall polysaccharide, particularly chitin, biosynthesis (Cabib and Farkas 1971).

Studies on protein traffic through the secretory pathway were more rudimentary. Two lines of investigation supported a role for secretory vesicles in the local growth of the bud cell wall. Secreted enzymes, such as the inducible acid phosphatase, could be visualized in a cluster of vesicles under the plasma membrane of an emerging bud (Linnemans et al. 1977). Newly secreted cell wall glycoproteins were visualized in the cell wall of the bud tip precisely where vesicles congregated in the cytoplasm (Tkacz and Lampen 1973). Although Golgi-like membranes were visualized, no role for the Golgi could be established other than by extension from studies on mammalian cells. The pathways of protein traffic to the vacuole and receptor or solute internalization by endocytosis had not been explored.

In an approximately 25-year period, starting in the late 1970s, tremendous progress—including genetic, molecular cloning, morphological, and biochemical analyses—transported yeast from a backwater of membrane traffic to a position as an essential element in the armamentarium of the cell biologist. Although many more could and should be credited, four key papers paved the way for this explosion. These four papers deal with the discovery of genes and proteins that form and target vesicles, the mechanics of vesicular traffic, and the genes and proteins that segregate membrane and soluble cargo proteins to ensure that organelles maintain a characteristic composition.

George Palade and his colleagues defined the broad contour of the secretory pathway by transmission electron microscopy of professional secretory cells such as are found in the pancreas. In 1974, when Palade was awarded the Nobel Prize for this pioneering effort, one could say that organelles of the secretion network communicate by vesicular intermediaries and yet the membranes of these organelles maintain a characteristic protein and, to some extent, lipid composition. How is this composition maintained in the face of the diffusional forces that would otherwise randomize proteins moving around in what amounts to a two-dimensional fluid? How are vesicles formed so as to capture only those cargo molecules that are designed to be transported? How do vesicles track to and merge with a particular compartment given the myriad alternatives available? Palade eloquently posed these questions, but no answers or even paths to answers were on the horizon (Palade 1975). Of course, for geneticists and biochemists, questions of specificity and mechanism are second nature. The solutions to the problems posed by Palade came with the introduction of experimental systems and approaches that permitted the nuts and bolts of the secretory pathway to be revealed.

REVEALING THE NUTS AND BOLTS OF THE SECRETORY PATHWAY

Novick et al. (1980) began with the assumption that secretion is essential for cell growth. Vesicles in the bud cytoplasm were predicted to convey plasma membrane proteins to the cell surface; thus, a failure in the creation or discharge of vesicles would block cell growth and perhaps lead to the accumulation of secretory organelles within the cell. This prediction was tested in a survey of a small collection of randomly selected temperature-sensitive isolates that were screened for pleiotropic defects in secretion (Novick and Schekman 1979). Wild-type cells maintain a small pool of molecular and organelle intermediates in secretion. However, in *sec1* mutant cells, a large increase in the intracellular level of invertase and acid phosphatase is accompanied by a corresponding accumulation of secretory vesicles. Concomitantly, the export of an induced form of sulfate permease is blocked. Accumulated secretory proteins are discharged at the cell surface when mutant cells are returned to a permissive temperature. Thus, the *sec1*-1 block introduces a reversible halt in secretion; we now know that the block is exerted at the end of the secretory pathway.

The next task was to find a selection procedure to enrich for more *sec* muants. *sec1* cells incubated at 37°C become phase refractile and arrest at random positions along the cell cycle without becoming larger. Susan Henry's lab reported that inositol auxotrophic cells cease net surface growth concomitant with a block in phospholipid synthesis, and as a result, inositol-starved cells experience an increase in buoyant density (Henry et al. 1977). Novick et al. (1980) found the same behavior for *sec1* mutant cells and used a density gradient to enrich for new *sec* mutants in cultures of mutagenized yeast cells. Among 70 isolates, 23 complementation groups were established that shared the pleiotropic accumulation of secretory enzymes and a defect in the export of the sulfate permease.

Representative alleles of each gene were examined by thin-section electron microscopy. Most genes defined one of three cytologic phenotypes: accumulation of endoplasmic reticulum (ER), accumulation of secretory vesicles, or accumulation of toroid-shaped membranes, referred to as Berkeley bodies, which subsequently were shown to be a dead-end form of Golgi membranes. Other mutants, such as *sec19*, which was subsequently shown to affect a GDI protein required to recycle GTP-binding proteins that act at multiple stages in the pathway, accumulated multiple organelles (Garrett et al. 1994).

The Novick et al. (1980) paper set the stage for a thorough genetic, molecular cloning, and biochemical analyses of the secretory pathway. The broad outline of the pathway was mapped by cytologic and genetic analyses of double-mutant strains, of glycoprotein intermediates of secretion, and of a branchpoint leading to the yeast vacuole. The *SEC* genes, and the many genes discovered subsequently and given different acronyms, ultimately revealed that the secretory pathway is functionally and structurally conserved in all eukaryotic cells (Fig. 1).

THE MECHANICS OF VESICULAR TRAFFIC

The first mechanistic insights to emerge from an analysis of the *sec* mutations came from an instructive sequence of the *SEC4* gene (Salminen and Novick 1987). Novick's group at Yale decided to focus on the late-acting *SEC* genes. *SEC4* was identified as a particularly appealing member of this set of approximately a dozen genes because of the interesting

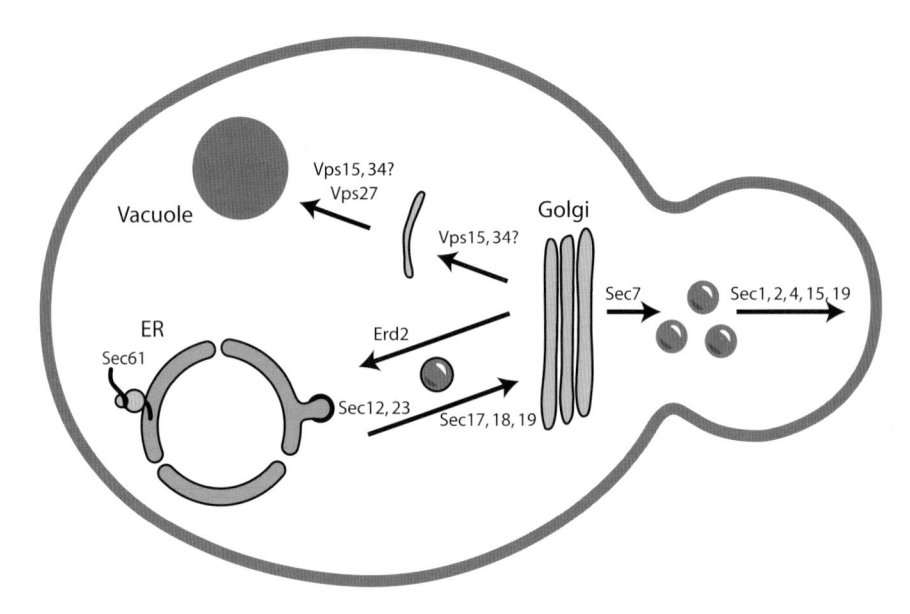

FIGURE 1. Membrane traffic, i.e., membrane budding, vesicular transport, and membrane fusion, is a highly dynamic process. The *SEC* genes discussed in the text are highlighted at the appropriate position in this cartoon of the secretory pathway.

genetic interactions that connect the gene and mutant alleles of *sec4* to mutations in other genes that affect this late step in the pathway. Salminen and Novick (1987) found that as little as a twofold elevation of *SEC4* gene copy at least partially phenotypically suppresses mutations in other genes such as *SEC15, SEC2*, and *SEC8*, but does not suppress mutations that block elsewhere in the secretory pathway. Likewise, certain combinations of mutations in the late-acting *SEC* genes produce a synthetic-lethal phenotype with a *sec4* allele. Thus, without any specific mechanistic understanding, it appeared that *SEC4* may encode a key regulator whose level could compensate for or compromise the action of crippled forms of gene products subject to control.

The *SEC4* sequence is 32% homologous to *ras*, and the residues of the GTP-binding domain of the Ras family of proteins are conserved in Sec4. This indicated a role for GTP and possibly GTP hydrolysis in a cycle involving Sec4 as a regulator of some aspect of secretory vesicle tracking, docking, or membrane fusion. Subsequent work, primarily from Novick's lab, showed that mutations in the GTP-binding domain which block hydrolysis render Sec4 dominantly defective in secretion (Walworth et al. 1989). Furthermore, a cycle of GTP hydrolysis and nucleotide exchange is indicated by the discovery of other gene products essential for secretion, such as Sec2, which is the nucleotide exchange catalyst for Sec4 (Walch-Solimena et al. 1997).

An even larger set of Sec proteins implicated in the last step of secretion form a complex, called the exocyst (TerBush et al. 1996), that interprets the GTP state of Sec4 and directs vesicles with bound Sec4-GTP to the bud site which appears to be marked by Sec3 (Finger et al. 1998; Guo et al. 1999). Genetic interactions identified by Salminen and Novick (1987) presaged the discovery of the biochemical interactions now known to be responsible for exocyst-mediated vesicle targeting.

Salminen and Novick's (1987) discovery also unleashed a cottage industry of effort to identify the numerous small GTP-binding proteins, so-called Rab proteins, that define the numerous avenues of transport in mammalian cells. More than three dozen Rab proteins and their even more numerous exchange (GEF), hydrolysis (GAP), and effector proteins have been described in *Caenorhabditis elegans, Drosophila*, and mammals. In many cases, these Rab proteins and their effectors facilitate specialized transport pathways not found in yeast.

A major branchpoint in the secretory pathway occurs when proteins destined to the lysosome-like vacuole are diverted from secretory proteins in the Golgi complex. Classic studies on this pathway in mammalian cells relied on the observation that lysosomal hydrolases share a common covalent modification, a mannose-6-phosphate signal (Hasilik and Neufeld 1980), of N-glycans that allows lysosomal and secretory proteins to be distinguished by a mannose-6-phosphate receptor (Gonzalez-Noriega et al. 1980). Much of the work on this pathway in mammals was focused on the elucidation of this signal and the sorting of the receptor, but little if anything was learned about the general machinery that conveys lysosomal proteins. Although *S. cerevisiae* does not employ the mannose-6-phosphate determinant for sorting to the vacuole, many features of the pathway appeared to be similar. Thus, a genetic approach to this process seemed most appealing (Stevens et al. 1982).

In the 1970s and 1980s, three groups developed a genetic analysis of vacuole protein biosynthesis. Elizabeth Jones developed a simple screening procedure to isolate mutants defective in the activity of vacuolar proteases (Jones 1977), and many of the genes she identified were shown subsequently to be involved in protein traffic to the vacuole. More direct approaches designed to enrich for vacuole protein sorting (*VPS*) mutants were developed by Tom Stevens and Scott Emr (Bankaitis et al. 1986; Rothman and Stevens 1986). In the intervening years, dozens of such genes have been characterized, and we now know a great deal about the sorting signals (Valls et al. 1987; Marcusson et al. 1994), a sorting receptor, and the many proteins that must interact to execute the sorting process. As a bonus, we can say with some assurance that the pathway is fundamentally conserved, as most of the *VPS* genes have mammalian equivalents that function in lysosomal biogenesis and endocytic vesicular traffic.

Nonetheless, although much can be said about the Vps proteins and their interactions, little is known about their catalytic role and virtually nothing can be explicated at a high level of mechanistic precision. One exception to this came from an important discovery reported by **Schu et al. (1993)**, who found that one of the VPS genes encodes a homolog of the catalytic subunit of bovine phosphatidylinositol (PI)-3 kinase.

In the course of cloning and molecular characterization, Scott Emr's group discovered that *VPS34* encodes a 110-kD protein with two regions of 33% sequence identity to a comparable carboxy-terminal domain of the bovine PI-3 kinase (Schu et al. 1993). Functional and genetic analyses demonstrated the catalytic identity of the yeast protein and the role of this enzyme reaction in the sorting of vacuolar proteins in vivo. Subsequent experiments in mammalian cells have documented a similar role for an enzyme of this specificity in the transport of lysosomal proteins and in traffic within the endosomal network.

Further molecular cloning and genetic analysis showed that Vps34 operates in a complex with a protein kinase encoded by the *VPS15* gene (Stack et al. 1993). The protein kinase activity of Vps15 is essential for expressing the lipid kinase activity of Vps34. Where exactly this complex acts to facilitate vesicular traffic is not known with any precision; however, a family of effector proteins containing a conserved sequence of approximately 70 amino acids, and referred to as the FYVE domain, appears to recognize PI-3-P (Burd and Emr 1998). Most importantly, several of the FYVE domain proteins are required for protein traffic to the vacuole (e.g., Vps27). These findings represent tantalizing but low-resolution images of the events in which the PI-3 kinase participates. Much remains to be done to pinpoint the exact step in which PI-3-P exerts its role in protein sorting (e.g., vesicle budding, targeting, or fusion).

MAINTAINING ORGANELLE COMPOSITION: HOW THE CELL AVOIDS SECRETING THE SECRETORY PATHWAY

One of the basic questions in the organization of the secretory pathway is how certain protein constituents are retained within an organelle in the face of a stream of protein traffic passing through on the way to subsequent stations. This problem seems particularly acute for soluble luminal resident proteins, several abundant representatives of which facilitate the folding of newly synthesized secretory and membrane proteins in the ER.

Three important breakthroughs were made by Hugh Pelham and his colleagues that led to our current understanding of how the pool of resident proteins is maintained. Munro and Pelham (1987) showed that luminal ER proteins share a carboxy-terminal tetrapeptide KDEL sequence, deletion of which leads to the slow but inexorable secretion of tailless forms of such chaperone proteins as BiP. These results suggested either that the KDEL sequence represents a retention signal that fastens BiP and other chaperones to some anchor protein in the ER (although how this anchor is itself retained is not specified by this view) or that the KDEL sequence represents a retrieval signal responsible for recovering ER luminal proteins from a distal compartment. Subsequent to the discovery of the KDEL signal on soluble luminal proteins, Cosson and Letourneur (1994) characterized a signal, KKXX, at the carboxyl terminus of type I membrane proteins responsible for their retention in the ER membrane. Deletion of the KKXX signal results in the export of tailless membrane proteins, and here again, the view was that this signal was required for retention or retrieval by processes formally analogous to those proposed for soluble luminal proteins.

Next, Pelham et al. (1988), Hardwick et al. (1990), and Semenza et al. (1990) developed a genetic approach to isolate genes essential for the retention or retrieval of proteins terminated by HDEL, the *S. cerevisiae* equivalent of the mammalian KDEL signal. Cells expressing a hybrid form of secretory invertase containing a carboxy-terminal HDEL were shown to accumulate invertase in the ER. Mutants that secrete hybrid invertase were identified with an activity assay suitable for large-scale screening of yeast colonies. Two genes, *ERD1* and *ERD2*, were selected, and *ERD2*, an essential gene, appeared to be the more interesting.

Given two genes that appear to be required for the retention of HDEL-terminated proteins, **Lewis et al. (1990)** devised an elegant test to demonstrate that *ERD2* exerts its role in a sequence-selective manner. They relied on the observation that *Kluyveromyces lactis* expresses ER luminal chaperones terminated in HDEL or DDEL retention sequences. *K. lactis* BiP, which terminates in the DDEL signal, fails to be retained when it is ectopically expressed in *S. cerevisiae*, suggesting that the *S. cerevisiae* retention/retrieval apparatus is more sequence-selective than the corresponding *K. lactis* system. Surrogate expression of the *K.*

lactis ERD2 and BiP genes in *S. cerevisiae* restores retention of *K. lactis* BiP. Thus, *ERD2* must contribute to the selective recognition of the retention/retrieval signal.

Subsequent evidence confirmed *ERD2* as the likely HDEL receptor in *S. cerevisiae*, and an ortholog in mammals as a KDEL receptor. However, the existence of such a receptor does not by itself distinguish the retention and retrieval models. For this, Lewis and Pelham (1992) tagged the mammalian Erd2p and showed that it localizes to post-ER compartments, the ER Golgi intermediate compartment, and the *cis*-Golgi cisternae, a localization most consistent with the view that KDEL/HDEL serves to permit escaped luminal proteins to be retrieved by a process of retrograde transport from post-ER compartments back to the ER.

The retrieval of KKXX-terminated ER membrane proteins is achieved by a parallel but independent process. Cosson and Letourneur (1994) found that a KKXX-terminated protein binds coatomer, a coat protein complex involved in vesicular traffic in the Golgi stack. Letourneur et al. (1994) used a KKXX-terminated form of Ste2, the α-factor receptor, to retain this protein within the cell, a condition that produces a sterile phenotype. Mutagenesis of such cells followed by a selection for mating proficiency yielded a set of conditional mutations that defined the subunits of the coatomer. From this, Letourneur et al. (1994) concluded that coatomer interacts with and retrieves KKXX-terminated proteins from a post-Golgi compartment. Indeed, it is now clear that coatomer is the principal vehicle for retrograde transport of soluble luminal and membrane resident proteins of the ER.

FINAL COMMENTS

Much of the current excitement in studies on protein transport in yeast focuses on the development of high-resolution mechanistic insights. Many of the proteins implicated in traffic have been isolated and atomic level structures have revealed intimate details of protein sorting, vesicle budding, targeting, and fusion. Nonetheless, surprising new limbs of the basic pathway continue to emerge at a purely descriptive level of analysis. Most recently, a long-standing question of the autonomous origin of peroxisomes has been challenged with the observation that a peroxisomal membrane protein required for peroxisome formation, Pex3, originates in the ER membrane (Hoepfner et al. 2005). Thus, peroxisomes may form by budding of vesicles from the ER distinct from the budding of secretory transport vesicles. With such basic features of membrane traffic only now emerging, one can safely assert that yeast cell biology is far from dead.

STUDY QUESTIONS

1. The first pleiotropic secretion mutant, *sec1*, was isolated from a random collection of 100 independently isolated temperature-sensitive growth mutants. Was this a lucky chance hit, or would one expect to have found this mutant in such a collection?

2. Why do wild-type yeast cells have a low level of molecular and organelle intermediates of the secretory pathway?

3. One of the original *SEC* genes defined by Novick et al. (1980) is not essential at room temperature. Deletion of *SEC22* produces a temperature-sensitive growth phenotype similar to that observed for the original temperature-sensitive isolate. How can this be?

4. Vesicles mediate traffic at many different stages in the secretory pathway. If so, why does the *sec4* mutation cause a block in secretion limited to post-Golgi secretory vesicles?

5. GTP-binding proteins such as Sec4 are recycled from a target membrane to the cytoplasm by the action of a GDP/GTP exchange inhibitory protein, GDI. Yeast has one such protein encoded by *SEC19*. Unlike *sec4* mutants, *sec19* mutants accumulate vesicles characteristic of several stages in the secretory pathway. Explain this apparent paradox.

6. The PI-3 kinase product of *VPS34* is responsible for some aspect of traffic of vacuolar precursor proteins from a *trans*-Golgi cisterna to the prevacuolar compartment, possibly the late endosome. Suggest molecular mechanisms by which the lipid PI-3-P could regulate the formation of a transport vesicle or the targeting and fusion of a transport vesicle.

7. A set of genes encoding yeast PI phosphatases has been cloned, no one of which is essential for PI turnover. However, deletion of all three genes produces a sickly growth phenotype and expands the pool of PI-3-P. Would you expect the triple-mutant strain to exaggerate or mitigate the vacuole sorting phenotype of a partial loss-of-function allele of *vps34*?

8. Deletion of the HDEL sequence of yeast BiP produces a truncated protein that is secreted, but with a rate of intercompartmental transport much slower than that found for typical secreted proteins. Offer two alternative explanations of the slow secretion phenotype or HDEL-less BiP.

9. You perform an experiment in which the HDEL sequence from an ER luminal chaperone and the KKXX sequence from a type I ER membrane protein are exchanged, such that a recombinant BiP containing carboxy-terminal KKXX is expressed in one strain and a recombinant type I membrane protein with a carboxy-terminal HDEL is expressed in another strain. Do you expect the BiP-KKXX chimera to be retained or secreted? If it is retained, will it respond to Erd2 protein or coatomer? Do you expect the type I membrane protein-HDEL chimera to be retained or transported to the cell surface? If it is retained, will it respond to the Erd2 protein or coatomer?

REFERENCES

Ballou C.E., Kern K.A., and Raschke W.C. 1973. Genetic control of yeast mamman structure: Complementation studies and properties of mannan mutants. *J. Biol. Chem.* **248:** 4667–4671.

Bankaitis V.A., Johnson L.M., and Emr S.D. 1986. Isolation of yeast mutants defective in protein sorting to the vacuole. *Proc. Natl. Acad. Sci.* **83:** 9075–9079.

Burd C.G. and Emr S.D. 1998. Phosphatidylinositol(3)-phosphate signaling mediated by specific binding to RING FYVE domains. *Mol. Cell* **2:** 157–162.

Cabib E. and Farkas V. 1971. The control of morphogenesis: An enzymatic mechanism for the initiation of septum formation in yeast. *Proc. Natl. Acad. Sci.* **68:** 2052–2056.

Cosson P. and Letourneur F. 1994. Coatomer interaction with di-lysine endoplasmic reticulum retention motifs. *Science* **263:** 1629–1631.

Criddle R.S. and Schatz G. 1969. Promitochondria of anaerobically grown yeast. I. Isolation and biochemical properties. *Biochemistry* **8:** 322–334.

Finger F.P., Hughes T.E., and Novick P.J. 1998. Sec3p is a spatial landmark for polarized secretion in budding yeast. *Cell* **92:** 559–571.

Garrett M.D., Zahner J.E., Cheney C.M., and Novick P.J. 1994. *GDI1* encodes a GDP dissociation inhibitor that plays an essential role in the yeast secretory pathway. *EMBO J.* **13:** 1718–1728.

Gonzalez-Noriega A., Grubb J.H., Talkad V., and Sly W.S. 1980. Chloroquine inhibits lysosomal enzyme pinocytosis and enhances lysosomal enzyme secretion by impairing receptor recycling. *J. Cell Biol.* **85:** 839–852.

Guo W., Roth D., Walch-Solimena C., and Novick P. 1999. The exocyst is an effector for Sec4p, targeting secretory vesicles to sites of exocytosis. *EMBO J.* **18:** 1071–1080.

Hardwick K.G., Lewis M.J., Semenza J., Dean N., and Pelham H.R.B. 1990. *ERD1*, a yeast gene required for the retention of luminal endoplasmic reticulum proteins, affects glycoprotein processing in the Golgi apparatus. *EMBO J.* **9:** 623–630.

Hasilik A. and Neufeld E.F. 1980. Biosynthesis of lysosomal enzymes in fibroblasts: Phosphorylation of mannose residues. *J. Biol. Chem.* **255:** 4946–4950.

Henry S.A., Atkinson K.D., Kolat A.I., and Culbertson M.R. 1977. Growth and metabolism of inositol-starved *Saccharomyces cerevisiae. J. Bacteriol.* **130:** 472–484.

Hoepfner D., Schildknegt D., Braakman I., Philippsen P., and Tabak H.F. 2005. Contribution of the endoplasmic reticulum to peroxisome formation. *Cell* **122:** 85–95.

Jones E.W. 1977. Proteinase mutants of *Saccharomyces cerevisiae. Genetics* **85:** 22–33.

Letourneur F., Gaynor E.C., Hennecke S., Demolliere G., Duden R., Emr S.D., Riezman H., and Cosson P. 1994. Coatomer is essential for retrieval of dilysine-tagged proteins to the endoplasmic reticulum. *Cell* **79:** 1199–1207.

Lewis M.J. and Pelham H.R.B. 1992. Ligand-induced redistribution of a human KDEL receptor from the Golgi complex to the endoplasmic reticulum. *Cell* **68:** 353–364.

Lewis M.J., Sweet D.J., and Pelham H.R.B. 1990. The *ERD2* gene determines the specificity of the luminal ER protein retention system. *Cell* **61:** 1359–1363.

Linnemans W.A.M., Boer P., and Elbers P.F. 1977. Localization of acid phosphatase in *Saccharomyces cerevisiae:* A clue to cell wall formation. *J. Bacteriol.* **131:** 638–644.

Marcusson E.B., Horazdovsky B.F., Cereghino J.L., Gharakhanian E., and Emr S.D. 1994. The sorting receptor for yeast carboxypeptidase Y is encoded by the *VPS10* gene. *Cell* **77:** 579–586.

Munro S. and Pelham H.R.B. 1987. A C-terminal signal prevents secretion of luminal ER proteins. *Cell* **48:** 899–907.

Neupert W., Sebald W., Schwab A.J., Massinger P., and Bucher T. 1969. Incorporation in vivo of ^{14}C-labelled amino acids into the proteins of mitochondrial ribosomes from *Neurospora crassa* sensitive to cycloheximide and insensitive to chloramphenicol. *Eur. J. Biochem.* **10:** 589–591.

Novick P. and Schekman R. 1979. Secretion and cell surface growth are blocked in a temperature-sensitive mutant of *Saccharomyces cerevisiae. Proc. Natl. Acad. Sci.* **76:** 1858–1862.

Novick P., Field C., and Schekman R. 1980. Identification of 23 complementation groups required for post-translational events in the yeast secretory pathway. *Cell* **21:** 205–215.

Palade G. 1975. Intracellular aspects of the process of protein synthesis. *Science* **189:** 347–358.

Pelham H.R.B., Hardwick K.G., and Lewis M.J. 1988. Sorting of soluble ER proteins in yeast. *EMBO J.* **7:** 1757–1762.

Rothman J.H. and Stevens T.H. 1986. Protein sorting in yeast: Mutants defective in vacuolar biogenesis mislocalize vacuolar proteins into the late secretory pathway. *Cell* **47:** 1041–1051.

Salminen A. and Novick P.J. 1987. A *ras*-like protein is required for a post-Golgi event in yeast secretion. *Cell* **49:** 527–538.

Schu P.V., Takegawa K., Fry M.J., Stack J.H., Waterfield M.D., and Emr S.D. 1993. Phosphatidylinositol 3-kinase encoded by yeast *VPS34* essential for protein sorting. *Science* **260:** 88–91.

Semenza J.C., Hardwick K.G., Dean N., and Pelham H.R.B. 1990. *ERD2*, a yeast gene required for the receptor-mediated retrieval of luminal ER proteins from the secretory pathway. *Cell* **61:** 1349–1357.

Stack J.H., Herman P.K., Schu P.V., and Emr S.D. 1993. A membrane-associated complex containing the Vps15 protein kinase and the vps34 PI 3-kinase is essential for protein sorting to the yeast lysosome-like vacuole. *EMBO J.* **12:** 2195–2204.

Stevens T.H., Esmon B., and Schekman R. 1982. Early stages of the yeast secretory pathway are required for transport of carboxypeptidase Y to the vacuole. *Cell* **30:** 439–448.

TerBush D.R., Maurice T., Roth D., and Novick P.J. 1996. The exocyst is a multiprotein complex required for exocytosis in *Saccharomyces cerevisiae*. *EMBO J.* **15:** 6483–6494.

Tkacz J.S. and Lampen J.O. 1973. Surface distribution of invertase on growing *Saccharomyces* cells. *J. Bacteriol.* **113:** 1073–1075.

Valls L.A., Hunter C.P., Rothman J.H., and Stevens T.H. 1987. Protein sorting in yeast: The localization determinant of yeast vacuolar carboxypeptidase Y resides in the propeptide. *Cell* **48:** 887–897.

Walch-Solimena C., Collins R.N., and Novick P.J. 1997. Sec2p mediates nucleotide exchange on Sec4p and is involved in polarized delivery of post-Golgi vesicles. *J. Cell Biol.* **137:** 1495–1509.

Walworth N.C., Goud B., Kabcenell A.K., and Novick P.J. 1989. Mutational analysis of *SEC4* suggests a cyclical mechanism for the regulation of vesicular traffic. *EMBO J.* **8:** 1685–1693.

13

Protein Translocation

Howard Riezman

Biochemistry Department
University of Geneva
CH-1211 Geneva 4, Switzerland

Nuclear localization, 255

Hall M.N., Hereford L., and Herskowitz I. 1984. Targeting of *E. coli* β-galactosidase to the nucleus in yeast. *Cell* **36:** 1057–1065.

ER translocation, 256

Deshaies R.J. and Schekman R. 1987. A yeast mutant defective at an early stage in import of secretory protein precursors into the endoplasmic reticulum. *J. Cell Biol.* **105:** 633–645.

Import into mitochondria, 257

Schleyer M. and Neupert W. 1985. Transport of proteins into mitochondria: Translocational intermediates spanning contact sites between outer and inner membranes. *Cell* **43:** 339–350.

Eilers M. and Schatz G. 1986. Binding of a specific ligand inhibits import of a purified precursor protein into mitochondria. *Nature* **322:** 228–232.

Note: The landmark papers listed above are those discussed in this chapter. Each landmark paper is preceded by the name of the section (with starting page number) where the paper is first discussed in detail.

ONE OF THE MAJOR DIFFERENCES BETWEEN PROKARYOTES and eukaryotes is the intracellular compartmentalization that occurs in eukaryotic cells. This creates organelles that allow intracellular specializations, but it also creates logistic problems concerning the targeting of proteins to the organelles. The intracellular compartments are bounded by membranes that allow the compartments to maintain their specific identities. This chapter deals with the targeting of proteins from the cytoplasm to specific organelles and the translocation of proteins across membranes that define these organelles.

One of the basic principles of protein targeting to an intracellular location was first postulated from studies on mammalian cells. It had been noticed that there are two populations of ribosomes: those in the cytosol and those attached to the endoplasmic reticulum (ER) membrane. Bound ribosomes were seen both in electron micrographs of the ER membrane (rough ER) and in biochemical studies (rough microsomes). The possibility was raised that these ribosomes could be involved in the synthesis of membrane proteins (Palade 1975). A tentative model for the role of ribosome-membrane interactions in protein translocation was

Note: Boldfaced references in the text denote landmark papers that are on the accompanying CD.

postulated by Blobel and Sabatini (1971). Milstein and colleagues (1972) then showed that the immunoglobulin light chain synthesized in vitro from mRNA was slightly larger than when synthesized in vivo. They postulated that the immunoglobulin light chain made in vitro could be a precursor, but did not provide evidence for a precursor-product relationship. Blobel and Dobberstein (1975a) then showed that in contrast to translation of purified mRNA, run-off translation of mRNA attached to membrane-bound polysomes yielded both processed and unprocessed light chains. This suggested that some of the light chains had already been proteolytically processed (presumably after crossing the membrane) before they were fully synthesized, consistent with a cotranslational mechanism of protein translocation. These data led to the postulation of the signal hypothesis. The basic features of the signal hypothesis, which are still valid today, were as follows: Proteins that are to be segregated across the ER membrane are synthesized as larger precursors with a 15–30-amino-acid extension at the amino terminus. When this sequence emerges from the ribosome, it directs the translation complex to the ER where ongoing protein synthesis propels the nascent protein through a transmembrane proteinaceous tunnel and into the ER lumen (see below). Once translocated into the ER lumen, the signal sequence is proteolytically removed from the protein. Several aspects of this hypothesis were immediately tested in a classic study involving reconstitution of protein transport across microsomal membranes. IgG light-chain mRNA was translated and the light chain was translocated across microsomal membranes into a space inaccessible to exogenously added protease. Translocation occurred obligatorily during translation, because completed precursors could not be translocated (Blobel and Dobberstein 1975b). The signal hypothesis was used as a general paradigm for the targeting of proteins from the cytoplasm to other organelles, including the semiautonomous organelles mitochondria and chloroplasts (Blobel 1980). Several groups set out at that time to identify signals for the different subcellular compartments.

At about the same time, studies of protein translocation across the *Escherichia coli* inner membrane showed that this event was also mediated by an ER-like signal sequence (Emr et al. 1980). However, there were also some important differences in protein translocation across the bacterial membrane from what had been observed with ER membranes. These differences sparked a lot of interest, controversy, and experiments. The classic studies reconstituting protein translocation into the ER membrane presented strong evidence that protein translocation was cotranslational, but translocation of M13 coat protein across bacterial membranes could occur posttranslationally (Ito et al. 1979). The very small size of the M13 coat protein was considered a reason for the posttranslational insertion properties of this protein. The coat protein is so small that its synthesis is essentially complete by the time the nascent amino terminus emerges from the ribosome. However, other studies in *E. coli* showed that larger proteins could also be translocated posttranslationally (Josefsson et al. 1982). Another difference was that translocation across the bacterial membrane required an electrochemical gradient (Date et al. 1980), whereas import into the ER did not. We will come back to these discrepancies later, but further experiments showed that what is common between co- and posttranslational import is as important as what is different.

Besides providing another model to study protein translocation, a most important feature of the work in bacterial systems was that it set the stage for many of the approaches that would be used in the future to study protein translocation in eukaryotes. Beckwith and colleagues introduced the use of gene fusions to define and test the function of signal sequences and to isolate mutants defective in protein secretion (Emr et al. 1981; Brickman et al. 1984). The initial studies in *E. coli* used fusions of the protein β-galactosidase to signal sequence containing proteins (Bassford et al. 1979; Emr et al. 1980). The fusion proteins were targeted to the export machinery of the bacterial membrane but were often arrested during the translocation process. In fact, arrested chimeric proteins interfered with the secretion of other proteins. The consequence of this was that the chimeric protein was lethal to the cell. This allowed the identification of two types of mutations: *cis* mutations in the signal sequence that abrogated targeting of the chimeric protein to the translocation machinery and extragenic mutations that defined the translocation machinery itself. Hybrid proteins were later introduced in yeast to identify nuclear localization signals and mitochondrial import signals. Gene fusions were also used to identify mutations in the ER translocation apparatus. Finally, mitochondrially targeted chimeric proteins that were arrested during the translocation process were instrumental in the studies that identified the mitochondrial import machinery. Therefore, the impact of the previous work from bacterial systems was vast.

NUCLEAR LOCALIZATION

The import of proteins into the nucleus differs from the above translocation processes because the proteins imported into the nucleus do not directly cross the lipid bilayers of the nuclear envelope, but pass through the nuclear pore, an aqueous channel through a large proteinaceous structure embedded in the nuclear envelope. Given the existence of the large nuclear pore complex, it was generally believed that all proteins freely diffused into the nucleoplasm with selective retention of nuclear proteins by binding to a nondiffusable substrate. Curiously, this retrospectively naïve model was accepted even though nuclear pores had been demonstrated to have an exclusion size limit of 70 kD, too small to accommodate diffusion of large nuclear proteins. In the early 1980s, the passive diffusion model was challenged and quickly replaced by the current model according to which nuclear proteins are selectively translocated across the nuclear envelope. The first major challenge to the passive diffusion model came from the Laskey lab. Dingwall et al. (1982) showed that a proteolytic fragment of nucleoplasmin, a large nuclear protein of the frog oocytes, contains a determinant that mediates nuclear entry, not nuclear retention. Such a determinant was a candidate for a nuclear localization signal (NLS) as stipulated by a selective uptake model. To identify a nuclear localization signal, **Hall et al. (1984)** used the hybrid protein approach that had proven to be so successful in the earlier studies on bacterial protein translocation. Hybrid proteins containing a variable amount of the regulatory protein Matα2 fused to a constant amount of active *E. coli* β-galactosidase were constructed. The fusion of the protein of interest, Matα2, to β-galactosidase had a particular advantage in this study. The large size of β-galactosidase (116 kD) precluded it from diffusing through the nuclear pore. Thus, nuclear localization of a Matα2–β-galactosidase hybrid would mean that the protein had been selectively translocated across the nuclear envelope. Furthermore, the β-galactosidase moiety provided an antigenic tag that allowed visualization of the hybrid protein by immunofluorescence on whole cells. Unlike proteins translocated into other organelles, nuclear proteins are not processed as part of the transport process. Assaying nuclear protein localization thus required direct visualization, rather than monitoring a molecular-weight change by one-dimensional gel electrophoresis. However, immunofluoresence had not yet been developed for yeast cells due to the yeast cell wall preventing entry of antibodies, and because yeast cells had been considered too small to be worthwhile subjects for subcellular visualization by light microscopy. Once the hybrid proteins had been constructed and a protocol for immunofluoresence on yeast cells had been worked out, it was demonstrated that β-galactosidase could indeed be targeted to the nucleus. The nuclear targeting of β-galactosidase was mediated by a short, basic amino acid sequence at the amino terminus of Matα2 (Hall et al. 1984) or by a second NLS in the internal region of Matα2 (Hall et al. 1990). This first identification of an NLS (Hall et al. 1984) established the selective uptake model of Dingwall et al. (1982), and quickly relegated the selective retention model to a faint memory. However, the impact of this paper goes beyond its characterization of a nuclear targeting signal. It was also among the first uses of gene fusion technology to gain new insights into protein targeting and membrane biogenesis in eukaryotic cells. Shortly thereafter, fusions were also used to study targeting to mitochondrial (Douglas et al. 1984) and ER membranes (Emr et al. 1984). Unfortunately, the import of the chimeric Matα2–β-galactosidase into the yeast nucleus did not lead to a phenotype that helped identify proteins involved in nuclear import, but it did serve to measure nuclear import. Most of the subsequent important discoveries from yeast concerning the mechanism of nuclear import came from other methods, especially the isolation of nuclear pore complexes (Hurt 1993; Strambio-de-Castillia et al. 1995).

Shortly after the identification of an NLS in yeast, work on SV40 large T-antigen mutations uncovered mutant T antigens that were defective in nuclear import (Kalderon et al. 1984a; Lanford and Butel 1984). Sequencing of these mutations identified a nuclear targeting signal composed largely of basic amino acids. The demonstration that a corresponding peptide could target an attached passenger protein to the nucleus confirmed the nuclear targeting function of the signal (Kalderon et al. 1984b; Goldfarb et al. 1986). Once the sequence of nucleoplasmin became available, it was shown to have two sequences that have homology with the SV40 large T-antigen NLS (Dingwall et al. 1987). Finally, studies visualizing nuclear import using electron microscopy showed that structures as large and inflexible as colloidal gold particles, when coated with a nuclear protein, can be imported into nuclei (Feldherr et al. 1984). This indicated that nuclear proteins do not need to be unfolded to cross the nuclear envelope and that the active transport channel across the nuclear pore complex is larger than the pore's passive diffusion channel. The nuclear pore complex contains a gated channel that is opened by an NLS.

ER TRANSLOCATION

As mentioned above, studies of protein translocation across bacterial membranes showed that translocation could be posttranslational, whereas translocation across dog pancreas microsomes was obligatorily cotranslational. This raised the possibility of a fundamental difference in mechanisms of protein translocation between prokaryotes and eukaryotes. This difference was rapidly dispelled, however, when prepro-α-factor was shown to be translocated across yeast membranes posttranslationally (Hansen et al. 1986; Rothblatt and Meyer 1986; Rothblatt et al. 1987). Therefore, it had to be concluded that both co- and posttranslational translocation were possible, depending on the particular protein. Indeed, some proteins, such as invertase, could not be posttranslationally translocated across yeast microsomes in vitro, but could be cotranslationally segregated in vivo (Hansen and Walter 1988). Therefore, by the identification of components involved in protein translocation into the ER in yeast, it would be possible to understand both co- and posttranslational translocation. What was needed was a selection scheme to isolate mutants defective in protein translocation into the ER. A strategy similar to that used in *E. coli* was attempted, but hybrid proteins containing an ER signal sequence fused to β-galactosidase were translocated into the ER without causing any apparent adverse effects (Emr et al. 1984). Therefore, a new strategy had to be devised. A clever new strategy, again involving the gene fusion approach, came from the Schekman lab. An endogenous cytoplasmic protein was coupled to an ER signal sequence. The cytoplasmic protein, encoding an essential enzyme in histidine biosynthesis, was efficiently targeted into the ER lumen where it was inactive. This allowed the isolation of mutants defective in protein translocation simply by selecting for growth of the appropriate histidine auxotroph on medium lacking histidine. An important aspect of this approach is that it allowed the isolation of conditional mutants altered in an essential process; a partial defect in translocation resulted in enough cytoplasmic enzyme for growth at the permissive temperature, whereas a complete or nearly complete defect at higher temperature resulted in lethality (**Deshaies and Schekman 1987**). The series of mutants isolated using this approach defined the components in the ER membrane that are required for co- and posttranslational translocation (Rothblatt et al. 1989). In particular, the major component of the ER translocation pore, Sec61, was identified using this method. Sec61, Sec62, and Sec63 were shown to form a protein complex (Deshaies et al. 1991). The identification of *SEC61* turned out to be a major breakthrough in the field, as it also led to the identification of the translocation pore in animal cells (Hartmann et al. 1994). Mammalian Sec62 and Sec63 also associate with Sec61 in mammalian cells (Meyer et al. 2000). The discovery of the nature of the translocation pore provided the final proof for the signal hypothesis. Studies on the yeast and mammalian Sec61 translocation pore eventually led to its structural characterization (Hanein et al. 1996; Menetret et al. 2000; Beckmann et al. 2001).

The availability of *sec61* mutant strains has also aided significantly in understanding the process of ER-associated protein degradation. Proteins that are inserted into the ER membrane or enter the ER lumen, but which are mutant proteins or do not fold properly, are exported from the ER back to the cytoplasm. In the cytoplasm, they are ubiquitinated and degraded by the proteasome. Mutant Sec61 in yeast was one of the first membrane protein substrates shown to be degraded by this pathway (Biederer et al. 1996). Importantly, *sec61* mutants, besides showing a defect in translocation into the ER, also show a defect in export of ER proteins that are substrates for degradation (Pilon et al. 1997; Plemper et al. 1997), implicating Sec61 in bidirectional transport (Romisch 1999).

IMPORT INTO MITOCHONDRIA

Yeast has had an important role in our understanding of protein translocation into nuclei and the ER, but studies in *Saccharomyces cerevisiae* (and to a somewhat lesser extent, *Neursopora crassa*) have dominated the field of mitochondrial protein import, leading to most of the major discoveries in this field. Therefore, because there are so many, it is quite difficult to choose landmark yeast papers in this field. Indeed, yeast has had a very prominent role in defining various aspects of the mitochondria (Ernster and Schatz 1981). It was shown early on that the vast majority of mitochondrial proteins are encoded in the nucleus, synthesized on cytoplasmic ribosomes, and then posttranslationally imported into the organelle (Harmey et al. 1977; Maccecchini et al. 1979). There were many suggestions that cotranslational import may also take place as cytoplasmic ribosomes were found bound to the surface of mitochondria (Kellems and Butow 1972; Kellems et al. 1974; Suissa and Schatz 1982), but it was clear that posttranslational import was possible for the majority of mitochondrial proteins.

Once nuclear genes encoding mitochondrial proteins were identified and cloned, it became possible to use a gene fusion strategy to characterize mitochondrial targeting sequences. Sequences of mitochondrial proteins were fused to β-galactosidase (Douglas et al. 1984; Emr et al. 1986), to dihydrofolate reductase (DHFR) (Hurt et al. 1984a,b), and to other proteins (Vestweber and Schatz 1988) for various purposes. Like proteins targeted into the ER, most mitochondrial proteins are targeted to the mitochondria by a cleavable, amino-terminal signal sequence (Maccecchini et al. 1979). Experiments using a gene fusion approach showed that the mitochondrial targeting sequence differs from the ER signal sequence. The ER signal sequence has a hydrophobic core, whereas the mitochondrial targeting sequence is characterized by the formation of an amphipathic helix with positively charged groups on one side of the helix and hydrophobic amino acids on the other (Allison and Schatz 1986; Roise et al. 1986). Gene fusions also turned out to be useful to produce pure substrates to analyze various aspects of mitochondrial import, including the energetics of the process (Pfanner and Neupert 1986; Eilers et al. 1987) and requirements for various protein factors and receptors (Pfanner et al. 1987). Mitochondrial signal sequences interact with a redundant set of receptors on the mitochondrial outer membrane to target proteins to the correct mitochondrial location (Kiebler et al. 1990; Baker and Schatz 1991). Some mitochondrial proteins do not contain cleavable signal sequences, but rather have internal signal sequences that are not removed (Hennig et al. 1983; Zwizinski and Neupert 1983).

One of the most striking differences between ER and mitochondrial import is that some ER proteins, unlike mitochondrial proteins, are obligatorily transported cotranslationally. An elaborate system recognizing a signal sequence on a nascent polypeptide and arresting translation until the translating complex is targeted to the ER for cotranslational translocation was described (Walter et al. 1984). These studies suggested that cells go to great lengths to ensure that protein translocation precedes protein folding, permitting translocation of an unfolded protein across the membrane. On the other hand, mitochondrial precursor proteins can be synthesized to completion in the cytoplasm and subsequently imported into mitochondria (Blobel et al. 1979; Reid and Schatz 1982; Teintze et al. 1982). These differences, and the fact that posttranslational translocation can also occur across the ER membrane in yeast, created great excitement at the time. The question arose as to whether proteins that are posttranslationally transported across membrane also traverse the membrane in an unfolded state and, if so, whether unfolding is required. To determine whether protein unfolding or a maintenance of an unfolded state could be necessary for posttranslational import into ER and mitochondria, the function of Hsp70 proteins in protein import was investigated, although at the time the precise function of Hsp70 in protein folding was not known. It is now known that Hsp70 is a chaperone that assists in the refolding of denatured proteins (Hartl and Martin 1995). Yeast has four genes encoding cytoplasmic Hsp70. Deletion of the two major genes, *SSA1* and *SSA2*, leads to a temperature-sensitive phenotype. Deletion of all four genes is lethal, thus necessitating construction of a conditional mutant (Werner-Washburne et al. 1987). It was found that depletion of Hsp70 has an adverse effect on import into both ER and mitochondria (Deshaies et al. 1988), suggesting a role for a protein chaperone in these processes. However, a more direct demonstration of a role for protein unfolding in mitochondrial import was soon to come.

The first demonstration that proteins traverse the two mitochondrial membranes in an unfolded state came from an elegant biochemical analysis of import by the Neupert lab (**Schleyer and Neupert 1985**). An import reaction was performed with isolated mitochondria and the precursor to the β subunit of the ATP synthase, in the presence of antibodies against the mature subunit. The bulky antibodies bound to a carboxy-terminal region of the β subunit and prevented its complete translocation into mitochondria. Importantly, the amino terminus of the precursor protein had reached the mitochondrial matrix, as indicated by the observation that it was processed by a protease in the mitochodrial matrix (Bohni et al. 1983; Schmidt et al. 1984), whereas the carboxyl terminus remained outside the mitochondrion as it was recognized by exogeneously added antibodies (Schleyer and Neupert 1985). This study proved that the precursor to the β subunit of the ATP synthase spanned both the inner and the outer mitochondrial membranes. The only way this could occur was if the precursor was highly extended and largely unfolded. Thus, this study showed that a precursor can span the membrane in a relatively unfolded state, but it did not show a requirement for unfolding, although this was suggested from the inhibition of import by antibodies.

The demonstration that protein unfolding is necessary for import into mitochondria came from a set of very elegant experiments described in another landmark paper (**Eilers and Schatz 1986**). Again, it involved a hybrid protein, this time of the mitochondrial targeting sequence of cytochrome oxidase subunit IV fused to DHFR. The antifolate drug methotrexate binds and stabilizes DHFR. Stabilization of the DHFR moiety of the hybrid protein with methotrexate inhibited import of the protein into mitochondria. At the same time, import of the authentic cytochrome *c* oxidase subunit IV was unaffected, ruling out a nonspecific effect of methotrexate. This provided the first direct evidence that unfolding of a mitochondrial precursor protein is required for its import.

The occurrence and requirement of protein unfolding for posttranslational import into mitochondria unified the mechanisms of co- and posttranslational import. Despite the great excitement at the time about the apparent differences between co- and posttranslational translocation, in both processes, a largely unfolded protein seems to be the translocation intermediate. In addition, in both processes, the unfolded protein traverses the membrane via a largely hydrophilic pore (Jensen and Kinnally 1997). This led investigators to look for the chaperones that keep precursor proteins in a soluble, unfolded state in the cytoplasm, and thus competent for subsequent import. As mentioned above, cytoplasmic Hsp70 is required for import of some precursors, but not all proteins require cytoplasmic chaperones for import (Gruhler et al. 1997). Chaperones were also proposed to act inside organelles where they help fold proteins once they are imported. The studies on chaperones inside mitochondria turned out to be particularly interesting.

An important conceptual aspect of cotranslational import was that protein synthesis could conceivably provide the energy required for the unidirectional translocation through an import or export channel. However, there were other possibile sources of energy for unidirectional transport. In mitochondria, one possible source of energy was the membrane potential that is required for translocation. Another was energy derived from protein refolding in the mitochondrial matrix. To test this latter possibility, the role of the mitochondrial Hsp70 protein (Ssc1) in protein import into the matrix compartment was tested. It was shown that Ssc1 was necessary for protein import and refolding inside the matrix (Kang et al. 1990; Ostermann et al. 1990; Scherer et al. 1990). Furthermore, an incompletely translocated precursor interacted with Ssc1, demonstrating that Ssc1 is engaged early in the translocation process and probably acts as a direct and crucial component of the translocation apparatus. Two basic mechanisms for the action of Ssc1 in import were proposed. In one model, Ssc1 drives translocation as a "by-product" of its protein-folding function. It would stimulate protein refolding on the matrix side of the mitochondria. Eventually, protein folding on the matrix side would favor translocation (Gaume et al. 1998). Another model suggests a more direct role for Ssc1. In this model, Ssc1 would act as an ATP-driven motor, latching onto the precursor in the matrix and translocating the precursor with a power stroke (Matouschek et al. 1997).

The finding that an extended, unfolded precursor protein is the substrate for import into mitochondria had tremendous impact on our understanding of protein translocation in general, but also provided a strategy to identify the first components of the mitochondrial import machinery. The reasoning was that if precursors span the mitochondrial inner and outer membranes and remain inside the import machinery, they should block the import of other precursor proteins. This was shown to be the case using an import-incompetent precursor protein with internal disulfide bridges (Vestweber and Schatz 1988), which then permitted the identification of one of the first components of the outer-membrane translocation site by cross-linking of a jammed precursor to its surrounding proteins. Tom40 (Isp42) was identified this way (Vestweber et al. 1989). Tom40 was later shown to be essential for import because depletion of this essential protein in vivo caused a block in import into mitochondria (Baker et al. 1990). Tom40 is part of a protein complex that also acts as a receptor that recognizes precursor proteins (Pfanner et al. 1991) and forms a cation-selective and voltage-gated channel in the mitochondrial outer membrane (Kunkele et al. 1998). Purification of the complex has allowed us to begin to understand its molecular architecture (Ahting et al. 1999), which can now also be compared to the Sec61 complex used in ER translocation. Both structures apparently have rather large cavities in the center of the complex, providing a large aqueous pore for the passage of proteins across membranes.

The Tom40 complex translocates precursors across the outer mitochondrial membrane, but some proteins must be translocated across the inner membrane as well. Experiments described above (Schleyer and Neupert 1985) showed that precursors can span both membranes simultaneously and are therefore most likely engaged in both translocation pores simultaneously. A component of the inner mitochondrial pore, Tim23, was identified (Bauer et al. 1996; Sirrenberg et al. 1996) and shown to be required for the multiple conductance channel (Lohret et al. 1997). Interestingly, Tom40 was found to cooperate functionally with the inner-membrane import machinery, in particular Tim23, and association of the outer- and inner-membrane pores could be seen during protein translocation (Sirrenberg et al. 1997; Rapaport et al. 1998; Schulke et al. 1999). Finally, Tim23 was shown to be anchored to the inner mitochondrial membrane by its carboxyl terminus and to have its amino terminus exposed on the outer surface of mitochondria, suggesting it tethers the inner- and outer-membrane translocases to each other (Donzeau et al. 2000). Tim23 interacts with another member of the Tim complex, Tim17, to form the protein import channel (Dekker et al. 1993; Ryan and Jensen 1993; Maarse et al. 1994; Ryan et al. 1994).

The Tim23 complex seems to be the major inner-membrane translocase for proteins being imported into the matrix compartment, but integral membrane proteins of the inner mitochondrial membrane are preferentially inserted into this membrane by another protein complex, called the Tim22 complex. This complex was first identified based on homology with Tim17 and Tim23 (Sirrenberg et al. 1996). Mutation of *TIM22* does not affect import in a general way, but has a specific effect on the import of inner-membrane proteins. Other components of this complex were then identified (Koehler 2000). In addition to the membrane components, soluble proteins of the intermembrane space are required for the import pathway into the mitochondrial inner membrane. These proteins, called the "tiny tims" due to their low molecular weight, have been proposed to act as chaperone-like molecules to guide precursors of inner-membrane proteins across the intermembrane space (Koehler 2000). Some of the major players of the mitochondrial import pathway discovered in yeast are shown in Figure 1.

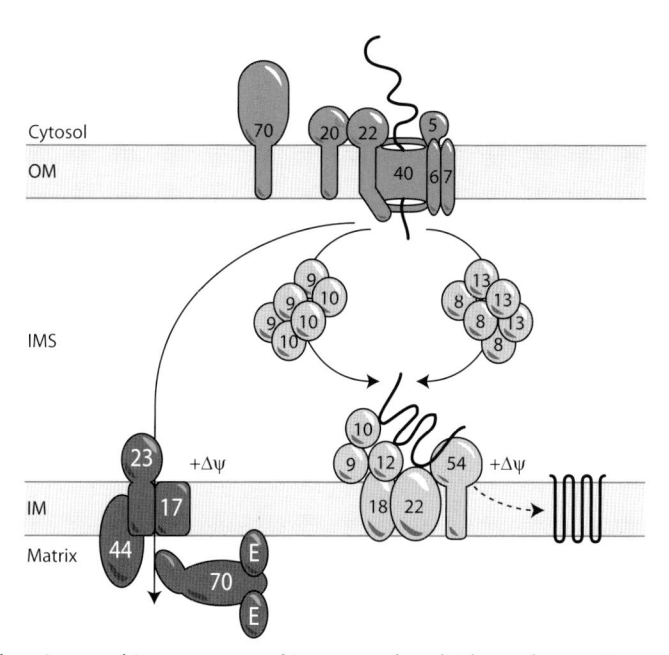

FIGURE 1. Translocation machinery across and into mitochondrial membranes. Precursors are recognized by receptors associated with the outer mitochondrial membrane translocase and are then shuttled to one of two different systems for further translocation. Translocation into the matrix compartment requires transfer of the precursor to the Tim23 complex. This complex can associate with the Tom complex. Mitochondrial Hsp70 requires ATP hydrolysis to translocate the precursor. An electrochemical gradient is also required. Transfer of inner-membrane protein precursors from the Tom complex to the Tim22 complex requires proteins of the inner-membrane space. Protein insertion into the inner membrane requires an electrochemical potential. The molecular weights indicated correspond also to the gene numbers of the components of the translocation machinery of the outer membrane (TOM) and inner membrane (TIM). This figure was kindly provided by Carla Koehler, University of California, Los Angeles.

SUMMARY

Yeast has contributed greatly to our understanding of protein translocation. Use of hybrid proteins and yeast genetics have been instrumental tools in coming to grips with how proteins cross membranes. In particular, the studies in yeast have allowed us to unify our concepts on how proteins cross lipid bilayers in a largely unfolded state and are refolded after translocation. Studies in yeast have been, by far, most important in our understanding of protein translocation into mitochondria. We have learned that nucleus-encoded mitochondrial proteins are often synthesized as larger precursor proteins with cleavable mitochondrial targeting sequences of an amphipathic nature. These signal sequences allow interaction with receptors on the mitochondrial surface and with the protein translocation machinery. The signal sequences probably have a role in gating the import channels. If the precursor protein had not been maintained in an unfolded state by chaperones, they must be unfolded for their subsequent translocation. Two forces could act to drive translocation of the precursor proteins, protein folding or an ATPase-dependent motor-like activity of a matrix chaperone. Import into mitochondria can require the translocation through the outer membrane, followed by insertion into or translocation through the inner membrane. Different protein complexes are involved in the two latter alternatives.

ACKNOWLEDGMENTS

The author thanks Carla Koehler for providing Figure 1, Jeff Schatz for his inspiration in the translocation field, and the Swiss National Science Foundation for support.

STUDY QUESTIONS

1. Even though we now know that proteins are unfolded for translocation across membranes, there must be a fundamental difference between the obligatory cotranslational mechanism common for the ER and import into mitochondria. What is this fundamental difference? Why do you suppose this difference has evolved?

2. Chaperones have an important role in import into ER and mitochondria. One of these roles is probably common and the other is different. Explain.

3. How would you use a gene fusion approach to determine whether proteins of peroxisomes must be unfolded for translocation?

4. Although genetics aided in the discovery of some components of the ER and mitochondrial translocation machinery, further progress and characterization of the translocation machinery depended largely on other methodology. What methodology was used? How was it combined with genetic techniques?

5. What supplies the energy for protein translocation across the ER membrane?

REFERENCES

Ahting U., Thun C., Hegerl R., Typke D., Nargang F.E., Neupert W., and Nussberger S. 1999. The TOM core complex: The general protein import pore of the outer membrane of mitochondria. *J. Cell Biol.* **147:** 959–968.

Allison D.S. and Schatz G. 1986. Artificial mitochondrial presequences. *Proc. Natl. Acad. Sci.* **83:** 9011–9015.
Baker K.P. and Schatz G. 1991. Mitochondrial proteins essential for viability mediate protein import into yeast mitochondria. *Nature* **349:** 205–208.
Baker K.P., Schaniel A., Vestweber D., and Schatz G. 1990. A yeast mitochondrial outer membrane protein essential for protein import and cell viability. *Nature* **348:** 605–609.
Bassford P.J., Jr., Silhavy T.J., and Beckwith J.R. 1979. Use of gene fusion to study secretion of maltose-binding protein into *Escherichia coli* periplasm. *J. Bacteriol.* **139:** 19–31.
Bauer M.F., Sirrenberg C., Neupert W., and Brunner M. 1996. Role of Tim23 as voltage sensor and presequence receptor in protein import into mitochondria. *Cell* **87:** 33–41.
Beckmann R., Spahn C.M., Eswar N., Helmers J., Penczek P.A., Sali A., Frank J., and Blobel G. 2001. Architecture of the protein-conducting channel associated with the translating 80S ribosome. *Cell* **107:** 361–372.
Biederer T., Volkwein C., and Sommer T. 1996. Degradation of subunits of the Sec61p complex, an integral component of the ER membrane, by the ubiquitin-proteasome pathway. *EMBO J.* **15:** 2069–2076.
Blobel G. 1980. Intracellular protein topogenesis. *Proc. Natl. Acad. Sci.* **77:** 1496–1500.
Blobel G. and Dobberstein B. 1975a. Transfer of proteins across membranes. I. Presence of proteolytically processed and unprocessed nascent immunoglobulin light chains on membrane-bound ribosomes of murine myeloma. *J. Cell Biol.* **67:** 835–851.
———. 1975b. Transfer to proteins across membranes. II. Reconstitution of functional rough microsomes from heterologous components. *J. Cell Biol.* **67:** 852–862.
Blobel G. and Sabatini D.D. 1971. Ribosome-membrane interaction in eukaryotic cells. In *Biomembranes* (ed. L.A. Manson), vol. 2, pp. 193–195. Plenum Press, New York.
Blobel G., Walter P., Chang C.N., Goldman B.M., Erickson A.H., and Lingappa V.R. 1979. Translocation of proteins across membranes: The signal hypothesis and beyond. *Symp. Soc. Exp. Biol.* **33:** 9–36.
Bohni P.C., Daum G., and Schatz G. 1983. Import of proteins into mitochondria. Partial purification of a matrix-located protease involved in cleavage of mitochondrial precursor polypeptides. *J. Biol. Chem.* **258:** 4937–4943.
Brickman E.R., Oliver D.B., Garwin J.L., Kumamoto C., and Beckwith J. 1984. The use of extragenic suppressors to define genes involved in protein export in *Escherichia coli*. *Mol. Gen. Genet.* **196:** 24–27.
Date T., Goodman J.M., and Wickner W.T. 1980. Procoat, the precursor of M13 coat protein, requires an electrochemical potential for membrane insertion. *Proc. Natl. Acad. Sci.* **77:** 4669–4673.
Dekker P.J., Keil P., Rassow J., Maarse A.C., Pfanner N., and Meijer M. 1993. Identification of MIM23, a putative component of the protein import machinery of the mitochondrial inner membrane. *FEBS Lett.* **330:** 66–70.
Deshaies R.J. and Schekman R. 1987. A yeast mutant defective at an early stage in import of secretory protein precursors into the endoplasmic reticulum. *J. Cell Biol.* **105:** 633–645.
Deshaies R.J., Sanders S.L., Feldheim D.A., and Schekman R. 1991. Assembly of yeast Sec proteins involved in translocation into the endoplasmic reticulum into a membrane-bound multisubunit complex. *Nature* **349:** 806–808.
Deshaies R.J., Koch B.D., Werner-Washburne M., Craig E.A., and Schekman R. 1988. A subfamily of stress proteins facilitates translocation of secretory and mitochondrial precursor polypeptides. *Nature* **332:** 800–805.
Dingwall C., Sharnick S.V., and Laskey R.A. 1982. A polypeptide domain that specifies migration of nucleoplasmin into the nucleus. *Cell* **30:** 449–458.
Dingwall C., Dilworth S.M., Black S.J., Kearsey S.E., Cox L.S., and Laskey R.A. 1987. Nucleoplasmin cDNA sequence reveals polyglutamic acid tracts and a cluster of sequences homologous to putative nuclear localization signals. *EMBO J.* **6:** 69–74.
Donzeau M., Kaldi K., Adam A., Paschen S., Wanner G., Guiard B., Bauer M.F., Neupert W., and Brunner M. 2000. Tim23 links the inner and outer mitochondrial membranes. *Cell* **101:** 401–412.
Douglas M.G., Geller B.L., and Emr S.D. 1984. Intracellular targeting and import of an F1-ATPase beta-subunit-beta-galactosidase hybrid protein into yeast mitochondria. *Proc. Natl. Acad. Sci.* **81:** 3983–3987.
Eilers M. and Schatz G. 1986. Binding of a specific ligand inhibits import of a purified precursor protein into mitochondria. *Nature* **322:** 228–232.
Eilers M., Oppliger W., and Schatz G. 1987. Both ATP and an energized inner membrane are required to import

a purified precursor protein into mitochondria. *EMBO J.* **6:** 1073–1077.
Emr S.D., Hall M.N., and Silhavy T.J. 1980. A mechanism of protein localization: The signal hypothesis and bacteria. *J. Cell Biol.* **86:** 701–711.
Emr S.D., Hanley-Way S., and Silhavy T.J. 1981. Suppressor mutations that restore export of a protein with a defective signal sequence. *Cell* **23:** 79–88.
Emr S.D., Schauer I., Hansen W., Esmon P., and Schekman R. 1984. Invertase beta-galactosidase hybrid proteins fail to be transported from the endoplasmic reticulum in *Saccharomyces cerevisiae*. *Mol. Cell. Biol.* **4:** 2347–2355.
Emr S.D., Vassarotti A., Garrett J., Geller B.L., Takeda M., and Douglas M.G. 1986. The amino terminus of the yeast F1-ATPase beta-subunit precursor functions as a mitochondrial import signal. *J. Cell Biol.* **102:** 523–533.
Ernster L. and Schatz G. 1981. Mitochondria: A historical review. *J. Cell Biol.* **91:** 227s–255s.
Feldherr C.M., Kallenbach E., and Schultz N. 1984. Movement of a karyophilic protein through the nuclear pores of oocytes. *J. Cell Biol.* **99:** 2216–2222.
Gaume B., Klaus C., Ungermann C., Guiard B., Neupert W., and Brunner M. 1998. Unfolding of preproteins upon import into mitochondria. *EMBO J.* **17:** 6497–6507.
Goldfarb D.S., Gariepy J., Schoolnik G., and Kornberg R.D. 1986. Synthetic peptides as nuclear localization signals. *Nature* **322:** 641–644.
Gruhler A., Arnold I., Seytter T., Guiard B., Schwarz E., Neupert W., and Stuart R.A. 1997. N-terminal hydrophobic sorting signals of preproteins confer mitochondrial hsp70 independence for import into mitochondria. *J. Biol. Chem.* **272:** 17410–17415.
Hall M.N., Craik C., and Hiraoka Y. 1990. Homeodomain of yeast repressor alpha 2 contains a nuclear localization signal. *Proc. Natl. Acad. Sci.* **87:** 6954–6958.
Hall M.N., Hereford L., and Herskowitz I. 1984. Targeting of *E. coli* β-galactosidase to the nucleus in yeast. *Cell* **36:** 1057–1065.
Hanein D., Matlack K.E., Jungnickel B., Plath K., Kalies K.U., Miller K.R., Rapoport T.A., and Akey C.W. 1996. Oligomeric rings of the Sec61p complex induced by ligands required for protein translocation. *Cell* **87:** 721–732.
Hansen W. and Walter P. 1988. Prepro-carboxypeptidase Y and a truncated form of pre-invertase, but not full-length pre-invertase, can be posttranslationally translocated across microsomal vesicle membranes from *Saccharomyces cerevisiae*. *J. Cell Biol.* **106:** 1075–1081.
Hansen W., Garcia P.D., and Walter P. 1986. In vitro protein translocation across the yeast endoplasmic reticulum: ATP-dependent posttranslational translocation of the prepro-alpha-factor. *Cell* **45:** 397–406.
Harmey M.A., Hallermayer G., Korb H., and Neupert W. 1977. Transport of cytoplasmically synthesized proteins into the mitochondria in a cell free system from *Neurospora crassa*. *Eur. J. Biochem.* **81:** 533–544.
Hartl F.U. and Martin J. 1995. Molecular chaperones in cellular protein folding. *Curr. Opin. Struct. Biol.* **5:** 92–102.
Hartmann E., Sommer T., Prehn S., Gorlich D., Jentsch S., and Rapoport T.A. 1994. Evolutionary conservation of components of the protein translocation complex. *Nature* **367:** 654–657.
Hennig B., Koehler H., and Neupert W. 1983. Receptor sites involved in posttranslational transport of apocytochrome c into mitochondria: Specificity, affinity, and number of sites. *Proc. Natl. Acad. Sci.* **80:** 4963–4967.
Hurt E.C. 1993. The nuclear pore complex. *FEBS Lett.* **325:** 76–80.
Hurt E.C., Pesold-Hurt B., and Schatz G. 1984a. The amino-terminal region of an imported mitochondrial precursor polypeptide can direct cytoplasmic dihydrofolate reductase into the mitochondrial matrix. *EMBO J.* **3:** 3149–3156.
———. 1984b. The cleavable prepiece of an imported mitochondrial protein is sufficient to direct cytosolic dihydrofolate reductase into the mitochondrial matrix. *FEBS Lett.* **178:** 306–310.
Ito K., Mandel G., and Wickner W. 1979. Soluble precursor of an integral membrane protein: Synthesis of procoat protein in *Escherichia coli* infected with bacteriophage M13. *Proc. Natl. Acad. Sci.* **76:** 1199–1203.
Jensen R.E. and Kinnally K.W. 1997. The mitochondrial protein import pathway: Are precursors imported through membrane channels? *J. Bioenerg. Biomembr.* **29:** 3–10.
Josefsson L.G., Hardy S., Harayama S., and Randall L. 1982. Studies on the export of the maltose-binding protein and the LamB protein. *Ann. Microbiol.* **133A:** 111–114.
Kalderon D., Richardson W.D., Markham A.F., and Smith A.E. 1984a. Sequence requirements for nuclear loca-

tion of simian virus-40 large T-antigen. *Nature* **311:** 33–38.
Kalderon D., Roberts B.L., Richardson W.D., and Smith A.E. 1984b. A short amino-acid sequence able to specify nuclear location. *Cell* **39:** 499–509.
Kang P.J., Ostermann J., Shilling J., Neupert W., Craig E.A., and Pfanner N. 1990. Requirement for hsp70 in the mitochondrial matrix for translocation and folding of precursor proteins. *Nature* **348:** 137–143.
Kellems R.E. and Butow R.A. 1972. Cytoplasmic-type 80 S ribosomes associated with yeast mitochondria. I. Evidence for ribosome binding sites on yeast mitochondria. *J. Biol. Chem.* **247:** 8043–8050.
Kellems R.E., Allison V.F., and Butow R.A. 1974. Cytoplasmic type 80 S ribosomes associated with yeast mitochondria. II. Evidence for the association of cytoplasmic ribosomes with the outer mitochondrial membrane in situ. *J. Biol. Chem.* **249:** 3297–3303.
Kiebler M., Pfaller R., Sollner T., Griffiths G., Horstmann H., Pfanner N., and Neupert W. 1990. Identification of a mitochondrial receptor complex required for recognition and membrane insertion of precursor proteins. *Nature* **348:** 610–616.
Koehler C.M. 2000. Protein translocation pathways of the mitochondrion. *FEBS Lett.* **476:** 27–31.
Kunkele K.P., Juin P., Pompa C., Nargang F.E., Henry J.P., Neupert W., Lill R., and Thieffry M. 1998. The isolated complex of the translocase of the outer membrane of mitochondria. Characterization of the cation-selective and voltage-gated preprotein-conducting pore. *J. Biol. Chem.* **273:** 31032–31039.
Lanford R.E. and Butel J.S. 1984. Construction and characterization of an SV40 mutant defective in nuclear transport of T-antigen. *Cell* **37:** 801–813.
Lohret T.A., Jensen R.E., and Kinnally K.W. 1997. Tim23, a protein import component of the mitochondrial inner membrane, is required for normal activity of the multiple conductance channel, MCC. *J. Cell Biol.* **137:** 377–386.
Maarse A.C., Blom J., Keil P., Pfanner N., and Meijer M. 1994. Identification of the essential yeast protein MIM17, an integral mitochondrial inner membrane protein involved in protein import. *FEBS Lett.* **349:** 215–221.
Maccecchini M.L., Rudin Y., Blobel G., and Schatz G. 1979. Import of proteins into mitochondria: Precursor forms of the extramitochondrially made F1-ATPase subunits in yeast. *Proc. Natl. Acad. Sci.* **76:** 343–347.
Matouschek A., Azem A., Ratliff K., Glick B.S., Schmid K., and Schatz G. 1997. Active unfolding of precursor proteins during mitochondrial protein import. *EMBO J.* **16:** 6727–6736.
Menetret J.F., Neuhof A., Morgan D.G., Plath K., Radermacher M., Rapoport T.A., and Akey C.W. 2000. The structure of ribosome-channel complexes engaged in protein translocation. *Mol. Cell* **6:** 1219–1232.
Meyer H.A., Grau H., Kraft R., Kostka S., Prehn S., Kalies K.U., and Hartmann E. 2000. Mammalian Sec61 is associated with Sec62 and Sec63. *J. Biol. Chem.* **275:** 14550–14557.
Milstein C., Brownlee G.G., Harrison T.M., and Mathews M.B. 1972. A possible precursor of immunoglobulin light chains. *Nat. New Biol.* **239:** 117–120.
Ostermann J., Voos W., Kang P.J., Craig E.A., Neupert W., and Pfanner N. 1990. Precursor proteins in transit through mitochondrial contact sites interact with hsp70 in the matrix. *FEBS Lett.* **277:** 281–284.
Palade G. 1975. Intracellular aspects of the process of protein synthesis. *Science* **189:** 347–358.
Pfanner N. and Neupert W. 1986. Transport of F1-ATPase subunit beta into mitochondria depends on both a membrane potential and nucleoside triphosphates. *FEBS Lett.* **209:** 152–156.
Pfanner N., Sollner T., and Neupert W. 1991. Mitochondrial import receptors for precursor proteins. *Trends Biochem. Sci.* **16:** 63–67.
Pfanner N., Muller H.K., Harmey M.A., and Neupert W. 1987. Mitochondrial protein import: Involvement of the mature part of a cleavable precursor protein in the binding to receptor sites. *EMBO J.* **6:** 3449–3454.
Pilon M., Schekman R., and Romisch K. 1997. Sec61p mediates export of a misfolded secretory protein from the endoplasmic reticulum to the cytosol for degradation. *EMBO J.* **16:** 4540–4548.
Plemper R.K., Bohmler S., Bordallo J., Sommer T., and Wolf D.H. 1997. Mutant analysis links the translocon and BiP to retrograde protein transport for ER degradation. *Nature* **388:** 891–895.
Rapaport D., Kunkele K.P., Dembowski M., Ahting U., Nargang F.E., Neupert W., and Lill R. 1998. Dynamics of the TOM complex of mitochondria during binding and translocation of preproteins. *Mol. Cell. Biol.* **18:** 5256–5262.
Reid G.A. and Schatz G. 1982. Import of proteins into mitochondria. Extramitochondrial pools and post-translational import of mitochondrial protein precursors in vivo. *J. Biol. Chem.* **257:** 13062–13067.

Roise D., Horvath S.J., Tomich J.M., Richards J.H., and Schatz G. 1986. A chemically synthesized pre-sequence of an imported mitochondrial protein can form an amphiphilic helix and perturb natural and artificial phospholipid bilayers. *EMBO J.* **5:** 1327–1334.

Romisch K. 1999. Surfing the Sec61 channel: Bidirectional protein translocation across the ER membrane. *J. Cell Sci.* **112:** 4185–4191.

Rothblatt J.A. and Meyer D.I. 1986. Secretion in yeast: Reconstitution of the translocation and glycosylation of alpha-factor and invertase in a homologous cell-free system. *Cell* **44:** 619–628.

Rothblatt J.A., Webb J.R., Ammerer G., and Meyer D.I. 1987. Secretion in yeast: Structural features influencing the post-translational translocation of prepro-alpha-factor in vitro. *EMBO J.* **6:** 3455–3463.

Rothblatt J.A., Deshaies R.J., Sanders S.L., Daum G., and Schekman R. 1989. Multiple genes are required for proper insertion of secretory proteins into the endoplasmic reticulum in yeast. *J. Cell Biol.* **109:** 2641–2652.

Ryan K.R. and Jensen R.E. 1993. Mas6p can be cross-linked to an arrested precursor and interacts with other proteins during mitochondrial protein import. *J. Biol. Chem.* **268:** 23743–23746.

Ryan K.R., Menold M.M., Garrett S., and Jensen R.E. 1994. SMS1, a high-copy suppressor of the yeast mas6 mutant, encodes an essential inner membrane protein required for mitochondrial protein import. *Mol. Biol. Cell* **5:** 529–538.

Scherer P.E., Krieg U.C., Hwang S.T., Vestweber D., and Schatz G. 1990. A precursor protein partly translocated into yeast mitochondria is bound to a 70 kd mitochondrial stress protein. *EMBO J.* **9:** 4315–4322.

Schleyer M. and Neupert W. 1985. Transport of proteins into mitochondria: Translocational intermediates spanning contact sites between outer and inner membranes. *Cell* **43:** 339–350.

Schmidt B., Wachter E., Sebald W., and Neupert W. 1984. Processing peptidase of *Neurospora* mitochondria. Two-step cleavage of imported ATPase subunit 9. *Eur. J. Biochem.* **144:** 581–588.

Schulke N., Sepuri N.B., Gordon D.M., Saxena S., Dancis A., and Pain D. 1999. A multisubunit complex of outer and inner mitochondrial membrane protein translocases stabilized in vivo by translocation intermediates. *J. Biol. Chem.* **274:** 22847–22854.

Sirrenberg C., Bauer M.F., Guiard B., Neupert W., and Brunner M. 1996. Import of carrier proteins into the mitochondrial inner membrane mediated by Tim22. *Nature* **384:** 582–585.

Sirrenberg C., Endres M., Becker K., Bauer M.F., Walther E., Neupert W., and Brunner M. 1997. Functional cooperation and stoichiometry of protein translocases of the outer and inner membranes of mitochondria. *J. Biol. Chem.* **272:** 29963–29966.

Strambio-de-Castillia C., Blobel G., and Rout M.P. 1995. Isolation and characterization of nuclear envelopes from the yeast *Saccharomyces*. *J. Cell Biol.* **131:** 19–31.

Suissa M. and Schatz G. 1982. Import of proteins into mitochondria. Translatable mRNAs for imported mitochondrial proteins are present in free as well as mitochondria-bound cytoplasmic polysomes. *J. Biol. Chem.* **257:** 13048–13055.

Teintze M., Slaughter M., Weiss H., and Neupert W. 1982. Biogenesis of mitochondrial ubiquinol:cytochrome c reductase (cytochrome bc1 complex). Precursor proteins and their transfer into mitochondria. *J. Biol. Chem.* **257:** 10364–10371.

Vestweber D. and Schatz G. 1988. A chimeric mitochondrial precursor protein with internal disulfide bridges blocks import of authentic precursors into mitochondria and allows quantitation of import sites. *J. Cell Biol.* **107:** 2037–2043.

Vestweber D., Brunner J., Baker A., and Schatz G. 1989. A 42K outer-membrane protein is a component of the yeast mitochondrial protein import site. *Nature* **341:** 205–209.

Walter P., Gilmore R., and Blobel G. 1984. Protein translocation across the endoplasmic reticulum. *Cell* **38:** 5–8.

Werner-Washburne M., Stone D.E., and Craig E.A. 1987. Complex interactions among members of an essential subfamily of hsp70 genes in *Saccharomyces cerevisiae*. *Mol. Cell. Biol.* **7:** 2568–2577.

Zwizinski C. and Neupert W. 1983. Precursor proteins are transported into mitochondria in the absence of proteolytic cleavage of the additional sequences. *J. Biol. Chem.* **258:** 13340–13346.

14

Ubiquitination and Protein Turnover

Mark Hochstrasser
Yale University
Department of Molecular Biophysics and Biochemistry
New Haven, Connecticut 06520-8114

Proteolytic targeting by specific protein signals, 269
Bachmair A., Finley D., and Varshavsky A. 1986. In vivo half-life of a protein is a function of its amino-terminal residue. *Science* **234:** 179–186.

Targeting of substrates across membranes to the proteasome, 273
Hiller M.M., Finger A., Schweiger M., and Wolf D.H. 1996. ER degradation of a misfolded luminal protein by the cytosolic ubiquitin-proteasome pathway. *Science* **273:** 1725–1728.

Beyond the proteasome: Ubiquitin as a signal for endocytosis, 274
Kölling R. and Hollenberg C.P. 1994. The ABC-transporter Ste6 accumulates in the plasma membrane in a ubiquitinated form in endocytosis mutants. *EMBO J.* **13:** 3261–3271.

Ubiquitin-dependent proteolysis and the cell cycle, 276
Schwob E., Böhm T., Mendenhall M.D., and Nasmyth K. 1994. The B-type cyclin kinase inhibitor p40^{SIC1} controls the G1 to S transition in *S. cerevisiae*. *Cell* **79:** 233–244.

Ubiquitin-like proteins: A paradigm extended, 277
Mizushima N., Noda T., Yoshimori T., Tanaka Y., Ishii T., George M.D., Klionsky D.J., Ohsumi M., and Ohsumi Y. 1998. A protein conjugation system essential for autophagy. *Nature* **395:** 395–398.

Note: The landmark papers listed above are those discussed in this chapter. Each landmark paper is preceded by the name of the section (with starting page number) where the paper is first discussed in detail.

UBIQUITIN IS A 76-RESIDUE POLYPEPTIDE WITH AN IMPACT ON cell physiology that belies its small size. It is found either free or covalently joined through its carboxyl terminus to a variety of cytoplasmic, nuclear, and membrane proteins. The best-defined function of ubiquitin, although by no means the only one, is to direct substrate proteins to their destruction by the 26S proteasome, a large multisubunit proteolytic complex. We are now also aware of a broader set of structurally related proteins—the ubiquitin-like proteins or Ubls—that are also covalently ligated to and removed from other macromolecules by specific enzymatic pathways. These enzymes are generally similar in both mechanism and sequence to those that act on ubiquitin. Although research on the ubiquitin-protein modification and degradation system began with studies on mammalian cells and cell extracts, it has been in the yeast *Saccharomyces cerevisiae* where many of the fundamental molecular attributes of the ubiquitin system have been determined and where its extraordinary functional diversity was first fully appreciated. The purpose of this retrospective is to highlight some of the key advances in our understanding of the ubiquitin system that were gained using *S. cerevisiae*.

A BRIEF HISTORY OF EARLY WORK ON UBIQUITIN LIGATION AND PROTEIN DEGRADATION

Reversible covalent attachment of one intracellular protein to another is now regarded as commonplace, but a quarter-century ago, this was not the case. Ubiquitin was first identified in 1975 in Gideon Goldstein's laboratory as a phylogenetically widespread protein with supposed immunomodulatory activity (it was original christened UBIP, for ubiquitous immunopoietic polypeptide) (Goldstein et al. 1975; Schlesinger et al. 1975). A year later, Harris Busch and colleagues (Olson et al. 1976) determined the peptide sequence of several stretches of a "histone-like" chromatin protein called A24 and found that the A24 sequences corresponded to segments of histone H2A and an unrelated protein that was subsequently recognized as ubiquitin (Hunt and Dayhoff 1977). It was Busch's group who made the remarkable finding that the two proteins were covalently connected between the carboxy-terminal carboxyl group of ubiquitin and a specific lysine side chain of the histone to form a Y-shaped molecule with an "isopeptide" linkage (Goldknopf and Busch 1977). Several years later, Avram Hershko and Irwin Rose and their colleagues independently discovered covalent protein modification by a heat-stable protein they termed APF-1 (for ATP-dependent proteolysis factor-1), which was required for protein degradation in a fractionated rabbit reticulocyte lysate (Ciechanover et al. 1980; Hershko et al. 1980). It was soon realized that APF-1 and ubiquitin were one and the same (Wilkinson et al. 1980). This was an important conclusion since it immediately suggested that the proteolytic pathway being studied in rabbit reticulocytes was widespread.

Biochemical characterization of the ubiquitin ligation system in reticulocyte extracts by Hershko, Rose, and co-workers quickly led to the basic outline of the pathway leading from ubiquitin activation to substrate ligation (Fig. 1) (Haas et al. 1982; Hershko et al. 1983). ATP-dependent ubiquitin ligation was demonstrated to be essential for protein degradation in these lysates. Early data also indicated that the conjugation of ubiquitin to proteins was reversed by specific enzymes (Matsui et al. 1982); however, it would be years before the gene for a deubiquitinating enzyme (DUB) was cloned—from *S. cerevisiae* (Miller et al. 1989). The proteolytic component of the ubiquitin-proteasome pathway initially proved to be more difficult to isolate and characterize than the ubiquitination enzymes (Hough et al. 1986). In retrospect, this is easy to understand: The protease, now generally called the 26S proteasome or simply the proteasome, is an approximately 2.5-MD complex of at least 32 distinct polypeptides that must be bathed in ATP in order to be maintained in a form capable of degrading ubiquitin-protein conjugates (DeMartino and Slaughter 1999).

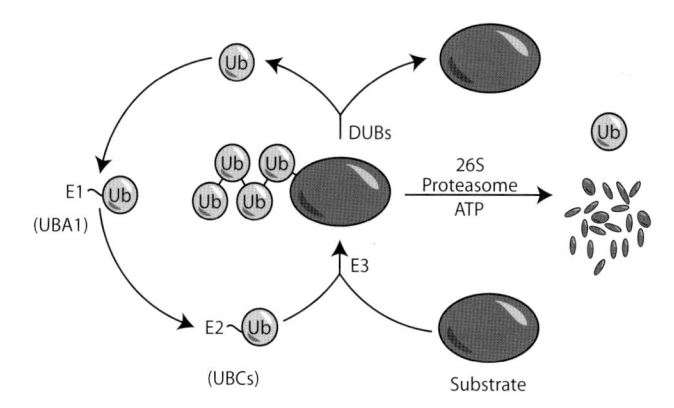

FIGURE 1. Outline of the ubiquitin-proteasome system.

Although the spotlight was focused on ubiquitin-protein ligation and its function in proteasome-mediated proteolysis, reports of ubiquitin-related polypeptides (Ubls) of uncertain function slowly began appearing in the literature. The first was an interferon-inducible protein, ISG15, which was discovered by Art Haas and colleagues to be conjugated, analogously to ubiquitin, to cellular proteins (Haas et al. 1987; Loeb and Haas 1992). With the massive genomic-scale DNA sequencing beginning in the 1990s, additional putative Ubls were soon uncovered. Subsequent analyses revealed that several were indeed also ligated to other proteins. At present, more than a dozen proven or putative Ubls, i.e., ubiquitin-related proteins that can be ligated to target molecules, are known (Hochstrasser 2000; Jentsch and Pyrowolakis 2000). Budding yeast has been a major experimental organism for the discovery and functional analysis of many of these Ubls, even though there are fewer such proteins in this species.

THE CONTRIBUTION OF YEAST GENETICS

Five studies using *S. cerevisiae* as the primary model have been picked based on their impact—not necessarily immediately felt—on our general understanding of the mechanisms or functions of the ubiquitin system. Although the studies in yeast were done after the original discovery of ubiquitin conjugation in other systems, it is fair to say that molecular genetic studies in yeast have been among the most seminal in advancing our understanding of ubiquitin-dependent molecular mechanisms and in vivo function. In an attempt to mirror the molecular diversity of the ubiquitin/Ubl system and its broad involvement in many different aspects of cell function, I have also tried in my selections to cover a wide array of physiological pathways on which the system impinges. These range from the regulation of cell differentiation and the cell cycle to the control of membrane protein endocytosis and autophagy.

PROTEOLYTIC TARGETING BY SPECIFIC PROTEIN SIGNALS

Following their groundbreaking discovery that ubiquitin-protein ligation is necessary for the bulk of intracellular protein degradation in mammalian cells and is essential for cell cycle progression (Ciechanover et al. 1984; Finley et al. 1984), Alex Varshavsky and colleagues turned to yeast to facilitate a deeper mechanistic analysis of these ubiquitin-dependent processes. A fundamental question that was unresolved at that time (and is still a very active area to this day) was exactly what structural features rendered a protein a target of ubiquitin-dependent proteolysis. By the mid 1980s, Hershko et al. had defined the key enzymatic steps required for protein ubiquitination (Hershko et al. 1983, 1986), but the exact substrate characteristics by which proteolytic targets were recognized were unknown. A variety of hypotheses had been put forward for the causes of such susceptibility, such as large protein size, aggregation, or amino acid oxidation. However, the data were largely correlative and no well-defined "signal" for degradation was generally accepted. An intriguing early observation made by Hershko and Rose was that a free α-amino group on the amino-terminal residue of a model substrate protein was necessary for its degradation in a fractionated reticulocyte lysate (Hershko et al. 1984). One interpretation of this result was that ubiquitin conjugation was initiated by ligation of ubiquitin to the α-amino group, followed by progressive ubiquitination of internal lysine side chains. At that time, most multiubiquitination of proteolytic substrates was thought to reflect the addition of single ubiquitin molecules to multiple lysines.

Bachmair et al. (1986) decided to test the idea of amino-terminal ubiquitination as an initiating signal for degradation by creating a translational fusion between yeast ubiquitin (the genes for which they had recently cloned) and the popular reporter protein, *Escherichia coli* β-galactosidase (β-gal). The prediction was that β-gal synthesized in yeast with an amino-terminal ubiquitin would be more rapidly degraded than β-gal without it. As it happened, the experiment could not be done as originally sketched because DUBs in the yeast cells rapidly clipped off the ubiquitin. In an effort to thwart this potent DUB activity, the authors decided to mutate the normally amino-terminal methionine residue of β-gal that linked it to the upstream ubiquitin moiety. By changing the methionine to other residues, they hoped to find a fusion in which the ubiquitin would persist on the amino terminus and possibly initiate proteolysis. Instead, the authors found that, with one exception, the yeast DUBs continued to cleave the ubiquitin from the altered β-gal residue. To their surprise, however, the steady-state level of the released β-gal varied enormously, depending on the identity of the amino-terminal residue. Pulse-chase experiments showed that these differences in levels were due to vastly different rates of proteolysis. For example, an arginine at the amino terminus resulted in a β-gal with a half-life of just a few minutes, whereas an otherwise identical protein with a methionine at this position persisted with a half-life of more than 20 hours. Bachmair et al. (1986) termed this relationship between amino-terminal amino acid identity and degradation rate the "N-End Rule."

In their paper, it was shown that the degradation rates also correlated with ubiquitination levels, with the β-gals that were most rapidly degraded being those most heavily ubiquitinated. Thus, ubiquitin-dependent degradation depended on an extremely simple determinant in the protein, namely, a particular type of amino acid residue at the very amino terminus. Of the 20 standard amino acids, 12 caused rapid degradation in yeast, 7 did not, and a proline residue at the junction between ubiquitin and β-gal inhibited deubiquitination. Thus, the ubiquitin–Pro–β-gal fusion represented the type of fusion Bachmair et al. (1986) had originally tried to create, and the fusion was far less metabolically stable than its unfused β-gal counterpart. However, given the rapid deubiquitination that occurs with most residues at the amino terminus of β-gal, it was not clear whether the α-amino group could serve as a posttranslational target for ubiquitin ligation, as originally envisaged. For the N-end rule substrates, it was later shown that specific internal lysines near the amino terminus were the sites of ubiquitin attachment; at least one of these lysines had to be present for degradation to take place (Bachmair and Varshavsky 1989; Chau et al. 1989). Thus, the amino-terminal degradation signal consisted of two determinants: a destabilizing amino-terminal residue and an accessible internal lysine residue.

The Bachmair et al. (1986) paper was important for many reasons. First, it offered the first evidence for a genetically encoded degradation determinant that could stimulate eukaryotic protein degradation. Ironically, demonstrating the physiological significance of the N-end recognition mechanism for degradation of naturally short-lived cellular substrates proved to be unusually difficult in comparison to most degradation determinants discovered subsequently. It was only recently, for instance, that an endogenous yeast protein was shown to be recognized by its amino-terminal residue (Rao et al. 2001). However, the greater significance of the Bachmair et al. (1986) paper was methodological: It provided a framework for the mechanistic dissection of substrate specificity in the ubiquitin system. Use of the ubiquitin–β-gal variants allowed the execution of a number of genetic screens for mutants defective in N-end rule substrate degradation. This has led to a thorough molecular genetic dissection of what is now generally referred to as the N-end rule pathway. Fruits of this work included the first cloning of an E3 ubiquitin-protein ligase (Bartel et al. 1990) and the first evidence for allosteric regulation of such an E3 (Turner et al. 2000). Interestingly, the E2 that works in this pathway was later shown to be Rad6 (Dohmen et al. 1991). The earlier discovery that Rad6, a protein crucial for DNA repair and induced mutagenesis, was a ubiquitin-conjugating enzyme was what had made it clear to all that the ubiquitin system indeed has a very broad range of functions in cellular physiology (Jentsch et al. 1987). Another key mechanistic insight made possible by use of the ubiquitin–β-gal proteins, as well as of another, non-N-end-based β-gal-derived substrate, was that ubiquitin-dependent protein degradation of multisubunit proteins is subunit-specific, i.e., degradation-signal-bearing subunits in a heteromeric complex could be destroyed selectively (Hochstrasser and Varshavsky 1990; Johnson et al. 1990). This property is now known to be important for removing inhibitory subunits from cell regulators such as the Cdk1 cell cycle kinase and the NFκB transcription factor (see Verma et al. 2001 and references therein).

Perhaps the most significant application of the ubiquitin–β-gal substrates was in a *tour-de-force* series of experiments by Vincent Chau et al. in which it was demonstrated that multiubiquitination of a protein substrate reflected the addition of a polymeric ubiquitin chain to a specific substrate lysine, rather than addition of single ubiquitin moieties to multiple lysines (Chau et al. 1989; Gregori et al. 1990). Importantly, the ubiquitins in the chain had a specific and uniform linkage to one another. By mutating the residue of ubiquitin involved in this isopeptide linkage, Chau et al. (1989) were able to show that multiubiquitination of a substrate was necessary for its efficient degradation in vitro. Multiubiquitin chains of the same linkage were soon confirmed to form on proteolytic substrates in vivo and were shown to be necessary for rapid substrate degradation (Hochstrasser et al. 1991). We now also know that ubiquitin chains with different ubiquitin-ubiquitin connectivities can be assembled and that these alternative topologies are associated with distinct functions (Arnason and Ellison 1994; Spence et al. 1995).

UBIQUITIN-DEPENDENT DEGRADATION OF NATURALLY SHORT-LIVED REGULATORS

Almost all of the early work on ubiquitin-dependent protein degradation was done with nonphysiological substrates. For example, the classic reticulocyte lysate studies used secreted proteins such as bovine serum albumin or hen egg-white lysozyme, which after they are synthesized would generally never come in contact with components of the ubiquitin-proteasome pathway. Early studies in yeast were also done with artificial test substrates, as noted above. Since the N-end rule degradation determinant could only be created at that time by the ubiquitin fusion technique, it was also by no means clear whether N-end recognition was ever normally used by cells to ubiquitinate proteins. Thus, even though it was apparent that the ubiquitin system was an essential part of many and probably all eukaryotes, it was not known how natural substrates of the system were recognized nor what ubiquitin pathway components were involved. The first natural substrate of the ubiquitin pathway discovered in yeast was Matα2, a transcriptional regulator important for cell-type determination (Hochstrasser and Varshavsky 1990; Hochstrasser et al. 1991; Chen et al. 1993). Chen et al. began with the observation that Matα2 is multiubiquitinated in vivo and proceeded to test strains deleted for the known E2 ubiquitin-conjugating enzymes (Ubcs) for defects in Matα2 degradation. At the time, 9 of the 11 yeast ubiquitin E2 isozymes had been identified by Stefan Jentsch and colleagues through systematic reverse-genetic screening (Jentsch 1992). Surprisingly, four of the nine E2/Ubc mutants displayed reduced rates of Matα2 degradation. The severity of the proteolytic defect was in all cases modest, with Matα2 stability increasing two- to threefold at most in the single mutants. However, when certain of the different mutations were combined, the rate of Matα2 degradation dropped dramatically. The genetic results were in the end very clear: Ubc6/Doa2 and Ubc7 worked in one ubiquitination pathway and the closely related Ubc4 and Ubc5 enzymes in a second one. Only when one or more mutations in both pathways were combined was a strong defect in Matα2 ubiquitination and proteolysis observed. Moreover, Ubc6 and Ubc7 appeared to be able to interact with each other based on yeast two-hybrid analysis; Ubc7-Ubc7 complexes were also suggested by these assays. Significantly, the Ubc6-Ubc7 ubiquitination pathway/complex was linked to a specific degradation signal in Matα2 called Deg1. Deg1 has at this point become one of the most thoroughly studied degradation signals or degrons, and attachment of the transplantable signal to various reporter proteins has allowed the identification and characterization of many components of the ubiquitin-proteasome system.

The Chen et al. (1993) study demonstrated for the first time that degradation signals in naturally short-lived regulatory factors can be recognized by specific ubiquitination enzymes and that a single substrate protein can be targeted—via distinct signals—by more than one ubiquitination pathway. Such signal-mediated degradation of regulatory proteins, often involving multiple mechanisms, has now been observed for numerous ubiquitin pathway substrates. These results were also important in that they suggested a basis for explaining the wide range of substrate specificities in the ubiquitin system, namely, that different E2s and presumably E3s could associate in diverse combinations to recognize particular degradation signals. This can greatly expand the repertoire of potential specificities. Distinct subunit compositions in multisubunit E3 complexes also contribute to the extension of substrate specificities, particularly in the SCF (Skp1-cullin-F-box protein) class of E3s.

TARGETING OF SUBSTRATES ACROSS MEMBRANES TO THE PROTEASOME

The existence of a quality control system in the endoplasmic reticulum (ER) that leads to rapid disposal of unassembled or misfolded ER proteins has been known for some time (Lippincott-Schwartz et al. 1988). This ER-associated degradation (ERAD) was initially assumed to involve ER-specific proteases because proteasomes and the ubiquitination enzymes are excluded from the ER lumen. The first hint that this assumption was incorrect came in 1995 when two groups published data with cultured mammalian cells that suggested the surprising idea that the transmembrane protein CFTR (cystic fibrosis transmembrane conductance regulator), when misfolded and retained in the ER rather than proceeding to the Golgi, is rapidly degraded at least in part through the ubiquitin-proteasome pathway (Jensen et al. 1995; Ward et al. 1995). However, about 60% of the polytopic CFTR protein faces the cytosol, so it was uncertain whether the entire molecule was degraded by the proteasome or only its cytosolically disposed segments. Most of the experiments relied on the use of inhibitors, so there also remained the possibility that the inhibitors affected multiple proteases and not just the proteasome.

Dieter Wolf and his colleagues had been conducting a wide range of studies on yeast proteases, including the vacuolar hydrolases and the cytosolic 20S proteasome core particle. Prior to the **Hiller et al. (1996)** paper, the Wolf group had identified mutant vacuolar hydrolase alleles that yielded proteins which, like misfolded mammalian CFTR, were retained in the ER and rapidly degraded, e.g., a carboxypeptidase Y point mutant called CPY* (Finger et al. 1993). Interestingly, it was concluded in this latter study that proteasomes were *not* involved in this ER-linked degradation, based on analysis of several partial loss-of-function 20S proteasome mutants. A subsequent mutant screen for ERAD mutants yielded four complementation groups called "*d*egradation in the *ER*" or *der* mutants (Knop et al. 1996). Unexpectedly, cloning of the *DER2* gene by Hiller et al. (1996) revealed that it was *UBC7*. This forced a reassessment of the potential contribution of the ubiquitin-proteasome pathway to ERAD. Hiller et al. (1996) analyzed multiple mutants of the proteasome and of the ubiquitin conjugation system. As with Matα2 degradation, both Ubc6 and Ubc7 were found to contribute to CPY* turnover, and the 26S proteasome was also required.

What is especially striking about CPY* is that it is a lumenal ER protein that is initially fully translocated into the ER and core-glycosylated (Plemper et al. 1999). It then must be retrotranslocated back across the ER membrane for degradation by the proteasome. Hiller et al. (1996) showed that glycosylated CPY* is ubiquitinated while still in the ER membrane. This depends in part on Ubc7, but other E2s are also involved (Friedlander et al. 2000). The principal E3 ligase was subsequently found to be Hrd1/Der3, a transmembrane protein of the ER which contains a RING domain (Hampton et al. 1996; Bordallo and Wolf 1999) that resides on cytosolic face of the ER (Gardner et al. 2000). RING motifs characterize the largest subclass of E3 enzymes. The most intriguing issues raised by this work, which have still not been fully resolved, are how the mutant proteins are "sensed" on the lumenal side of the ER membrane and what translocation machinery allows retrotranslocation to the cytosol. The Sec61 translocon, which is responsible for most anterograde transport across the ER membrane, has been implicated in retrotranslocation as well.

The Hiller et al. (1996) paper, together with an independent study on a virally induced "dislocation" of the major histocompatibility complex (MHC) class I molecule from the ER (Wiertz et al. 1996), provided the first compelling evidence for retrotanslocation of large proteins out of the ER. From the latter study, it was initially believed that ubiquitination of the substrate would not be required for degradation by the proteasome. Later work showed that as with CPY* degradation, ubiquitination was involved. These results were completely unexpected at the time. It is now clear that many different membrane and secreted proteins are subject to quality controls or checkpoints in the ER, with the major disposal pathway working through substrate ejection back into the cytoplasm and degradation by the proteasome (Bonifacino and Weissman 1998). In some cases, the "extraction" of membrane proteins from the ER depends on proteasome activity (Mayer et al. 1998), whereas in other examples, the protein can be fully retrotranslocated into the cytosol with little or no proteasome catalytic function (Wiertz et al. 1996). ERAD has become a major focus of research within the ubiquitin field because of the biomedically relevant proteins, such as CFTR and hydroxymethlyglutaryl–coenzyme A (HMG-CoA) reductase, that are degraded at the ER and because of the interesting mechanistic questions that arise with this branch of the ubiquitin system. There is also a close connection between ERAD and the unfolded protein response (UPR), a homeostatic mechanism for preventing overaccumulation of misfolded or misassembled proteins in the ER (for review, see Hampton 2000). Neither ERAD nor the UPR is required for normal growth under most conditions, but simultaneous loss of both mechanisms is strongly deleterious. The Hiller et al. (1996) study was at the vanguard of research into the role of ubiquitin-dependent ERAD in ER quality control, and CPY* continues to be one of the most useful substrates for characterization of these phenomena.

BEYOND THE PROTEASOME: UBIQUITIN AS A SIGNAL FOR ENDOCYTOSIS

In the late 1980s and early 1990s, the proteasome had been considered to be the principal if not the only destination for all rapidly degraded ubiquitin-tagged proteins. Indeed, work in yeast figured prominently in establishing this connection in vivo (see, e.g., Heinemeyer et al. 1991; Seufert and Jentsch 1992). On the other hand, reports of ubiquitin modification of cell surface proteins had appeared sporadically in the literature, but whether this led to degradation of these membrane proteins at all, let alone by the proteasome, was not known.

An early analysis of ligand-induced degradation of the platelet-derived growth factor receptor (PDGF-R) in porcine cells had implicated receptor ubiquitination in the degradation of the ligand-receptor complex (Mori et al. 1992). A large carboxy-terminal deletion of PDGF-R blocked its ubiquitination and this correlated with a modest inhibition of degradation. These data were suggestive, but the correlation did not prove a causal role for ubiquitin in receptor degradation, nor did the data indicate at what step in the receptor down-regulation pathway that ubiquitination could be exerting its effect. The study by **Kölling and Hollenberg (1994)** provided the first strong evidence that ubiquitination of plasma membrane proteins not only was important for their degradation, but also probably worked by stimulating the endocytosis of these proteins and/or their subsequent trafficking into the vacuole (the yeast equivalent of the vertebrate lysosome). This represented a major shift in how we could view the function of ubiquitin in proteolysis, and, it is fair to say, not all of us fully assimilated it at the time.

Kölling and Hollenberg (1994) were studying the biogenesis and fate of an ABC transporter protein called Ste6, which promotes secretion of the **a**-factor pheromone, a small lipidated peptide. Somewhat surprisingly, they found that the major fraction of this multimembrane-spanning protein was localized to the Golgi at steady state, rather than at the cell surface, and that it was continuously degraded, with a half-life of approximately 13 minutes. To verify that Ste6 did in fact traffic to the cell surface, they examined its distribution in endocytosis mutants using both immunofluorescence localization and subcellular fractionation. Ste6 accumulated at the plasma membrane in these mutants. Interestingly, it was noticed that the plasma membrane fraction of Ste6 was modified in some way to yield forms that migrated more slowly on SDS-polyacrylamide gels. Kölling and Hollenberg (1994) tested the idea that these forms might be ubiquitinated species of Ste6, and the results were positive, suggesting that Ste6 was ubiquitinated at the cell membrane.

To determine whether Ste6 ubiquitination contributed to its degradation, several ubiquitin-conjugating enzyme mutants were tested. In cells deleted for both *UBC4* and *UBC5*, which encode E2 proteins that are 92% identical, Ste6 was stabilized approximately threefold. At the same time, Kölling and Hollenberg (1994) found that degradation of Ste6 was almost completely blocked in mutants lacking vacuolar protease activity; similar results were reported independently by Susan Michaelis's group (Berkower et al. 1994). The simplest model to account for these data was that Ste6 was ubiquitinated at the cell surface, thereby triggering its uptake into endocytic vesicles that traveled through the endocytic pathway to the vacuole where Ste6 was degraded by resident proteases.

Nevertheless, there were gaps in these initial studies, most of which were addressed in a follow-up report (Kölling and Losko 1997). One was the lack of any test of known proteasome mutants for impairment of cell membrane protein degradation. The prediction would have been that these mutations would have little or no effect on Ste6 degradation (as was later shown). Another missing link was the absence of direct evidence for a role of Ste6 ubiquitination per se in its endocytosis or trafficking to the vacuole. The *trans*-acting mutations in *UBC4/5* might have affected Ste6 degradation indirectly, e.g., through a failure to ubiquitinate some endocytic factor. However, subsequent analysis (Kölling and Losko 1997) provided evidence correlating a reduction of Ste6 ubiquitination through mutations in Ste6 itself to reduced endocytosis and degradation.

These issues were also addressed in an elegant paper published in 1996 by Hicke and Riezman (Hicke and Riezman 1996) using ligand-induced uptake and degradation of the Ste2 α-factor receptor as a model. More recent studies indicate that attachment of a single ubiquitin molecule to a cell surface protein, at least in yeast, is usually sufficient to generate an endocytic signal (Terrell et al. 1998; Roth and Davis 2000). Virtually all tested yeast plasma membrane proteins are now known to be endocytosed through a ubiquitin-dependent mechanism (Shaw et al. 2001). This appears to be true for a number of mammalian proteins too, such as the PDGF-R and the growth hormone receptor, but the precise role(s) of ubiquitin in these cases is not as well-defined. Moreover, the proteasome does appear to participate in the trafficking of these proteins to the lysosome, but not in the direct degradation of the cell surface receptor itself (Longva et al. 2002).

UBIQUITIN-DEPENDENT PROTEOLYSIS AND THE CELL CYCLE

If one wished to point to one area of research that most galvanized general interest in the ubiquitin system, it would have to be in the study of the cell cycle. A mitotic cell cycle arrest was the most prominent trait noted for the first ubiquitin system mutant ever described (Matsumoto et al. 1983; Ciechanover et al. 1984). Moreover, the cyclin proteins, which must be degraded for cells to exit mitosis, were among the first substrates suggested to be degraded by the ubiquitin pathway, based on an in vitro *Xenopus* system that faithfully recapitulated key aspects of the embryonic cell cycle (Glotzer et al. 1991).

Cell cycle research in *S. cerevisiae*, on the other hand, has had a long and well-chronicled history, particularly the isolation by Lee Hartwell of mutants conditionally defective for cell cycle progression (see Chapter 6). Among these mutants was *cdc34*, which was later shown to have a defect in an E2 ubiquitin-conjugating enzyme (Goebl et al. 1988). This mutant was specifically defective in the G_1-to-S cell cycle transition, accumulating at the restrictive temperature with duplicated spindle-pole bodies and with multiple elongated buds, but with only a single, unreplicated nucleus. The data suggested that a cell cycle regulatory protein(s) needed to be ubiquitinated and probably degraded for DNA replication to ensue. This example is illustrative of virtually all the earlier literature on the link between protein ubiquitination and cell cycle progression: Ubiquitination enzymes had in some cases been identified, as had proteins degraded in a cell-cycle-dependent fashion, but there was no well-established case for the ubiquitin-dependent degradation of a specific protein driving a particular cell cycle transition in vivo. For instance, cyclin-B degradation had initially been assumed to be important for the metaphase-to-anaphase transition in both vertebrates and yeast, but nondestructible versions of the cyclin were later demonstrated to support normal progression into anaphase (Holloway et al. 1993; Surana et al. 1993). Using relatively simple genetic experiments in *S. cerevisiae*, **Schwob et al. (1994)** demonstrated a clear link between a particular ubiquitination pathway and a specific substrate, Sic1, and showed that Sic1 degradation was critical for G_1-to-S progression. This work in turn led to an intense search for all the relevant ubiquitination factors needed for Sic1 modification, leading to the discovery of the SCF class (Skp1/Cdc53/F-box protein complex; Skowyra et al. 1997) of E3 ligases.

The cyclins are activating subunits of the central cell cycle kinase Cdc28 (also called Cdk1). Schwob et al. (1994) initially set out to examine the contribution of B-type cyclins to the initiation of S phase. *S. cerevisiae* expresses six such cyclins, and by inactivating them all simultaneously, Schwob et al. (1994) were able to show that they were essential for S-phase entry, as well as for preventing rebudding and hyperpolarized bud growth. Tellingly, the terminal-arrest phenotype of the sextuple cyclin mutant looked very similar to that of *cdc34* mutants. At the same time, the authors noted that forced early G_1 expression of Clb5, a B-type cyclin which is normally induced only in late G_1, did not change the kinetics of replication initiation or Cdc28–cyclin B kinase activation, suggesting that cells contain an inhibitor of this activity. This inference was confirmed by cell-extract-mixing experiments, which demonstrated that cells in G_1 but not other cell cycle stages contained a kinase-inhibiting activity.

It was at this point that some earlier biochemical work became crucial. A G_1-phase Cdc28 inhibitor had been identified previously by the Mendenhall laboratory, which had managed to purify enough of the scarce protein to obtain peptide sequences, which allowed cloning of the corresponding gene, *SIC1* (Mendenhall 1993; Nugroho and Mendenhall 1994). Sic1 was a prime candidate for the kinase inhibitor implicated by the cell-extract-mixing results, and when Schwob et al. (1994) repeated these experiments using G_1 cell extracts from *sic1Δ* cells, the Cdc28-inhibitory activity had disappeared. Recombinant, bacterially expressed Sic1 was shown to bind the Clb5-Cdc28 kinase complex and to inhibit its kinase activity. In vegetatively growing cells, Sic1 accumulated specifically in G_1 and then was eliminated shortly before the onset of S phase. This degradation required the Cdc34 ubiquitin-conjugating enzyme. Finally, in the most important experiment of their paper, Schwob et al. (1994) made a *cdc34-ts sic1Δ* double mutant and found that the usual cell cycle arrest in G_1 of the *cdc34* single mutant at nonpermissive temperature was completely bypassed. Therefore, Sic1 was a critical target of the Cdc34 ubiquitination pathway, the elimination of which allowed the activated Cdc28–cyclin B kinase complex to induce entry into S phase.

Later studies have confirmed these insights and added the important discovery that a multisubunit E3 ubiquitin ligase, called SCF^{Cdc4}, functions with the Cdc34 E2 to ubiquitinate Sic1. The SCF ligases include a variable subunit that contains a so-called F-box, a degenerate motif that functions as a Skp1-interaction domain (Bai et al. 1996), and this large group of E3s has now been shown to be extremely widespread among eukaryotes, with many substrates of biomedical importance. Interestingly, whereas the Cdc4 and Cdc53 (cullin) subunits of the SCF^{Cdc4} E3 were initially uncovered by the same genetic screens used to identify Cdc34, Skp1 is multifunctional and was first found in a study of the yeast kinetochore (Connelly and Hieter 1996).

A significant effect of the Schwob et al. (1994) work was to redirect attempts at purifying the ubiquitination enzymes required for the G_1-S transition, which were originally focused on the Cln-class G_1 cyclin substrates, to the Sic1 substrate (Deshaies 1999). Sic1 would become the first natural substrate whose ubiquitination was reconstituted in vitro with purified recombinant proteins (Feldman et al. 1997; Skowyra et al. 1997). It is of interest that some of the recombinant proteins were synthesized in insect cells, resulting in the inadvertant copurification of a host protein. This host protein was later shown to replace an essential SCF subunit, Rbx1, which contains a RING domain (see Tyers and Willems 1999). These studies were among the earliest to recognize the general importance of RING motifs as a signature for a broad class of E3 ligases. Finally, the Schwob et al. (1994) study also suggested that degradation of Sic1 might require its prior phosphorylation by the Cdc28 kinase. This was confirmed in later work, culminating in an elegant recent study in which it was shown that multiple sites on Sic1 needed to be phosphorylated for its efficient ubiquitination, which allows Sic1 degradation to be exquisitely timed for the appropriate point in the cell cycle (Nash et al. 2001).

UBIQUITIN-LIKE PROTEINS: A PARADIGM EXTENDED

Among the most exciting developments in the ubiquitin field over the past years has been the realization that ubiquitin is not unique in its ability to modify other proteins but is a member of a larger class of intracellular protein modifiers. Known Ubls include SUMO/Smt3, a protein important for nucleocytoplasmic trafficking, cell cycle progression, and chromosome stability; NEDD8/Rub1, a modifier that regulates SCF ligases (see previous section); and ISG15, an interferon-induced antiviral factor in mammals (for review, see

Hochstrasser 2000; Jentsch and Pyrowolakis 2000). The functions of the Ubl modifiers are very different from those of ubiquitin. Perhaps the clearest examples of this have come from molecular studies of autophagy in *S. cerevisiae*, which have also generated some of the biggest surprises about the diversity of Ubl pathways and their molecular components.

Autophagy (or more specifically, macroautophagy) is a process wherein regions of cytoplasm are engulfed by double-membrane sheets to form autophagosomes; the outer membrane of the autophagosome then fuses with the vacuole/lysosome, releasing the autophagic bodies to the vacuolar interior, where their contents are enzymatically digested. Autophagy is generally induced by nutrient limitation, although a specific variant of this pathway operates under normal growth conditions for the biosynthetic delivery of certain vacuolar enzymes (Scott et al. 1996). Work from several laboratories, particularly that of Yoshinori Ohsumi, had implicated at least 16 genes in the process of autophagosome formation in *S. cerevisiae*, but the sequences of most of the cloned genes yielded few clues to their function. Unexpectedly, when **Mizushima et al. (1998)** began looking at the protein products of the autophagy genes, they discovered that one of them, Atg12, was covalently linked in vivo to another one, Atg5. A third factor, Atg7, was necessary for this ligation and bore weak similarity to E1-like enzymes. (Note: A recent nomenclature revision has replaced the "Apg" protein designations used in this paper with "Atg," which will be used here.) These data immediately suggested to Mizushima et al. (1998) that they were looking at a ubiquitin-like modification system, and in short order they were able to demonstrate an amide linkage between the carboxy-terminal glycine of Atg12 and a specific lysine of Atg5, similar to the linkage of ubiquitin to substrates. An additional autophagy factor, Atg10, was required for Atg12-Atg5 conjugation and behaved like an E2 enzyme, i.e., it formed an Atg7-dependent thioester with Atg12 (despite lacking detectable sequence similarity to the E2s for ubiquitin or other known Ubls). The Atg12-Atg5 conjugate interacts noncovalently with a third protein, Atg16, to form a 350-kD oligomer that may act as a coat to drive autophagosome formation (Kuma et al. 2002).

A number of discoveries with the Atg12 ligation pathway have had a major impact on our current view of the ubiquitin sytem. First, the Atg12 factor is essentially unrecognizable as an Ubl based on primary sequence; it does not even have the usual pair of glycines at its carboxyl terminus (but it does have one terminal glycine that is essential for ligation). Hence, it is possible that additional Ubls, which have escaped detection by standard similarity-search algorithms, remain to be uncovered. Atg12 and the Atg8 (see below) are modestly related in sequence (Hochstrasser 2000), and a human homolog of Atg8 was shown to have the ubiquitin fold (Paz et al. 2000). Second, Atg12 appears to have only one target, Atg5, and its ligation to Atg5 is likely to be irreversible. These are features that are not shared with ubiquitin or the other tested Ubl ligation systems. Finally, Atg12 is activated by an E1-like protein that was recently shown to activate another Ubl called Atg8, which is also an autophagy factor (Ichimura et al. 2000). The Atg8 ligation pathway uses a distinct E2, Apg3, which has extremely weak similarity to the Atg10 E2 but no detectable similarity to any other E2-like factors. It is possible that Atg10 and Apg3 share three-dimensional structural similarity with the canonical E2s, but this has not yet been determined. A most intriguing property of these two Ubl systems is that a single E1 passes an activated Ubl only to the cognate E2 for that Ubl. How this dual specificity is maintained remains a fascinating unanswered question.

The Atg8 ligation system, whose discovery owes a substantial debt to the Atg12 studies by Mizushima et al. (1998), has another remarkable feature: Its natural targets are membrane lipids rather than proteins (Ichimura et al. 2000). In particular, Atg8 is conjugated reversibly to phosphatidylethanolamine by an amide linkage to the ethanolamine moiety. At present, it is not known whether Atg8 functions as a carrier for specific lipids during formation of autophagosome precursors or, conversely, whether lipid conjugation allows cyclical attachment of Atg8 to these precursors in order to facilitate autophagosome formation in some way, e.g., to recruit other autophagy factors. Finally, it is worth noting that the Atg8 and Atg12 ligation systems have been conserved during eukaryotic evolution, and their discovery by Ohsumi and colleagues should greatly aid the molecular dissection of autophagy in mammals and other multicellular eukaryotes, where the phenomenon was originally described but where mechanistic studies had been almost nonexistent.

CONCLUSION

Our understanding of the ubiquitin system has expanded tremendously in recent years, and there can be little argument that analyses in *S. cerevisiae* have often led the way. The earlier studies discussed above provided many of the mutants, genes, biochemical tools, and analytical approaches that continue to help ongoing work in this burgeoning field. In addition to the classical genetic and biochemical strategies that have been brought to bear on problems in the ubiquitin system, "postgenomic" approaches will begin to have an impact on the field as well. Among these will be the systematic identification of ubiquitin and Ubl conjugates in yeast under various conditions, particularly by mass spectrometry techniques, and genome/proteome-scale evaluations of changes in mRNA and protein levels induced by loss or overproduction of various ubiquitin system components. New landmarks are sure to follow.

ACKNOWLEDGMENTS

I wish to thank Nick Davis, Randy Hampton, Daniel Klionsky, Jeff Laney, Mike Mendenhall, Susan Michaelis, Cecile Pickart, Mark Solomon, and Alex Varshavsky for comments on the manuscript and/or advice about the choice of papers. Work from my laboratory has been supported by grants from the National Institutes of Health.

STUDY QUESTIONS

1. If you noticed a sequence in the protein databases that showed significant similarity to ubiquitin, what features might suggest that the protein is a Ubl, i.e., can be ligated to another molecule?

2. Supposing that you identified a candidate Ubl in the yeast proteome, how would you determine whether it was in fact conjugated to other molecules by a ubiquitin-like pathway?

3. If mutations in several ubiquitin ligation pathway enzymes were found to impair the degradation of the protein you are studying, what genetic approaches can be used to

determine if these enzymes work together or independently in this degradation? What biochemical approaches could be employed to substantiate your conclusion?

4. Virtually every cellular protein, if misfolded or misassembled, could in principle be a target for ubiquitination and proteasomal degradation. What substrate features might be used by the ubiquitin system to discriminate a misfolded/misassembled protein from a correctly folded one?

5. For those proteins synthesized with a penultimate residue that would be "destabilizing" by the yeast N-end rule if at the very amino terminus, methionine aminopeptidase will not remove the terminal methionine. What would therefore be necessary for an intracellular protein to be a substrate of the N-end rule pathway? What conditions would need to be met for this relationship between MAP activity and the N-end rule to be useful to the cell for identifying normally secreted or compartmentalized proteins that mislocalize to the cytoplasm?

6. The Sec61 translocon or the ER has been suggested to participate in both anterograde and retrograde transport of proteins across the ER membrane. What kinds of mutations in the translocon would be necessary to make a strong claim for this idea based on genetic experiments? How would you go about isolating such mutations?

7. Although ubiquitination of plasma membrane proteins appears to trigger their endocytosis, a complicating factor in this interpretation is the ability of endocytosed proteins either to recycle to the cell surface or to continue through the endosomal pathway to the vacuole. What alternative interpretation of the function of receptor ubiquitination can be suggested in this light? How would you distinguish this from a direct role in the internalization step? For a discussion of this issue, see Losko et al. (2001).

8. It appears that the autophagy Ubl called Atg12 has only one intracellular substrate, Atg5, to which it becomes irreversibly ligated. What might be the advantage of creating the Atg12-Atg5 conjugate by such a posttranslational mechanism rather than synthesizing Atg12 and Atg5 as a fusion protein? How might you test your hypothesis?

9. The Atg7 E1-like enzyme transfers Atg12 to the Atg10 E2-like protein and transfers Atg8 to the distantly related Apg3 E2-like protein. Propose a mechanism that could account for this unusual dual specificity. What experiments would you do to dissect the regulated E1-to-E2 transthiolation mechanism?

10. In the Schwob et al. (1994) paper, no evidence for Sic1 ubiquitination by the Cdc34 E2 was presented. Suggest another formal explanation for the suppression by *sic1Δ* of the G_1-to-S phase arrest in *cdc34-2* cells. Sic1 persists in synchronized *cdc34-2* cultures for far longer than in congenic wild-type cells. Give two possible mechanisms by which inactivation of Cdc34 might cause this.

REFERENCES

Arnason T. and Ellison M.J. 1994. Stress resistance in *Saccharomyces cerevisiae* is strongly correlated with assembly of a novel type of multiubiquitin chain. *Mol. Cell. Biol.* **14:** 7876–7883.

Bachmair A. and Varshavsky A. 1989. The degradation signal in a short-lived protein. *Cell* **56:** 1019–1032.

Bachmair A., Finley D., and Varshavsky A. 1986. In vivo half-life of a protein is a function of its amino-terminal residue. *Science* **234:** 179–186.

Bai C., Sen P., Hofmann K., Ma L., Goebl M., Harper J.W., and Elledge S.J. 1996. SKP1 connects cell cycle regulators to the ubiquitin proteolysis machinery through a novel motif, the F-box. *Cell* **86:** 263–274.

Bartel B., Wunning I., and Varshavsky A. 1990. The recognition component of the N-end rule pathway. *EMBO J.* **9:** 3179–3189.

Berkower C., Loayza D., and Michaelis S. 1994. Metabolic instability and constitutive endocytosis of *STE6*, the α-factor transporter of *Saccharomyces cerevisiae*. *Mol. Biol. Cell* **5:** 1185–1198.

Bonifacino J.S. and Weissman A.M. 1998. Ubiquitin and the control of protein fate in the secretory and endocytic pathways. *Annu. Rev. Cell Dev. Biol.* **14:** 19–57.

Bordallo J. and Wolf D.H. 1999. A RING-H2 finger motif is essential for the function of Der3/Hrd1 in endoplasmic reticulum associated protein degradation in the yeast *Saccharomyces cerevisiae*. *FEBS Lett.* **448:** 244–248.

Chau V., Tobias J.W., Bachmair A., Marriott D., Ecker D.J., Gonda D.K., and Varshavsky A. 1989. A multiubiquitin chain is confined to specific lysine in a targeted short-lived protein. *Science* **243:** 1576–1583.

Chen P., Johnson P., Sommer T., Jentsch S., and Hochstrasser M. 1993. Multiple ubiquitin-conjugating enzymes participate in the in vivo degradation of the yeast MATα2 repressor. *Cell* **74:** 357–369.

Ciechanover A., Finley D., and Varshavsky A. 1984. Ubiquitin dependence of selective protein degradation demonstrated in the mammalian cell cycle mutant ts85. *Cell* **37:** 57–66.

Ciechanover A., Heller H., Elias S., Haas A.L., and Hershko A. 1980. ATP-dependent conjugation of reticulocyte proteins with the polypeptide required for protein degradation. *Proc. Natl. Acad. Sci.* **77:** 1365–1368.

Connelly C. and Hieter P. 1996. Budding yeast *SKP1* encodes an evolutionarily conserved kinetochore protein required for cell cycle progression. *Cell* **86:** 275–285.

DeMartino G.N. and Slaughter C.A. 1999. The proteasome, a novel protease regulated by multiple mechanisms. *J. Biol. Chem.* **274:** 22123–22126.

Deshaies R.J. 1999. SCF and Cullin/Ring H2-based ubiquitin ligases. *Annu. Rev. Cell Dev. Biol.* **15:** 435–467.

Dohmen R.J., Madura K., Bartel B., and Varshavsky A. 1991. The N-end rule is mediated by the UBC2(RAD6) ubiquitin-conjugating enzyme. *Proc. Natl. Acad. Sci.* **88:** 7351–7355.

Feldman R.M., Correll C.C., Kaplan K.B., and Deshaies R.J. 1997. A complex of Cdc4p, Skp1p, and Cdc53p/cullin catalyzes ubiquitination of the phosphorylated CDK inhibitor Sic1p. *Cell* **91:** 221–230.

Finger A., Knop M., and Wolf D.H. 1993. Analysis of two mutated vacuolar proteins reveals a degradation pathway in the endoplasmic reticulum or a related compartment of yeast. *Eur. J. Biochem.* **218:** 565–574.

Finley D., Ciechanover A., and Varshavsky A. 1984. Thermolability of ubiquitin-activating enzyme from the mammalian cell cycle mutant ts85. *Cell* **37:** 43–55.

Friedlander R., Jarosch E., Urban J., Volkwein C., and Sommer T. 2000. A regulatory link between ER-associated protein degradation and the unfolded-protein response. *Nat. Cell Biol.* **2:** 379–384.

Gardner R.G., Swarbrick G.M., Bays N.W., Cronin S.R., Wilhovsky S., Seelig L., Kim C., and Hampton R.Y. 2000. Endoplasmic reticulum degradation requires lumen to cytosol signaling. Transmembrane control of Hrd1p by Hrd3p. *J. Cell Biol.* **151:** 69–82.

Glotzer M., Murray A.W., and Kirschner M.W. 1991. Cyclin is degraded by the ubiquitin pathway. *Nature* **349:** 132–138.

Goebl M.G., Yochem J., Jentsch S., McGrath J.P., Varshavsky A., and Byers B. 1988. The yeast cell cycle gene CDC34 encodes a ubiquitin-conjugating enzyme. *Science* **241:** 1331–1335.

Goldknopf I.L. and Busch H. 1977. Isopeptide linkage between nonhistone and histone 2A polypeptides of chromosomal conjugate-protein A24. *Proc. Natl. Acad. Sci.* **74:** 864–868.

Goldstein G., Scheid M., Hammerling U., Schlesinger D.H., Niall H.D., and Boyse E.A. 1975. Isolation of a polypeptide that has lymphocyte-differentiating properties and is probably represented universally in living cells. *Proc. Natl. Acad. Sci.* **72:** 11–15.

Gregori L., Poosch M.S., Cousins G., and Chau V. 1990. A uniform isopeptide-linked multiubiquitin chain is sufficient to target substrate for degradation in ubiquitin-mediated proteolysis. *J. Biol. Chem.* **265:** 8354–8357.

Haas A.L., Ahrens P., Bright P.M., and Ankel H. 1987. Interferon induces a 15-kilodalton protein exhibiting marked homology to ubiquitin. *J. Biol. Chem.* **262:** 11315–11323.

Haas A.L., Warms J.V., Hershko A., and Rose I.A. 1982. Ubiquitin-activating enzyme. Mechanism and role in protein-ubiquitin conjugation. *J. Biol. Chem.* **257:** 2543–2548.

Hampton R.Y. 2000. ER stress response: Getting the UPR hand on misfolded proteins. *Curr. Biol.* **10:** R518–R521.

Hampton R.Y., Gardner R.G., and Rine J. 1996. Role of 26S proteasome and HRD genes in the degradation of 3-hydroxy-3-methylglutaryl-CoA reductase, an integral endoplasmic reticulum membrane protein. *Mol. Biol. Cell* **7:** 2029–2044.

Heinemeyer W., Kleinschmidt J.A., Saidowsky J., Escher C., and Wolf D.H. 1991. Proteinase yscE, the yeast proteasome/multicatalytic-multifunctional proteinase: Mutants unravel its function in stress induced proteolysis and uncover its necessity for cell survival. *EMBO J.* **10:** 555–562.

Hershko A., Heller H., Elias S., and Ciechanover A. 1983. Components of ubiquitin-protein ligase system. Resolution, affinity purification, and role in protein breakdown. *J. Biol. Chem.* **258:** 8206–8214.

Hershko A., Heller H., Eytan E., and Reiss Y. 1986. The protein substrate binding site of the ubiquitin-protein ligase system. *J. Biol. Chem.* **261:** 11992–11999.

Hershko A., Ciechanover A., Heller H., Haas A.L., and Rose I.A. 1980. Proposed role of ATP in protein breakdown: Conjugation of protein with multiple chains of the polypeptide of ATP-dependent proteolysis. *Proc. Natl. Acad. Sci.* **77:** 1783–1786.

Hershko A., Heller H., Eytan E., Kaklij G., and Rose I.A. 1984. Role of the alpha-amino group of protein in ubiquitin-mediated protein breakdown. *Proc. Natl. Acad. Sci.* **81:** 7021–7025.

Hicke L. and Riezman H. 1996. Ubiquitination of a yeast plasma membrane receptor signals its ligand-stimulated endocytosis. *Cell* **84:** 277–287.

Hiller M.M., Finger A., Schweiger M., and Wolf D.H. 1996. ER degradation of a misfolded luminal protein by the cytosolic ubiquitin-proteasome pathway. *Science* **273:** 1725–1728.

Hochstrasser M. 2000. Evolution and function of ubiquitin-like protein-conjugation systems. *Nat. Cell Biol.* **2:** E153–E157.

Hochstrasser M. and Varshavsky A. 1990. In vivo degradation of a transcriptional regulator: The yeast α2 repressor. *Cell* **61:** 697–708.

Hochstrasser M., Ellison M.J., Chau V., and Varshavsky A. 1991. The short-lived MATα2 transcriptional regulator is ubiquitinated in vivo. *Proc. Natl. Acad. Sci.* **88:** 4606–4610.

Holloway S.L., Glotzer M., King R.W., and Murray A.W. 1993. Anaphase is initiated by proteolysis rather than by the inactivation of maturation-promoting factor. *Cell* **73:** 1393–1402.

Hough R., Pratt G., and Rechsteiner M. 1986. Ubiquitin-lysozyme conjugates: Identification and characterization of an ATP-dependent protease from rabbit reticulocyte lysates. *J. Biol. Chem.* **261:** 2400–2408.

Hunt L.T. and Dayhoff M.O. 1977. Amino-terminal sequence identity of ubiquitin and the nonhistone component of nuclear protein A24. *Biochem. Biophys. Res. Commun.* **74:** 650–655.

Ichimura Y., Kirisako T., Takao T., Satomi Y., Shimonishi Y., Ishihara N., Mizushima N., Tanida I., Kominami E., Ohsumi M., Noda T., and Ohsumi Y. 2000. A ubiquitin-like system mediates protein lipidation. *Nature* **408:** 488–492.

Jensen T.J., Loo M.A., Pind S., Williams D.B., Goldberg A.L., and Riordan J.R. 1995. Multiple proteolytic systems, including the proteasome, contribute to CFTR processing. *Cell* **83:** 129–135.

Jentsch S. 1992. The ubiquitin conjugation system. *Annu. Rev. Genet.* **26:** 177–205.

Jentsch S. and Pyrowolakis G. 2000. Ubiquitin and its kin: How close are the family ties? *Trends Cell Biol.* **10:** 335–342.

Jentsch S., McGrath J.P., and Varshavsky A. 1987. The yeast DNA repair gene RAD6 encodes a ubiquitin-conjugating enzyme. *Nature* **329:** 131–134.

Johnson E.S., Gonda D.K., and Varshavsky A. 1990. cis-trans recognition and subunit-specific degradation of short-lived proteins. *Nature* **346:** 287–291.

Knop M., Finger A., Braun T., Hellmuth K., and Wolf D.H. 1996. Der1, a novel protein specifically required for endoplasmic reticulum degradation in yeast. *EMBO J.* **15:** 753–763.

Kölling R. and Hollenberg C.P. 1994. The ABC-transporter Ste6 accumulates in the plasma membrane in a ubiquitinated form in endocytosis mutants. *EMBO J.* **13:** 3261–3271.

Kölling R. and Losko S. 1997. The linker region of the ABC-transporter Ste6 mediates ubiquitination and fast turnover of the protein. *EMBO J.* **16:** 2251–2261.

Kuma A., Mizushima N., Ishihara N., and Ohsumi Y. 2002. Formation of the ~350 kD Apg12-Apg5-Apg16 multimeric complex, mediated by Apg16 oligomerization, is essential for autophagy in yeast. *J. Biol. Chem.* **277:** 18619–18625.

Lippincott-Schwartz J., Bonifacino J.S., Yuan L.C., and Klausner R.D. 1988. Degradation from the endoplasmic reticulum: Disposing of newly synthesized proteins. *Cell* **54:** 209–220.

Loeb K.R. and Haas A.L. 1992. The interferon-inducible 15-kDa ubiquitin homolog conjugates to intracellular proteins. *J. Biol. Chem.* **267:** 7806–7813.

Longva K.E., Blystad F.D., Stang E., Larsen A.M., Johannessen L.E., and Madshus I.H. 2002. Ubiquitination and proteasomal activity is required for transport of the EGF receptor to inner membranes of multivesicular bodies. *J. Cell Biol.* **156:** 843–854.

Losko S., Kopp F., Kranz A., and Kolling R. 2001. Uptake of the ATP-binding cassette (ABC) transporter Ste6 into the yeast vacuole is blocked in the *doa4* mutant. *Mol. Biol. Cell* **12:** 1047–1059.

Matsui S., Sandberg A.A., Negoro S., Seon B.K., and Goldstein G. 1982. Isopeptidase: A novel eukaryotic enzyme that cleaves isopeptide bonds. *Proc. Natl. Acad. Sci.* **79:** 1535–1539.

Matsumoto Y., Yasuda H., Marunouchi T., and Yamada M. 1983. Decrease in uH2A (protein A24) of a mouse temperature-sensitive mutant. *FEBS Lett.* **151:** 139–142.

Mayer T.U., Braun T., and Jentsch S. 1998. Role of the proteasome in membrane extraction of a short-lived ER-transmembrane protein. *EMBO J.* **17:** 3251–3257.

Mendenhall M.D. 1993. An inhibitor of p34CDC28 protein kinase activity from *Saccharomyces cerevisiae*. *Science* **259:** 216–219.

Miller H.I., Henzel W.J., Ridgway J.B., Kuang W.-J., Chisholm V., and Liu C.C. 1989. Cloning and expression of a yeast ubiquitin-protein cleaving activity in *Escherichia coli*. *Bio/Technology* **7:** 698–709.

Mizushima N., Noda T., Yoshimori T., Tanaka Y., Ishii T., George M.D., Klionsky D.J., Ohsumi M., and Ohsumi Y. 1998. A protein conjugation system essential for autophagy. *Nature* **395:** 395–398.

Mori S., Heldin C.H., and Claessonwelsh L. 1992. Ligand-induced polyubiquitination of the platelet-derived growth factor beta-receptor. *J. Biol. Chem.* **267:** 6429–6434.

Nash P., Tang X., Orlicky S., Chen Q., Gertler F.B., Mendenhall M.D., Sicheri F., Pawson T., and Tyers M. 2001. Multisite phosphorylation of a CDK inhibitor sets a threshold for the onset of DNA replication. *Nature* **414:** 514–521.

Nugroho T.T. and Mendenhall M.D. 1994. An inhibitor of yeast cyclin-dependent protein kinase plays an important role in ensuring the genomic integrity of daughter cells. *Mol. Cell. Biol.* **14:** 3320–3328.

Olson M.O., Goldknopf I.L., Guetzow K.A., James G.T., Hawkins T.C., Mays-Rothberg C.J., and Busch H. 1976. The NH2- and COOH-terminal amino acid sequence of nuclear protein A24. *J. Biol. Chem.* **251:** 5901–5903.

Paz Y., Elazar Z., and Fass D. 2000. Structure of GATE-16, membrane transport modulator and mammalian ortholog of autophagocytosis factor Aut7p. *J. Biol. Chem.* **275:** 25445–25450.

Plemper R.K., Deak P.M., Otto R.T., and Wolf D.H. 1999. Re-entering the translocon from the lumenal side of the endoplasmic reticulum. Studies on mutated carboxypeptidase yscY species. *FEBS Lett.* **443:** 241–245.

Rao H., Uhlmann F., Nasmyth K., and Varshavsky A. 2001. Degradation of a cohesin subunit by the N-end rule pathway is essential for chromosome stability. *Nature* **410:** 955–959.

Roth A.F. and Davis N.G. 2000. Ubiquitination of the PEST-like endocytosis signal of the yeast α-factor receptor. *J. Biol. Chem.* **275:** 8143–8153.

Schlesinger D.H., Goldstein G., and Niall H.D. 1975. The complete amino acid sequence of ubiquitin, an adenylate cyclase stimulating polypeptide probably universal in living cells. *Biochemistry* **14:** 2214–2218.

Schwob E., Böhm T., Mendenhall M.D., and Nasmyth K. 1994. The B-type cyclin kinase inhibitor p40^{SIC1} controls the G1 to S transition in *S. cerevisiae*. *Cell* **79:** 233–244.

Scott S.V., Hefner-Gravink A., Morano K.A., Noda T., Ohsumi Y., and Klionsky D.J. 1996. Cytoplasm-to-vacuole targeting and autophagy employ the same machinery to deliver proteins to the yeast vacuole. *Proc. Natl. Acad. Sci.* **93:** 12304–12308.

Seufert W. and Jentsch S. 1992. In vivo function of the proteasome in the ubiquitin pathway. *EMBO J.* **11:** 3077–3080.

Shaw J.D., Cummings K.B., Huyer G., Michaelis S., and Wendland B. 2001. Yeast as a model system for studying endocytosis. *Exp. Cell Res.* **271:** 1–9.

Skowyra D., Craig K.L., Tyers M., Elledge S.J., and Harper J.W. 1997. F-box proteins are receptors that recruit phosphorylated substrates to the SCF ubiquitin-ligase complex. *Cell* **91:** 209–219.

Spence J., Sadis S., Haas A.L., and Finley D. 1995. A ubiquitin mutant with specific defects in DNA repair and multiubiquitination. *Mol. Cell. Biol.* **15:** 1265–1273.

Surana U., Amon A., Dowzer C., McGrew J., Byers B., and Nasmyth K. 1993. Destruction of the CDC28/CLB mitotic kinase is not required for the metaphase to anaphase transition in budding yeast. *EMBO J.* **12:** 1969–1978.

Terrell J., Shih S., Dunn R., and Hicke L. 1998. A function for monoubiquitination in the internalization of a G protein-coupled receptor. *Mol. Cell* **1:** 193–202.

Turner G.C., Du F., and Varshavsky A. 2000. Peptides accelerate their uptake by activating a ubiquitin-dependent proteolytic pathway. *Nature* **405:** 579–583.

Tyers M. and Willems A.R. 1999. One ring to rule a superfamily of E3 ubiquitin ligases. *Science* **284:** 601–604.

Verma R., McDonald H., Yates J.R., III, and Deshaies R.J. 2001. Selective degradation of ubiquitinated Sic1 by purified 26S proteasome yields active S phase cyclin-Cdk. *Mol. Cell* **8:** 439–448.

Ward C.L., Omura S., and Kopito R.R. 1995. Degradation of CFTR by the ubiquitin-proteasome pathway. *Cell* **83:** 121–127.

Wiertz E.J., Jones T.R., Sun L., Bogyo M., Geuze H.J., and Ploegh H.L. 1996. The human cytomegalovirus US11 gene product dislocates MHC class I heavy chains from the endoplasmic reticulum to the cytosol. *Cell* **84:** 769–779.

Wilkinson K.D., Urban M.K., and Haas A.L. 1980. Ubiquitin is the ATP-dependent proteolysis factor I of rabbit reticulocytes. *J. Biol. Chem.* **255:** 7529–7532.

15

Genomics

Mark Johnston
Department of Genetics
Washington University Medical School
St. Louis, Missouri 63110

Philip Hieter
Michael Smith Laboratories
University of British Columbia
Vancouver, British Columbia V6T 1Z4, Canada

Genetic mapping, 287
Petes T.D. and Botstein D. 1977. Simple Mendelian inheritance of the reiterated ribosomal DNA of yeast. *Proc. Natl. Acad. Sci.* **74:** 5091–5095.

Physical mapping, 288
Olson M.V., Dutchik J.E., Graham M.Y., Brodeur G.M., Helms C., Frank M., MacCollin M., Scheinman R., and Frank T. 1986. Random-clone strategy for genomic restriction mapping in yeast. *Proc. Natl. Acad. Sci.* **83:** 7826–7830.

Genome sequencing, 291
Oliver S.G., van der Aart Q.J.M., Agostoni-Carbone M.L., Aigle M., Alberghina L., Alexandraki D., Antoine G., Anwar R., Ballesta J.P.G., Benit P., et al. 1992. The complete DNA sequence of yeast chromosome III. *Nature* **357:** 38–46.

SNP mapping, 293
Winzeler E.A., Richards D.R., Conway A.R., Goldstein A.L., Kalman S., McCullough M.J., McCusker J.H., Stevens D.A., Wodicka L., Lockhart D.J., and Davis R.W. 1998. Direct allelic variation scanning of the yeast genome. *Science* **281:** 1194–1197.

Functional genomics, 294
Giaever G., Chu A.M., Ni L., Connelly C., Riles L., Véronneau S., Dow S., Lucau-Danila A., Anderson K., André B., et al. 2002. Functional profiling of the *Saccharomyces cerevisiae* genome. *Nature* **418:** 387–391.

Note: The landmark papers listed above are those discussed in this chapter. Each landmark paper is preceded by the name of the section (with starting page number) where the paper is first discussed in detail.

WHAT IS GENOMICS? THE ORTHODOX VIEW IS THAT IT IS SIMPLY THE STUDY of genomes: enumerating their components and determining their organization. The major product of the genome projects, after all, has been genome sequences, and the complete gene lists of organisms. By this relatively prosaic definition, one might say that the field is approaching its end. But looking back to the recent past and foreseeing into the near future, we see that genomics has become a respectable scientific discipline with a life of its own. Today it is fashionable to perform experiments on entire genomes; tomorrow it will be essential.

Genomics is really about resource and tool development (Johnston and Fields 2000). By providing the information resource of the genome sequence of an organism, and the tools to enable productive use of this information (e.g., cloning vectors, novel genetic methods, and DNA microarrays), genome science has had a major impact on biology—an impact some say (perhaps hyperbolically) may be as big as that made by Darwin or by Watson and Crick. Inasmuch as most major advances in biology are, arguably, driven by technical advances, genomics is currently a highly important area of biology.

Compared to other landmarks presented in this book, yeast genomics is one of the newest areas of yeast biology. Some trace its beginnings in the yeast field to the late 1980s, when Andre Goffeau and his collaborators in Europe embarked on a bold project to determine the DNA sequence of the yeast genome (Goffeau and Vassarotti 1991). Others trace it to the late 1970s, when Maynard Olson had the foresight to see the need for a physical map of the yeast genome, and the audacity to attempt to produce one. Still others might say it began in the middle 1970s, with Hereford and Rosbash's analysis of the yeast transcriptome (although they did not call it that) (Hereford and Rosbash 1977). But, since the overriding goal of genomics is the characterization of a genome—the identification and mapping of all its genes—it is clear that the genomics of yeast originated in the early 1950s, when Bob Mortimer and Don Hawthorne began to develop a comprehensive genetic map of the yeast genome (Mortimer and Hawthorne 1966). (By this criterion, Bridges, who produced a cytogenetic map of the chromosomes of the fruit fly in the first decades of the last century, was the true father of the field of Genomics.)

All of these pioneers took a similar approach that at the time was rather unorthodox: Instead of focusing on any particular group of genes or biological process, they took on the whole genome. Their goal was not to understand how genes and their protein products function in any particular biological process, but, in the first instance, simply to identify *all* the genes of the organism and understand how they and their surrounding DNA sequences are organized. For this, they were often reviled, as prophets often are. By eschewing an interest in specific biological processes, they opened themselves up to attack from those who use the name "technologist" as a pejorative. We clearly remember people saying "he's *just* a mapper" or "*just* a sequencer" about some of the first genome scientists. "What has she learned about how things work?" "Who cares where all the *Hin*dIII and *Eco*RI sites in the yeast genome are?" "Where's the biology?"

By now, nearly everybody (even most of those early critics) realizes that the genome pioneers contributed much more than they would have had they focused their energies on particular biological processes. They provided all of us with the information and tools necessary to make progress in the particular biological processes that the rest of us study. They have made our work easier, and extended our reach greatly. By enabling others to make discoveries, the influence of the genome pioneers has spread into every corner of biology. In a very real sense, they contributed to all recent discoveries.

GENETIC MAPPING

The ability to clone and specifically detect fragments of complex genomes, gained in the 1970s, enabled the isolation of genes that, when defective, lead to disease. But, before a disease gene can be cloned, its location in the genome has to be known. The paper by **Petes and Botstein (1977)** is a landmark because it showed a way to do this. It led directly to an explosion of human disease gene identification and had a major impact on our understanding and diagnosis of human disease. This paper is an excellent example of how basic research can, quite unexpectedly, lead to major breakthroughs that have practical applications.

In the early 1970s, people using DNA restriction enzymes and DNA ligase were in the vanguard of yeast genetics, and they were busy learning how to use these reagents to clone genes. The first papers reporting the isolation of genes appeared in the middle of the decade, and everybody was off to the races. Among the first genes to be cloned were the ones encoding ribosomal RNA (Philippsen et al. 1978), because they were the easiest to isolate: They are already partially purified in preparations of genomic DNA because they are enriched in the genome relative to most other genes, since each cell contains about 100 copies of the rRNA-encoding genes. Tom Petes had his hands on these genes and noticed that some of the yeast strains he was working with differed in the number of fragments that appeared after digestion of the rDNA with *Eco*RI.

He and David Botstein realized that this difference in the DNA of the strains could be used to genetically map the rDNA and that this would enable them to resolve the "controversy" that raged over whether yeast carried its rRNA genes at several different locations in the genome (as is the case with most other species that had been analyzed) or at one location (Petes and Botstein 1977). Petes simply scored the segregation of this trait in spores that emanated from sporulation of a diploid that exhibited both patterns of restriction fragments. He found that the two patterns of fragments segregated as alleles, which proved that the rDNA is present at only one location (more precisely, they could only be certain that no more than 10% of the rDNA genes are at a second chromosomal location). Petes went on to map this location to the left arm of chromosome 12 (Petes 1979a,b). In parallel work, Maynard Olson and his colleagues mapped tRNA-encoding genes in a similar way (Goodman et al. 1977).

The paper by Petes and Botstein (1977) is significant not because of what it contributed to the field of yeast genetics, but because of the seminal idea it contributed to the field of genetics: that one could use differences (polymorphisms) in DNA sequence as "DNA-sequence-based markers" to map genes. The first incarnation of this approach—RFLP (restriction fragment length polymorphism) mapping—enabled the first genetic mapping and positional cloning of human disease genes (Knowlton et al. 1985; Monaco et al. 1985; Tsui et al. 1985; Wexler et al. 1985). RFLPs were eventually superseded for genetic mapping by DNA differences that were easier to score and more informative (SSLPs, VNTRs, STRs), but Petes and Botstein's formulation and implementation of the idea of linking detectable differences in DNA sequence in individuals to differences in their phenotype were a real breakthrough.

The idea did not come to them immediately, though. For a long time, they thought the contribution of their work was limited to the narrow issue of the genetic map location of yeast rDNA. The idea that RFLPs could be used to genetically map traits in humans came to David Botstein during discussion of a graduate student's talk at a meeting held in 1978 at the Alta ski resort in Utah (to review the University of Utah's Genetics Training Grant). The student was struggling to genetically map a human disease (Kravitz et al. 1979), and it quickly became clear that his problem was the dearth of genetic markers available for mapping. During the heated discussion that followed the student's talk, it dawned on Botstein and Ron Davis that DNA sequence differences like the ones Petes used to map the yeast rDNA genes should be spread throughout the genome. Continued discussion in the ski lodge's bar that evening quickly solidified the idea that these sequence differences could be used for genetic mapping in humans, and it was formally described a short time later (Botstein et al. 1980). People who attended that meeting say that the significance of the idea was immediately obvious. Indeed, it was rapidly implemented by human geneticists, allowing them to quickly construct a genetic map based on DNA-based markers (Drayna et al. 1984; Donis-Keller et al. 1987) and to clone many human disease genes. Yeast geneticists were gratified (but not surprised) that research on the organism had a significant impact on the field of human genetics.

PHYSICAL MAPPING

In 1986, Olson and co-workers reported their progress on producing a physical map of the yeast genome (**Olson et al. 1986**). This is perhaps an odd paper to be anointed a landmark in the field. It does not shed any light on a biological process; the subject matter—mapping restriction enzyme cutting sites in the yeast genome—seems pedestrian, and it does not even tell a complete story. Nevertheless, the origins of the Human Genome Project, which some people believe is the most significant accomplishment in modern biological science, can be traced back to this paper.

By the late 1970s, recombinant DNA had captured the imagination of experimental biologists, and the *avant garde* of molecular biology was busy using DNA restriction enzymes (discovered just 10 years earlier) to map and clone genes. Methods for determining the sequence of DNA (Maxam and Gilbert 1977; Sanger et al. 1977a) had just been invented and were beginning to be disseminated to a few laboratories. Everyone was glad to have these powerful new methods for isolating and analyzing genes, because they promised great breakthroughs, or at least a significant increase in productivity. There was a great urgency among molecular biologists to acquire these revolutionary techniques and apply them to their favorite biological questions. The excitement in the field was palpable. But the focus remained on individual genes and processes. No one was thinking of applying these methods to the analysis of entire genomes.

No one, that is, except Maynard Olson and John Sulston. Both of these scientists were trained as chemists (Olson as an inorganic chemist, Sulston as an organic chemist), but turned to genetics and molecular biology (undoubtedly because of the palpable excitement in these fields). While a postdoctoral fellow in Ben Hall's lab at the University of Washington in

Seattle, Olson was one of the early discoverers of introns (Goodman et al. 1977). He joined the faculty of Washington University in St. Louis in 1978 as an assistant professor in the Department of Genetics, where he set out to clone and map the entire yeast genome. To most yeast geneticists, this endeavor seemed of little potential value, and they often reviled Olson for pursuing it. A few years later, at the MRC laboratories in Cambridge, England, John Sulston had just completed the remarkable feat of determining the lineage of all 1090 cells of the roundworm *Caenorhabditis elegans* (for which he was awarded the Nobel Prize in 2002). Looking for another challenge, Sulston set himself the goal of cloning and mapping the approximately 100-Mb genome of *C. elegans*.

By 1986, Olson and Sulston were far enough along in their projects to warrant publication of a description of their approach and a report of their progress, and their manuscripts appeared back-to-back (Coulson 1986; Olson et al. 1986). These reports were timely, because by then debates on the possibility and desirability of determining the sequence of the human genome were just beginning to take place. It was another 3–4 years before the full impact of Olson and Sulston's papers began to be felt. By 1989, enough people had bought into the idea of whole-genome sequencing to enable launching of the Human Genome Project, and the model organisms that were to be the first expeditions into this frontier were being chosen. A necessity for a genome sequencing project was a set of mapped, "sequence-ready" clones, and Olson and Sulston's 1986 papers showed that these were well on their way to being available for yeast and worms. Because of the availability of these clones, the first eukaryotic genome sequences to be determined were of yeast (1996) and *C. elegans* (1998). About half of the yeast genome sequence was determined using Olson's clones; the clones for the other half came from a parallel mapping project initiated about 10 years after Olson began to map the yeast genome (Dujon 1993).

Sulston and Olson's work paved the way for sequencing the human genome. Determining the sequence of mapped clones turned out to be essential for obtaining the complete sequence (Waterston et al. 2002), and the methods used to produce those maps were basically those developed by Sulston and Olson. The method was simple: Produce a large number of clones of fragments of the target genome, determine the size of the subfragments produced by digestion of these clones with restriction enzymes (Olson used *Eco*RI and *Hin*dIII), and then determine which clones overlap based on the fragments they share. If enough clones were analyzed, there should be sufficient overlap among them to be able to "tile" a unique map (Fig. 1). The biggest challenge was to develop algorithms that could sort through the many possibilities for clone overlaps and decide on the one correct map. Most of Olson's paper describes how he met this challenge. However, even the best algorithm cannot deduce a map with bad data, so accurately measuring DNA fragment sizes, day after day, was a significant technical challenge that is not discussed in the paper.

Olson went on to make seminal technical and intellectual contributions in genome sequencing. Among these were the development of yeast artificial chromosome (YAC) cloning vectors (Burke et al. 1987) by David Burke, a graduate student in Olson's lab, who recognized the necessity of cloning and mapping large DNA fragments to provide long-range continuity to the yeast genome physical map. Another important development was the use of the newly developed polymerase chain reaction (PCR) as a genome-mapping tool (representing the genome map as a set of ordered sequence-tagged sites, or STSs), which had a major impact on the development of physical maps of the human genome (Green et al. 1991).

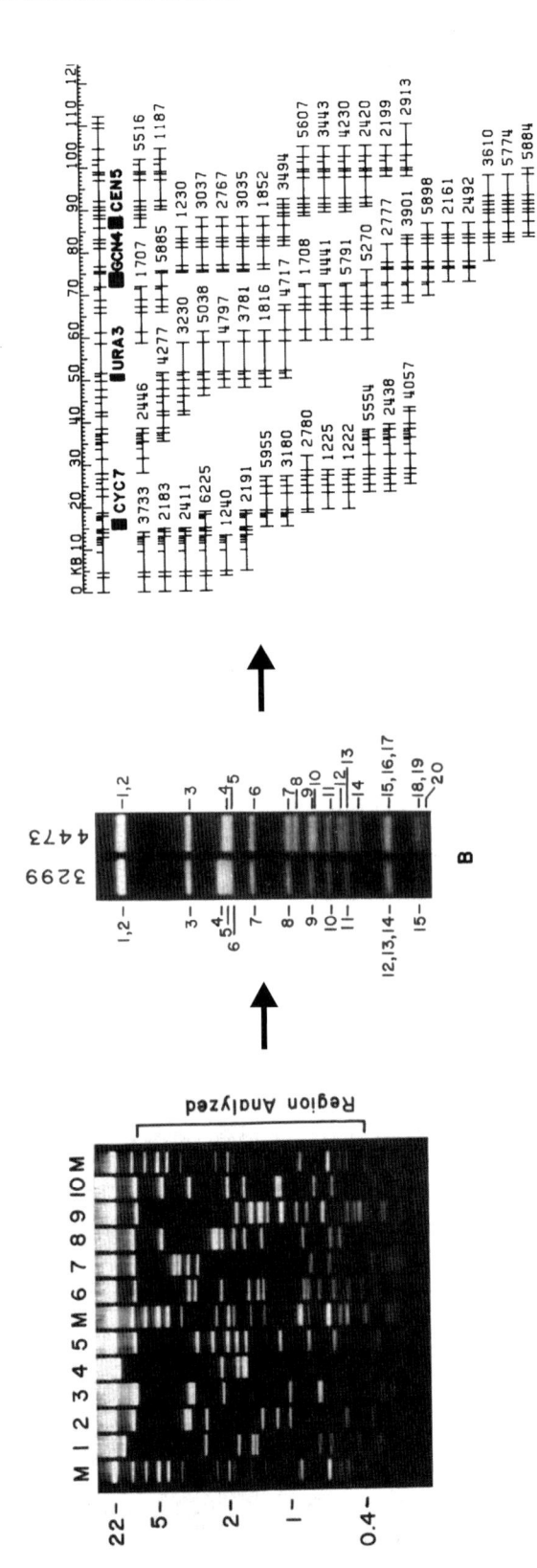

FIGURE 1. (*Left panel*) Agarose gel of the type used to collect the fragment-size data from *Eco*RI/*Hin*dIII (RH)-digested clones of yeast genomic DNA. (*Middle panel*) RH fragments of two bacteriophage λ clones, with the fragments numbered starting with the largest. (*Right panel*) Physical map of a region of chromosome V that results from analysis of these data. The order of the RH sites denoted by vertical lines that extend above and below the horizontal lines is known; the relative order of the RH sites denoted by lines that extend only above the horizontal line is unknown. (Modified from Olson et al. 1986.)

The paper by Olson and collaborators and its companion authored by John Sulston and his colleagues were the first to describe whole-genome analysis, and thus helped spawn a new era of biology. These papers slowly stimulated others to think in terms of entire genomes rather than individual genes.

GENOME SEQUENCING

By the late 1980s, a few brash (some said reckless, others said foolish) molecular biologists were considering the possibility of determining the DNA sequence of entire genomes. The sequence of a few relatively small viral genomes (~50 kb or less, see Table 1) had been determined, but scaling up to the much larger cellular genomes was a daunting prospect. Some people said it could not be done; others said it should not be done. Even those who were optimistic about the prospects for success and enthusiastic about the potential scientific return on an investment in genome sequencing thought it would take a long time. No one could know how quickly it would be accomplished.

The bacterium *Escherichia coli* and the yeast *Saccharomyces cerevisiae* were natural targets for the first genome sequencing efforts, largely because the relatively small size of their genomes made them the most realistic targets for a genome sequencing effort, but also because of their long history of study, which provided a rich knowledge base of the biology of these organisms. Yeast was a particularly attractive target because, being a eukaryote, its genome sequence would reveal the basic set of genes required for organization and function of a eukaryotic cell.

While most people in the young field of genome sequencing were still debating the method of approach and trying to develop new (hopefully revolutionary) techniques for DNA sequencing, Andre Goffeau and Steve Oliver and their colleagues in Europe were quietly beginning to determine the sequence of the yeast genome. The entire 14-Mb

TABLE 1. Nucleotide sequencing timeline

1965	Yeast tRNA sequence	Holley et al. (1965)
1973–1975	*E. coli lacO*	Maxam and Gilbert (1973); Dickson et al. (1975)
1977	Chemical DNA sequencing	Maxam and Gilbert (1977)
1977	Dideoxy DNA sequencing	Sanger et al. (1977a)
1977	Yeast 5S rRNA	Valenzuela et al. (1977)
1977	Yeast tRNA	Goodman et al. (1977)
1977	Bacteriophage ΦX174 (5.375 kb)	Sanger et al. (1977b)
1978	Yeast *CYC1* gene	Montgomery et al. (1978)
1982	Bacteriophage λ (48 kb)	Sanger et al. (1982)
1995	*Haemophilus influenzae* (1,830 kb)	Fleischmann et al. (1995)
1996	*Saccharomyces cerevisiae* (12,068 kb)	Goffeau et al. (1996)
1997	*Escherichia coli* (4,639 kb)	Blattner et al. (1997)
1998	*Caenorhabditis elegans* (97,000 kb)	Consortium (1998)
2000	*Drosophila melanogaster* (120,000 kb)	Adams et al. (2000)
2001	*Homo sapiens* (3,000,000 kb)	Lander et al. (2001); Venter et al. (2001)

genome was, of course, too big to tackle all at once, so they chose to focus on an individual chromosome. They chose chromosome III for two reasons. First, it was probably the best-studied chromosome, mostly because it contains genes responsible for determining the mating type of cells (at the *MAT* locus in the middle of the chromosome) and genes that enable cells to switch their mating type (the "silent" copies of the *MAT* genes—*HML* and *HMR*—located toward each chromosome end), genes that were (and still are) of great interest. More important, an ordered set of clones of DNA fragments spanning most of chromosome III were available, because of a unique property of the chromosome: It occasionally converts into a circular chromosome by recombination between the *HML* and *HMR* loci (Strathern et al. 1979; Haber et al. 1980). This "ring chromosome" can be separated from the 15 other chromosomes by gel electrophoresis and isolated intact. A library of clones of this ring chromosome had been made and mapped (Newlon et al. 1991) and provided most of the needed sequence-ready clones for the project.

But determining the sequence of even just one small chromosome was a huge undertaking, so Goffeau and Oliver recruited several of their European colleagues studying yeast to participate in the project. The clones were distributed to 35 laboratories throughout Europe, where their sequence was determined, and then sent to a central database (MIPS) for assembly and analysis. The results were exciting, and they electrified the audience that first heard about them at the Yeast Meeting held in San Francisco in May, 1991 (**Oliver et al. 1992**). Many people in the audience sensed that they were witnessing a defining moment in the field: It was clear from Steve Oliver's presentation that the sequence of the entire genome was within reach!

Perhaps the most surprising result was that most of the genes on chromosome III had not been identified by yeast geneticists: only 34 of the 182 predicted genes on this chromosome had been described (Oliver et al. 1992). Many yeast geneticists had long suspected that their field was overcrowded, but this fear evaporated with the realization that this army of scientists had missed most of the genes in the organism. The sequencing project had provided a large treasure chest of new genes to analyze.

But there was still a long way to go before the sequence of the entire yeast genome would be in hand. It had taken more than 2 years to complete about 2.5% of the yeast genome sequence. Even with anticipated increases in efficiency, the project was expected to extend into the next century (Anderson 2000), and nobody had the patience for that. The European sequencers' success encouraged the large-scale sequencing centers in St. Louis (run by Bob Waterston) and Cambridge, England (run by John Sulston) to join the project. Waterston and Sulston were on a course to finish the worm genome sequence well before the sequence of the yeast genome was to be completed, a situation that was less than ideal. The worm sequencers needed to know the basic set of genes in a eukaryotic cell, to help them annotate the worm genome sequence. They were also planning on using YAC clones of worm DNA for some of their sequencing, and they needed to be able to recognize the yeast sequence so they could throw it out before trying to assemble the worm sequence. (They ended up doing little sequencing of YACs.)

The entry of the two large sequencing centers, with their large stables of automated sequencing machines, promised to speed up the project. Indeed, papers describing the sequence of other yeast chromosomes appeared with increasing regularity, so that by early 1995 papers describing the sequence of a yeast chromosome were no longer *de rigueur*. By early 1996, only 4 years after the publication of the sequence of the first yeast chromosome, the sequence of the entire genome was in hand. This landmark event was announced on April 24, 1996 at joint press conferences in Europe and the United States; an excellent paper describing the results was published a short time later (Goffeau et al. 1996), followed by an entire issue of the journal *Nature* devoted to a description of the yeast genome sequence and its analysis (Yeast Genome Directory 1997). And only 4 years later, the draft sequence of the human genome sequence was completed (Lander et al. 2001; Venter et al. 2001)!

The European sequencing network ended up determining more than half of the yeast genome sequence, the two large sequencing centers in St. Louis and Cambridge contributed about one third of the sequence, and groups in Canada, Japan, and at Stanford University contributed the rest. But Oliver and Goffeau and their colleagues led the way. It is clear that they were experimenting with sequencing methods on chromosome III, since several different sequencing approaches were employed. (Chromosome III was later resequenced, because the quality, although good at the time—at about 1 mistake per 1000 base pairs, it was approximately fourfold more accurate than the yeast sequences in the public databases at the time—was not up to the standards met at the end of the project: No more than 1 mistake per 5000 base pairs.) They, too, turned more and more to automated sequencing as that methodology matured. Everybody now agrees that distributing the work among many labs was an approach that was too inefficient to last, but the well-organized network of European researchers turned out to be a model for the collaborative effort that was needed to determine the sequence of the human genome, as Oliver et al. suggested it would. In fact, it was critical for the project, because it provided the first success and provided confidence that the task could be completed. The rest is, as they say, history.

SNP MAPPING

One of the most exciting things about Genomics is that one can see the potential for practical applications that promise to bring great advances in medicine. Genome scientists are already saying that the day is near when we will all carry our genotype on a card in our wallets or purses, and the recent determination of the human genome sequence brings that day even closer. The hope is that with this information in our back pocket, we will be able to predict our susceptibility to many diseases, which should enable us to minimize our disease risks by modifying our behavior accordingly, and perhaps enable custom-designed treatments. But how will we determine our genotypes? Once again, the pioneering work was carried out on yeast, with the mapping of single-nucleotide polymorphisms (SNPs) by **Winzeler et al. (1998)**.

Genes have long been mapped by a laborious, chromosome-by-chromosome search. One identifies a heritable difference in phenotype, then maps that difference either genetically (by looking among progeny of a cross of the appropriate individuals for its linkage to other heritable differences in phenotype) or physically (by looking for a difference in DNA sequence that is linked to a heritable trait). In yeast, one must do many crosses to many strains with different mutations to find linkage of two traits, and even then, linkage is often not detected (because of the high rate of genetic recombination in yeast). Mapping a mutation even in an organism as manipulatable as yeast is often frustrating; it is even more laborious in humans, because of several limitations (e.g., the crosses cannot be designed, and the number of progeny that can be scored is usually limited). An even more daunting aspect of identifying gene variants that predispose individuals to disease is that inheritance is often multigenic in nature.

DNA microarrays ("gene chips") promise to simplify gene mapping. In one experiment performed on a space no larger than a fingernail, it should be possible to map differences in an individual's DNA (i.e., "genotype" the individual). Winzeler et al. (1998) did just that for yeast. They used an array of oligonucleotides (an Affymetrix chip) to identify differences in the DNA sequences of two closely related yeast strains. They probed an Affymetrix chip that contains oligonucleotides representing the sequence of a common lab strain (S288c, the strain whose genome was sequenced) with DNA from a wild isolate of *S. cerevisiae* (a pathogenic strain isolated from an immunocompromised patient). The DNA sequences of the genomes of these two strains are over 99% identical (one nucleotide difference approximately every 160 bp). For most of the oligonucleotides on the chip, the sequence of the wild strain is identical to that of the lab strain, and therefore the probe (DNA from the wild strain) hybridizes to them. However, oligonucleotides on the chip that have sequences of regions of the genome that differ in the two yeast strains hybridize significantly less well to the probe (sometimes not at all). This latter type of oligonucleotides, whose sequences and positions on the chip are known, marks the position of the sequence differences, i.e., alleles, between the two strains. Since the genome sequence is known, the precise location of these sequence differences is also known (to a level of resolution of the length of the oligonucleotide probe—20 nucleotides).

Winzeler and co-workers mapped 3714 sequence differences between the two yeast strains. This gives an average genetic resolution of one allele every 1 cM and an average physical resolution of one allele approximately every 3.5 kb. They used these alleles to map, simultaneously, five independently segregating traits that differ between these two strains. This was accomplished by analyzing the segregation of all 3714 markers in ten meiotic segregants that had inherited the five traits. They confirmed the chromosomal location of mutations in four genes known to cause phenotypic differences (*MAT, LYS2, LYS5,* and *HO*). They also identified a 57-kb chromosomal region that harbored a gene for resistance to cycloheximide. They noticed a gene (*PDR5*) within this region of the genome likely to be responsible for this phenotype (a "candidate gene") and validated its role in the phenotype. They also mapped the location of all 97 DNA crossover events that occurred in a single meiosis (Fig. 2). Thus, this method proved to be highly efficient for mapping mutations and identifying phenotypes associated to particular genes.

Optimists would say that it is just a short leap to application of this method to humans and that our genotypes are nearly in our wallets. But more realistic people point out that the human genome is much more complex than the yeast genome, so reliably detecting sequence differences in the human genome by hybridization of genomic DNA to short oligonucleotides remains a significant challenge.

FUNCTIONAL GENOMICS

The ability to view the entire genome began to change the way yeast geneticists approached tackling biological problems. It provided them with the potential to test *all* genes in their experiments. If one wanted to identify genes involved in a particular cellular process, why throw oneself at the mercy of fickle fate and be satisfied with what comes out

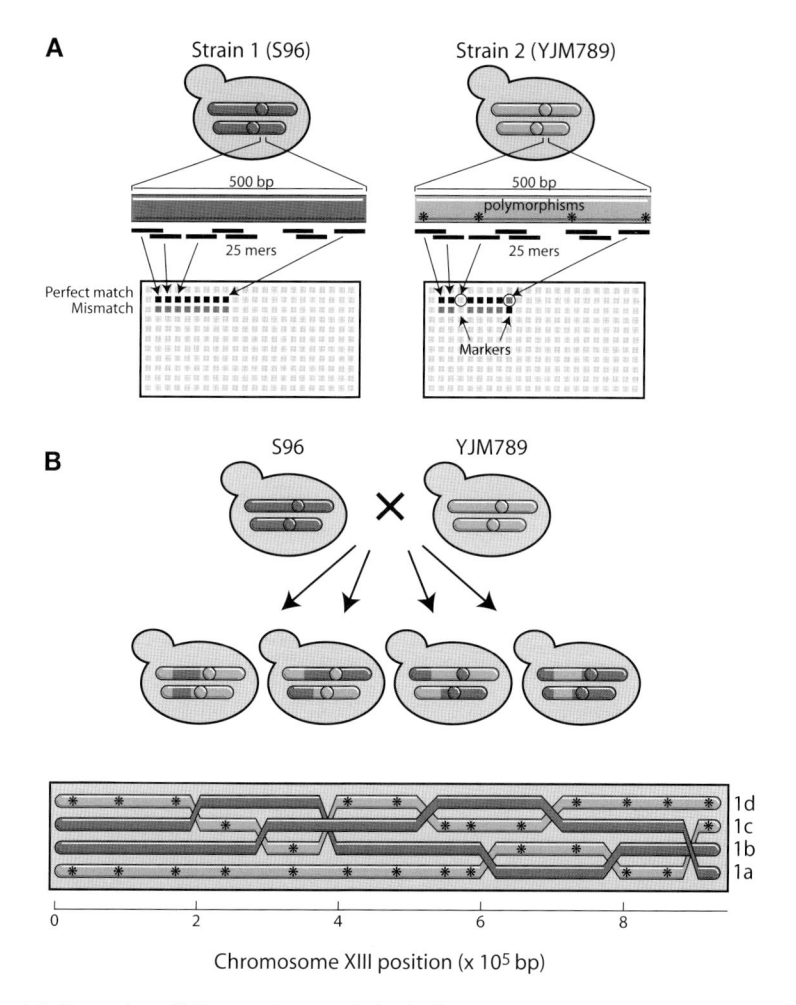

FIGURE 2. (*A*) Detecting allelic variation with high-density DNA arrays. An array of 25-bp oligonucleotide probes (an Affymetrix "gene chip") was probed with labeled genomic DNA from two different strains. One strain (S96) matches the probe sequences perfectly (because the array was designed using this genome sequence); the other strain has one single-nucleotide polymorphism approximately every 160 bp. The array also contains probes with a single-base mismatch in the central position of the oligonucleotide (Mismatch) to serve as background and nonspecific hybridization controls. Probes complementary to YJM789 DNA fragments that carry single-nucleotide polymorphisms (∗) yield a decreased signal intensity relative to the S96 signal. These signal differences between S96 and YJM789 DNA probes are the alleles that are scored. (*B*) Schematic presentation of the inheritance of the single-nucleotide polymorphisms on chromosome XIII in one tetrad from a cross between YJM789 and S96. The location of all the crossover events in this meiosis can be mapped. (Redrawn, with permission, from Winzeler et al. 1998 [©AAAS].)

of a random mutagenesis experiment? Why not systematically test *all* of the genes for their role in the process? If one is going to measure gene expression in response to a particular perturbation, why not do it for *all* genes in the genome?

Even before determination of the yeast sequence was completed, investigators were thinking of how best to use the data and were beginning to develop the genetic tools that would be necessary. Mike Snyder and his colleagues were the first to take this on in a big way (Burns et al. 1994) with a large-scale random mutagenesis with *lacZ* insertions to identify genes expressed under different conditions and to determine the subcellular locations of the encoded gene products. Then, under the leadership of Steve Oliver, the network of European labs sequencing the yeast genome reorganized into a "functional analysis network" (EUROFAN) (Oliver 1996), whose goal was to systematically generate and analyze mutants defective in each gene of the organism. Like with the sequencing, they were quickly joined by their North American colleagues in a collaborative effort to achieve the task. That international project culminated with the report of the complete set of nearly 6000 gene deletion mutants (**Giaever et al. 2002**).

The main goal of the project was simply to produce and distribute to the scientific community a resource: the complete collection of gene deletion mutants. This "YKO collection" (Giaver et al. 2002) quickly found its way into many labs around the world, and several reports of its use to analyze different cellular processes soon appeared (some before this paper describing the YKO collection was published!). But the YKO collection was also set up to enable testing, in a single experiment, the growth of all 6000 mutants under a large number of different conditions. This was enabled by Ron Davis's clever idea to tag each deletion mutant with a different 20-nucleotide sequence (incorporated into the oligonucleotides used to make the deletions) (Shoemaker et al. 1996). The presence of this sequence could be detected, after the appropriate polymerase chain reactions, by its hybridization to a DNA microarray of all 6000 sequence tags. The tags corresponding to each mutant in a mixture of equal numbers of the 6000 mutants should hybridize to their complementary sequences on the array with equal intensity. If this mixture of mutants is cultured overnight, those missing genes required for growth under that condition will be underrepresented in the resultant culture, and consequently the intensity of hybridization of their sequence tags to the array will be reduced. The intensity of hybridization of a tag to the array can be quantified, allowing estimation of the contribution of its gene to fitness under the growth condition.

In this way, Giaever et al. (2002) identified all of the genes required for maximal growth under several conditions. It is perhaps somewhat surprising that they identified several genes not previously known to be required for growth on galactose, despite this being one of the best studied phenotypes of yeast. The most provocative result reported in this paper is that regulation of gene expression does not seem to be very well correlated with gene function; i.e., expression of many of the genes that are required for maximal growth under a condition is not regulated by that condition, and, conversely, many genes whose expression is regulated by a particular condition are not required for growth under that condition.

The report by Giaever et al. (2002) was chosen rather arbitrarily as a landmark paper. It serves as a representative of many other equally seminal papers that constitute the early canon of Functional Genomics. Reports of genome-wide surveys of gene expression are arguably the papers that pioneered the field of Functional Genomics (DeRisi et al. 1997; Lashkari et al. 1997; Velculescu et al. 1997; Wodicka et al. 1997). Other seminal papers are those that reported genome-wide surveys of protein-protein interactions (Ito et al. 2000; Uetz et al. 2000; Gavin et al. 2002; Ho et al. 2002), protein-DNA interactions (Ren et al. 2000), gene-gene interactions (Tong et al. 2001), and gene-metabolite interactions (Raamsdonk et al. 2001). Even biochemical experiments are now done on a genome-wide scale (Martzen et al. 1999; Zhu et al. 2001). With the sequence of the yeast genome in our computers, and wonderful tools and reagents on our lab benches, there is no going back to business as usual: We have entered the golden age of yeast genetics (Johnston 2000).

STUDY QUESTIONS

1. Why could Petes and Botstein (1977) not be certain that all of the rDNA is at a single locus?

2. What result led Petes and Botstein (1977) to conclude that the rDNA locus is not close to a centromere?

3. What is the limit of resolution of Olson's "RH map" (Olson et al. 1986) of the yeast genome? What determines this limit?

4. Some regions of Olson's map (Olson et al. 1986) include several RH fragments whose order is not known. Why are they unordered?

5. What is the difference between a genetic map and a physical map of a chromosome? How can the two maps be integrated?

6. What is the difference between "top-down" and "bottom-up" genome mapping?

7. Only yeast genes larger than 100 codons were annotated in the genome sequence. Why? Some of the annotated genes are probably false (i.e., not real genes). How might they be identified?

8. What are some major differences between yeast and human genome organization?

9. Why do you think the method of Winzeler et al. (1998) has not yet been applied to humans?

10. When screening the yeast gene deletion collection for phenotypes, it is best to use the homozygous diploids. Why?

11. Can you come up with a reasonable explanation for why genes whose expression is regulated by a particular treatment have no effect on the response of cells to that treatment?

REFERENCES

Anderson A. 2000. Yeast Genome Project: 300,000 and counting. *Science* **256**: 462.

Adams M.D., Celniker S.E., Holt R.A., Evans C.A., Gocayne J.D., Amanatides P.G., Scherer S.E., Li P.W., Hoskins R.A., Galle R.F., George R.A., Lewis S.E., Richards S., Ashburner M., Henderson S.N., Sutton G.G., Wortman J.R., Yandell M.D., Zhang Q., Chen L.X., Brandon R.C., Rogers Y.H., Blazej R.G., Champe M., Pfeiffer B.D., et al. 2000. The genome sequence of *Drosophila melanogaster*. *Science* **287**: 2185–2195.

Blattner F.R., Plunkett G., III, Bloch C.A., Perna N.T., Burland V., Riley M., Collado-Vides J., Glasner J.D., Rode C.K., Mayhew G.F., Gregor J., Davis N.W., Kirkpatrick H.A., Goeden M.A., Rose D.J., Mau B., and Shao Y. 1997. The complete genome sequence of *Escherichia coli* K-12. *Science* **277**: 1453–1474.

Botstein D., White R.L., Skolnick M., and Davis R.W. 1980. Construction of a genetic linkage map in man using restriction fragment length polymorphisms. *Am. J. Hum. Genet.* **32**: 314–331.

Burke D.T., Carle G.F., and Olson M.V. 1987. Cloning of large segments of exogenous DNA into yeast by means of artificial chromosome vectors. *Science* **236:** 806–812.

Burns N., Grimwade B., Ross-Macdonald P.B., Choi E.Y., Finberg K., Roeder G.S., and Snyder M. 1994. Large-scale analysis of gene expression, protein localization, and gene disruption in *Saccharomyces cerevisiae*. *Genes Dev.* **8:** 1087–1105.

Consortium (The *C. elegans* Sequencing Consortium). 1998. Genome sequence of the nematode *C. elegans:* A platform for investigating biology. *Science* **282:** 2012–2018.

Coulson A., Sulston J.E, Brenner S., and Karn J. 1986. Towards a physical map of the genome of the nematode *Caenorhabditis elegans*. *Proc. Natl. Acad. Sci.* **83:** 7821–7825.

DeRisi J.L., Iyer V.R., and Brown P.O. 1997. Exploring the metabolic and genetic control of gene expression on a genomic scale. *Science* **278:** 680–686.

Dickson R.C., Abelson J., Barnes W.M., and Reznikoff W.S. 1975. Genetic regulation: The Lac control region. *Science* **87:** 27–35.

Donis-Keller H., Green P., Helms C., Cartinhour S., Weiffenbach B., Stephens K., Keith T.P., Bowden D.W., Smith D.R., and Lander E.S. 1987. A genetic linkage map of the human genome. *Cell* **51:** 319–337.

Drayna D., Davies K., Hartley D., Mandel J.L., Camerino G., Williamson R., and White R. 1984. Genetic mapping of the human X chromosome by using restriction fragment length polymorphisms. *Proc. Natl. Acad. Sci.* **81:** 2836–2839.

Dujon B. 1993. Mapping and sequencing the nuclear genome of the yeast *Saccharomyces cerevisiae:* Strategies and results of the European enterprise. *Cold Spring Harbor Symp. Quant. Biol.* **58:** 357–366.

Fleischmann R.D., Adams M.D., White O., Clayton R.A., Kirkness E.F., Kerlavage A.R., Bult C.J., Tomb J.F., Dougherty B.A., Merrick J.M., McKenney K., Sutton G.G., FitzHugh W., Fields C.A., Gocayne J.D., Scott J.D., Shirley R., Liu L.I., Glodek A., Kelley J.M., Weidman J.F., Phillips C.A., Spriggs T., Hedblom E., Cotton M.D., et al. 1995. Whole-genome random sequencing and assembly of *Haemophilus influenzae* Rd. *Science* **269:** 496–512.

Gavin A.C., Bosche M., Krause R., Grandi P., Marzioch M., Bauer A., Schultz J., Rick J.M., Michon A.M., and Cruciat C.M. 2002. Functional organization of the yeast proteome by systematic analysis of protein complexes. *Nature* **415:** 141–147.

Giaever G., Chu A.M., Ni L., Connelly C., Riles L., Véronneau S., Dow S., Lucau-Danila A., Anderson K., André B., Arkin A.P., Astromoff A., El-Bakkoury M., Bangham R., Benito R., Brachat S., Campanaro S., Curtiss M., Davis K., Deutschbauer A., Entian K.D., Flaherty P., Foury F., Garfinkel D.J., Gerstein M., et al. 2002. Functional profiling of the *Saccharomyces cerevisiae* genome. *Nature* **418:** 387–391.

Gilbert W. and Maxam A. 1973. The nucleotide sequence of the lac operator. *Proc. Natl. Acad. Sci.* **70:** 3581–3584.

Goffeau A. and Vassarotti A. 1991. The European project for sequencing the yeast genome. *Res. Microbiol.* **142:** 901–903.

Goffeau A., Barrell B.G., Bussey H., Davis R.W., Dujon B., Feldmann H., Galibert F., Hoheisel J.D., Jacq C., and Johnston M. 1996. Life with 6000 genes. *Science* **274:** 546–567.

Goodman H.M., Olson M.V., and Hall B.D. 1977. Nucleotide sequence of a mutant eukaryotic gene: The yeast tyrosine-inserting ochre suppressor SUP4-o. *Proc. Natl. Acad. Sci.* **74:** 5453–5457.

Green E.D., Mohr R.M., Idol J.R., Jones M., Buckingham J.M., Deaven L.L., Moyzis R.K., and Olson M.V. 1991. Systematic generation of sequence-tagged sites for physical mapping of human chromosomes: Application to the mapping of human chromosome 7 using yeast artificial chromosomes. *Genomics* **11:** 548–564.

Haber J.E., Rogers D.T., and McCusker J.H. 1980. Homothallic conversions of yeast mating-type genes occur by intrachromosomal recombination. *Cell* **22:** 277–289.

Hereford L.M. and Rosbash M. 1977. Number and distribution of polyadenylated RNA sequences in yeast. *Cell* **10:** 453–462.

Ho Y., Gruhler A., Heilbut A., Bader G.D., Moore L., Adams S.L., Millar A., Taylor P., Bennett K., and Boutilier K. 2002. Systematic identification of protein complexes in *Saccharomyces cerevisiae* by mass spectrometry. *Nature* **415:** 180–183.

Holley R.W., Apgar J., Everett G.A., Madison J.T., Marquisee M., Merrill S.H., Penswick J.R., and Zamir A. 1965. Structure of a ribonucleic acid. *Science* **147:** 1462–1465.

Ito T., Tashiro K., Muta S., Ozawa R., Chiba T., Nishizawa M., Yamamoto K., Kuhara S., and Sakaki Y. 2000.

Toward a protein-protein interaction map of the budding yeast: A comprehensive system to examine two-hybrid interactions in all possible combinations between the yeast proteins. *Proc. Natl. Acad. Sci.* **97:** 1143–1147.

Johnston M. 2000. The yeast genome: On the road to the Golden Age. *Curr. Opin. Genet. Dev.* **10:** 617–623.

Johnston M. and Fields S. 2000. Grass-roots genomics. *Nat. Genet.* **24:** 5–6.

Knowlton R.G., Cohen-Haguenauer O., Van Cong N., Frezal J., Brown V.A., Barker D., Braman J.C., Schumm J.W., Tsui L.C., and Buchwald M. 1985. A polymorphic DNA marker linked to cystic fibrosis is located on chromosome 7. *Nature* **318:** 380–382.

Kravitz K., Skolnick M., Cannings C., Carmelli D., Baty B., Amos B., Johnson A., Mendell N., Edwards C., and Cartwright G. 1979. Genetic linkage between hereditary hemochromatosis and HLA. *Am. J. Hum. Genet.* **31:** 601–619.

Lander E.S., Linton L.M., Birren B., Nusbaum C., Zody M.C., Baldwin J., Devon K., Dewar K., Doyle M., FitzHugh W., Funke R., Gage D., Harris K., Heaford AS., Howland J., Kann L., Lehoczky J., LeVine R., McEwan P., McKernan K., Meldrim J., Mesirov J.P., Miranda C., Morris W., Naylor J., et al.; International Human Genome Sequencing Consortium. 2001. Initial sequencing and analysis of the human genome. *Nature* **409:** 860–921.

Lashkari D.A., DeRisi J.L., McCusker J.H., Namath A.F., Gentile C., Hwang S.Y., Brown P.O., and Davis R.W. 1997. Yeast microarrays for genome wide parallel genetic and gene expression analysis. *Proc. Natl. Acad. Sci.* **94:** 13057–13062.

Martzen M.R., McCraith S.M., Spinelli S.L., Torres F.M., Fields S., Grayhack E.J., and Phizicky E.M. 1999. A biochemical genomics approach for identifying genes by the activity of their products. *Science* **286:** 1153–1155.

Maxam A.M. and Gilbert W. 1977. A new method for sequencing DNA. *Proc. Natl. Acad. Sci.* **74:** 560–564.

Monaco A.P., Bertelson C.J., Middlesworth W., Colletti C.A., Aldridge J., Fischbeck K.H., Bartlett R., Pericak-Vance M.A., Roses A.D., and Kunkel L.M. 1985. Detection of deletions spanning the Duchenne muscular dystrophy locus using a tightly linked DNA segment. *Nature* **316:** 842–845.

Montgomery D.L., Hall B.D., Gillam S., and Smith M. 1978. Identification and isolation of the yeast cytochrome *c* gene. *Cell* **14:** 673–680.

Mortimer R.K. and Hawthorne D.C. 1966. Genetic mapping in *Saccharomyces*. *Genetics* **53:** 165–173.

Newlon C.S., Lipchitz L.R., Collins I., Deshpande A., Devenish R.J., Green R.P., Klein H.L., Palzkill T.G., Ren R.B., and Synn S. 1991. Analysis of a circular derivative of *Saccharomyces cerevisiae* chromosome III: A physical map and identification and location of ARS elements. *Genetics* **129:** 343–357.

Oliver S. 1996. A network approach to the systematic analysis of yeast gene function. *Trends Genet.* **12:** 241–242.

Oliver S.G., van der Aart Q.J.M., Agostoni-Carbone M.L., Aigle M., Alberghina L., Alexandraki D., Antoine G., Anwar R., Ballesta J.P.G., Benit P., Berben G., Bergantino E., Biteau N., Bolle P.A., Bolotin-Fukuhara M., Brown A., Brown A.J.P., Buhler J.M., Carcano C., Carignani G., Cederberg H., Chanet R., Contreras R., Crouzet M., Daignan-Fornier B., et al. 1992. The complete DNA sequence of yeast chromosome III. *Nature* **357:** 38–46.

Olson M.V., Dutchik J.E., Graham M.Y., Brodeur G.M., Helms C., Frank M., MacCollin M., Scheinman R., and Frank T. 1986. Random-clone strategy for genomic restriction mapping in yeast. *Proc. Natl. Acad. Sci.* **83:** 7826–7830.

Petes T.D. 1979a. Meiotic mapping of yeast ribosomal deoxyribonucleic acid on chromosome XII. *J. Bacteriol.* **138:** 185–192.

———. 1979b. Yeast ribosomal DNA genes are located on chromosome XII. *Proc. Natl. Acad. Sci.* **76:** 410–414.

Petes T.D. and Botstein D. 1977. Simple Mendelian inheritance of the reiterated ribosomal DNA of yeast. *Proc. Natl. Acad. Sci.* **74:** 5091–5095.

Philippsen P., Kramer R.A., and Davis R.W. 1978. Cloning of the yeast ribosomal DNA repeat unit in SstI and HindIII lambda vectors using genetic and physical size selections. *J. Mol. Biol.* **123:** 371–386.

Raamsdonk L.M., Teusink B., Broadhurst D., Zhang N., Hayes A., Walsh M.C., Berden J.A., Brindle K.M., Kell D.B., and Rowland J.J. 2001. A functional genomics strategy that uses metabolome data to reveal the phenotype of silent mutations. *Nat. Biotechnol.* **19:** 45–50.

Ren B., Robert F., Wyrick J.J., Aparicio O., Jennings E.G., Simon I., Zeitlinger J., Schreiber J., Hannett N., and Kanin E. 2000. Genome-wide location and function of DNA binding proteins. *Science* **290:** 2306–2309.

Sanger F., Nicklen S., and Coulson A.R. 1977a. DNA sequencing with chain-terminating inhibitors. *Proc. Natl. Acad. Sci.* **74:** 5463–5467.

Sanger F., Coulson A.R., Hong G.F., Hill D.F., and Petersen G.B. 1982. Nucleotide sequence of bacteriophage lambda DNA. *J. Mol. Biol.* **162:** 729–773.

Sanger F., Air G.M., Barrell B.G., Brown N.L., Coulson A.R., Fiddes C.A., Hutchison C.A., Slocombe P.M., and Smith M. 1977b. Nucleotide sequence of bacteriophage phi X174 DNA. *Nature* **265:** 687–695.

Shoemaker D.D., Lashkari D.A., Morris D., Mittmann M., and Davis R.W. 1996. Quantitative phenotypic analysis of yeast deletion mutants using a highly parallel molecular bar-coding strategy. *Nat. Genet.* **14:** 450–456.

Strathern J.N., Newlon C.S., Herskowitz I., and Hicks J.B. 1979. Isolation of a circular derivative of yeast chromosome III: Implications for the mechanism of mating type interconversion. *Cell* **18:** 309–319.

Tong A.H., Evangelista M., Parsons A.B., Xu H., Bader G.D., Page N., Robinson M., Raghibizadeh S., Hogue C.W., and Bussey H. 2001. Systematic genetic analysis with ordered arrays of yeast deletion mutants. *Science* **294:** 2364–2368.

Tsui L.C., Buchwald M., Barker D., Braman J.C., Knowlton R., Schumm J.W., Eiberg H., Mohr J., Kennedy D., and Plavsic N. 1985. Cystic fibrosis locus defined by a genetically linked polymorphic DNA marker. *Science* **230:** 1054–1057.

Uetz P., Giot L., Cagney G., Mansfield T.A., Judson R.S., Knight J.R., Lockshon D., Narayan V., Srinivasan M., and Pochart P. 2000. A comprehensive analysis of protein-protein interactions in *Saccharomyces cerevisiae*. *Nature* **403:** 623–627.

Valenzuela P., Bell G.I., Masiarz F.R., DeGennaro L.J., and Rutter W.J. 1977. Nucleotide sequence of the yeast 5S ribosomal RNA gene and adjacent putative control regions. *Nature* **267:** 641–643.

Velculescu V.E., Zhang L., Zhou W., Vogelstein J., Basrai M.A., Bassett D.E., Jr., Hieter P., Vogelstein B., and Kinzler K.W. 1997. Characterization of the yeast transcriptome. *Cell* **88:** 243–251.

Venter J.C., Adams M.D., Myers E.W., Li P.W., Mural R.J., Sutton G.G., Smith H.O., Yandell M., Evans C.A., Holt R.A., Gocayne J.D., Amanatides P., Ballew R.M., Huson D.H., Wortman J.R., Zhang Q., Kodira C.D., Zheng X.H., Chen L., Skupski M., Subramanian G., Thomas P.D., Zhang J., Gabor Miklos G.L., Nelson C., et al. 2001. The sequence of the human genome. *Science* **291:** 1304–1351.

Wexler N.S., Conneally P.M., Housman D., and Gusella J.F. 1985. A DNA polymorphism for Huntington's disease marks the future. *Arch. Neurol.* **42:** 20–24.

Winzeler E.A., Richards D.R., Conway A.R., Goldstein A.L., Kalman S., McCullough M.J., McCusker J.H., Stevens D.A., Wodicka L., Lockhart D.J., and Davis R.W. 1998. Direct allelic variation scanning of the yeast genome. *Science* **281:** 1194–1197.

Wodicka L., Dong H., Mittmann M., Ho M.H., and Lockhart D.J. 1997. Genome-wide expression monitoring in *Saccharomyces cerevisiae*. *Nat. Biotechnol.* **15:** 1359–1367.

Yeast Genome Directory. 1997. The yeast genome directory. *Nature* (suppl. 6632) **387:** 5–105.

Zhu H., Bilgin M., Bangham R., Hall D., Casamayor A., Bertone P., Lan N., Jansen R., Bidlingmaier S., and Houfek T. 2001. Global analysis of protein activities using proteome chips. *Science* **293:** 2101–2105.

Index

A

α1-α2 hypothesis, 144–145
ACS. *See* ARS consensus sequence
Actin cytoskeleton
    Arp2/3 complex, 220
    cables and patches, 219
    cytoplasmic microtubule interactions, 217
    depolarization, 220
    fluorescence studies, 218
    mutants, 218–219
    myosins, 219
ade2, 20–21
Anaphase-promoting complex (APC/C)
    cell cycle control, 113–116
    components, 114
    discovery, 113
    Pds1 inactivation, 114–115
APC/C. *See* Anaphase-promoting complex
APF-1, 268
ARG4, 39–40
Arp2/3 complex, 220
ARS consensus sequence (ACS), replicators, 2, 51–54
Atg ligation systems, 278–279
AUG. *See* Translation initiation
Axl1, 223–224

B

BAR1, 197
BCY1, 130, 133
Bem1, 225
BNI1, 220
Bud site selection
    Bud proteins, 223
    cyclin-dependent kinase roles
        Cdc24, 222
        Cdc28, 223
        Cdc42, 222, 225–227
    landmarks, 223–224
    polarity-establishing proteins, 224–225
    Rho GTPases, 226–227
    scaffold proteins, 225

C

Carboxypeptidase Y, degradation studies, 273–274
Cassette hypothesis, mobile mating type, 148–152
CBF2, 59
CBF3, 59–60
Cdks. *See* Cyclin-dependent kinases
CEN3, 58
Centromere
    cloning, 58
    definition, 49, 57
    kinetochore proteins, 59–60
    sister centromere cohesion regulation in MI, 171–173
    structure, 58–59
Chiasmata, 168
ChIP. *See* Chromatin immunoprecipitation
Chromatin
    gene regulation, 74
    structure, 73–74
Chromatin immunoprecipitation (ChIP), 75
Chromosome cycle
    anaphase-promoting complex and sister chromatid separation, 112–116
    cyclin-dependent kinases
        mitotic triggers, 116–119
        oscillation halting mechanisms, 119–121
    DNA replication initiation, 110–112, 118
    overview, 109–110
Chromosome replication
    centromeres, 57–60
    replicators
        discovery, 50–51
        initiator protein identification, 52–54
        origin mapping, 51–52
    telomeres, 54–57
CLN2, 165
Cln3, 136
cMTs. *See* Cytoplasmic microtubules
Cohesin, 113, 115–116
CPA1, translational control, 98

CYC1
    first AUG rule formulation in translation, 88–90
    transcriptional regulation studies, 71
Cyclin-dependent kinases (Cdks)
    bud site selection
        Cdc24, 222
        Cdc28, 223
        Cdc42, 222, 225–227
    Cdc28
        degradation, 276–27
        mutants and mating, 197
        pachytene checkpoint role, 181–182
    DNA replication initiation, 112
    mitotic triggers, 116–119
    mutants, 117
    oscillation halting mechanisms, 119–121
Cyclosome. *See* Anaphase-promoting complex
CYR1, 130
Cytoplasmic inheritance
    cytoduction test, 12
    discovery, 12
    evidence, 12
Cytoplasmic microtubules (cMTs)
    actin cytoskeleton interactions, 217
    cortical factor attachment, 217
    dynamic instability, 217–218
    functional overview, 216
    motor proteins, 217
    spindle asymmetry establishment, 218
Cytoskeleton
    actin cytoskeleton, 218–220
    bud site selection and polarization signaling, 222–227
    cytoplasmic microtubules, 216–218
    early studies, 212–213
    mitotic spindle, 213–214
    septins, 220–222
    spindle-pole body, 214–215

D

DDEL signal, 248
DER2, 273
DHFR. *See* Dihydrofolate reductase
Dihydrofolate reductase (DHFR), fusion protein studies of mitochondrial import, 258–259
DMC1, 45
DNA microarray
    gene mapping, 294
    sequence tags, 296
DNA replication, initiation, 110–112, 118
DNA sequencing
    genome sequencing, 291–293

    milestones, 291
Double-strand-break repair (DSBR) model
    branched DNA intermediate identification, 42–43
    inconsistencies, 43–44
    initiation site mapping, 39–40
    meiotic double-strand break characterization, 41–42
    mutant studies, 45
    origins, 37–38
    overview, 37–39
    Spo11 role, 42
    trans-acting factors, 41
    visualization, 40–41
DSBR model. *See* Double-strand-break repair model
Dynein, 217

E

eIF2, phosphorylation, 94–95, 97–98
eIF4
    components, 99
    function, 99–101
    Pab1 interactions, 101–102
Endoplasmic reticulum translocation
    cotranslational translocation, 256
    fusion protein studies, 257
    mutants, 257
    ribosome-membrane interactions, 253–254
    signal sequences, 248–249, 254
Endosomal reticulum-associated degradation (ERAD), 273–274
ERAD. *See* Endosomal reticulum-associated degradation
Escherichia coli, protein translocation studies, 254–255
EST1, 56–57

F

FEAR system, meiosis exit, 175
Filamentation, control mechanisms, 152–153
First AUG rule. *See* Translation initiation
Formins, 220
Future landmark papers, identification, 5–7

G

GAL genes
    regulation studies, 68–71
    upstream activation site, 72
GCN4, translational control, 95–98
GCN5, 76
Genome, diploid yeast features, 13
Genomics
    functional genomics, 294–297
    genetic mapping, 287–288
    genome sequencing, 291–293

overview, 286–287
physical mapping, 288–291
single nucleotide polymorphism mapping, 293–294
GPCR. *See* G-protein-coupled receptor
G-protein-coupled receptor (GPCR), *STE* genes and signaling, 198–199
Growth cycle, coupling to cell cycle, 128–130

H

HDEL signal, 248–249
HIS4, 40–41, 91
Histones, mutation studies, 75
HML, 3, 38, 74, 150–152
HMR, 3, 38, 74, 150–152
HO, 149–152
Holliday model, homologous recombination, 34–37
Homologous recombination (HR)
    double-strand-break repair model
        branched DNA intermediate identification, 42–43
        inconsistencies, 43–44
        initiation site mapping, 39–40
        meiotic double-strand break characterization, 41–42
        mutant studies, 45
        origins, 37–38
        overview, 37–39
        Spo11 role, 42
        trans-acting factors, 41
        visualization, 40–41
    early models, 34–37
    event type, 34–35
    overview, 33–34
HR. *See* Homologous recombination
Hsp70, protein translocation role, 258–260

I

IME1, 161, 165–167, 179–180
IME2, 161, 167
Internal ribosome entry site (IRES), 92
IRES. *See* Internal ribosome entry site
ISG15, 277

K

KAR1, cytoduction, 12
KDEL signal, 248–249
Killer viruses, Mendelian gene interactions, 20
Kinesins, 214, 217
KKXX signal, 248–249

L

Latrunculin-A, actin depolymerization, 220
LEU2, 40–41

M

Mad2, 182
Mam1, 174
MAPK. *See* Mitogen-activated protein kinase
MAPs. *See* Microtubule-associated proteins
Mating type
    α1-α2 hypothesis, 144–145
    α2-mediated repression of **a**-genes, 147
    cassette hypothesis and mobile mating type, 148–152
    complementation by mutation, 143–144
    MAT locus
        models of control, 142–143
        mutants, 195
        protein degradation signals, 272
        transcripts and upstream activation sequences, 145–147
    Mcm1-α complexes in transcriptional activation, 147–148
    pheromones, 142
    regulatory genes, 143
    Ssn6-Tup1 in repression, 148
    STE genes in pheromone response. *See STE* genes
    types, 194
Mcm1, α complexes in transcriptional activation, 147–148
Mediator complex, discovery, 76–77
Meiosis
    commitment, 176–177
    context of studies, 158
    dependency relationships of events, 177–178
    exit, 174–175
    function, 157–158
    gene identification
        bypassing normal controls in meiotic initiation, 161
        recessive mutants in ascus production, 160–161
        single division and diploid versus haploid mutants, 161–163
    homolog separation in MI
        sister centromere cohesion regulation, 171–173
        sister kinetochore coorientation and monopolin, 173–174
    initiation
        cell-type control of entry, 165
        Ime1 role, 166–167

Meiosis (*continued*)
  Ime2 role, 167
  nutritional control, 165–166
  synaptonemal complex, 167–171
  landmarks
    cytology, 163–164
    overview, 159
    recombination, segregation, and spore formation, 164–165
    return to growth, 164
  Ndt80 regulation, 181
  pachytene checkpoint, 181–182
  progress in understanding, 159–160
  spindle checkpoint, 182–183
  transcriptional program, 178–181
  yeast advantages as model system, 159
MEN system, meiosis exit, 175
MER2, 169
Microtubule. *See* Cytoplasmic microtubules
Microtubule-associated proteins (MAPs), 214
Mitochondrial DNA
  null mutants, 19
  oxidative phosphorylation and respiratory chain, mutants, 17–18
  viability genes, 19
Mitochondrial import
  cotranslational translocation, 258
  dihydrofolate reductase fusion protein studies, 258–259
  energetics, 259–260
  protein unfolding studies, 259
  Tim22 complex, 261
  Tim23 complex, 260–261
  Tom40 role, 260–261
Mitochondrial markers, recombination, 16–17
Mitogen-activated protein kinase (MAPK)
  Ras signaling, 131
  STE genes, 199–201
Mitotic spindle
  electron microscopy, 213
  functions, 213
  kinesins, 214
Monopolin, 174

N

Ndt80, meiotic regulation, 181
NEDD8, 277
N-end rule pathway, 271
Nfs⁻ phenotype, 17–18
Nuclear localization
  deubiquitinating enzyme cloning, 268
  fusion protein studies, 255–256
  signal discovery, 255
  SV40 T antigen mutants, 256

O

ORC. *See* Origin-recognition complex
Origin-recognition complex (ORC), 111
Oxidative phosphorylation, mutants, 17–18

P

Pab1, translation initiation role, 88, 100–103
PCR. *See* Polymerase chain reaction
PDGFR, 274–275
PDR5, 294
Pds1, 114–115
Petites
  backcrosses, 15
  discovery, 14
  induction, 16
Pex3, 249
PGK1, 58
PHO5, chromatin structure and transcriptional control, 74
PKA. *See* Protein kinase A
Polarization, signaling, 222–227
Polymerase chain reaction (PCR), 289
Prions, 22–24
Promoter, early studies, 71–72
Proteasome
  early studies, 268–269
  substrate targeting across membranes, 273–274
Protein degradation. *See* Proteasome; Ubiquitination
Protein kinase A (PKA)
  control of cell growth response to nutrients, 132–133
  meiosis entry control, 166
Protein translocation. *See* Endoplasmic reticulum translocation; Mitochondrial import; Nuclear localization
[*PSI+*]
  discovery and function, 20
  prion activity, 22–24

R

Rab proteins, 247
RAD50, 41–42
Ras
  control of cell growth response to nutrients, 130–132
  filamentation control, 153
  mitogen-activated protein kinase signaling, 131
  oncogenic mutations, 4
rDNA, mapping, 287

Rec8, 171–174
Replicator
  definition, 49
  discovery, 50–51
  initiator protein identification, 52–54
  origin mapping, 51–52
Respiratory chain, mutants, 17–18
Restriction fragment length polymorphism (RFLP), 287–288
RFLP. *See* Restriction fragment length polymorphism
Rho GTPases, bud site selection, 226–227
Rim proteins, meiotic regulation, 180–181
RME1, 161, 165–167
RNA polymerase II, Mediator complex discovery, 76–77
Rsr1, 222–223
Rub1, 277

S

Saccharomyces Genome Database (SGD), 7
SC. *See* Synaptonemal complex
SCF complex, cell cycle regulation, 276–277
Secretory pathway
  avoidance mechanisms and organelle maintenance, 248–249
  early studies, 243–244
  electron microscopy, 244
  mutants, 245, 247
  SEC genes, 245–247
  vesicular traffic mechanics, 245–248
  Vps proteins, 247–248
Securin, 113, 116
Separase, 113, 115–116
Septins, 220–222
Sexual cycle, *Saccharomyces*, 14
SGD. *See Saccharomyces* Genome Database
Sgo1, 173–174
Sharing, resources and reagents, 7
Sic1, 277
Single nucleotide polymorphism (SNP), mapping, 293–294
SIR2, 5
Sister chromatids, separation, 112–116
Smt3, 277
Snf1
  control of cell growth response to nutrients, 134
  filamentation control, 153
SNP. *See* Single nucleotide polymorphism
Sok2, 166
SOS system, cyclin-dependent kinase regulation, 120–121

SPB. *See* Spindle-pole body
Spindle-pole body (SPB)
  asymmetry establishment, 218
  electron microscopy, 214–215
  purification, 214
  separation, 213
  structure, 215
SPO genes
  recessive mutants in ascus production, 160–161
  Spo13, 173–175
  Spo11 catalysis of double-strand breaks, 42
Spore formation
  assays, 164–165
  exit, 175–176
  gene identification
    bypassing normal controls in meiotic initiation, 161
    recessive mutants in ascus production, 160–161
    single division and diploid versus haploid mutants, 161–163
Ssn6, 148
STE genes
  G-protein-coupled receptor signaling, 198–199
  mitogen-activated protein kinases, 199–201
  nonmating mutants in identification, 195–198
  protein types and functions, 196
  protein ubiquitination, 275
  Ste5 scaffold function, 201–203
SUC2, regulation studies, 75
Sui- mutants, 93–94
SUMO, 277
SUP35, prion activity, 22–24
Super suppressors
  discovery, 20
  SUQ5, 20–21
Swi/Snf complex, transcriptional regulation, 75–76
Synaptonemal complex (SC)
  assembly, 170
  chiasmata role, 168
  components, 169–170
  cytology of meiosis, 163–164
  elements, 167–168
  functions, 159, 170–171
  recombination initiation, 168–169
  Zip1 functions, 170

T

Technology, development cycle, 2
Telomere
  cloning, 55
  definition, 49